STUDENT SOLUTIONS M

to accompany

Cutnell & Johnson

Physics

Young and Stadler

Tenth Edition

David Young

Shane Stadler

Louisiana State University

WILEY

ASSOCIATE DEVELOPMENT EDITOR	Alyson Rentrop
SENIOR PRODUCTION EDITOR	Elizabeth Swain
COVER PHOTO	© Samran wonglakorn/Shutterstock

Founded in 1807, John Wiley & Sons, Inc. has been a valued source of knowledge and understanding for more than 200 years, helping people around the world meet their needs and fulfi ll their aspirations. Our company is built on a foundation of principles that include responsibility to the communities we serve and where we live and work. In 2008, we launched a Corporate Citizenship Initiative, a global eff ort to address the environmental, social, economic, and ethical challenges we face in our business. Among the issues we are addressing are carbon impact, paper specifi cations and procurement, ethical conduct within our business and among our vendors, and community and charitable support. For more information, please visit our Web site: www.wiley.com/go/citizenship.

ISBN: 978-1-118-83690-3

Printed in the United States of America

10 9 8 7 6 5 4 3 2 1

PREFACE

This volume contains the complete solutions to those Problems in the text that are marked with an SSM icon. There are about 600 such Problems, and they are found at the end of each chapter in the text.

The solutions are worked out with great care, and all the steps are included. Most of the solutions are comprised of two parts, a REASONING part followed by a SOLUTION part. In the REASONING part we explain what motivates our procedure for solving the problem, before any algebraic or numerical work is done. During the SOLUTION part, numerical calculations are performed, and the answer to the problem is obtained.

We welcome any suggestions that you may have for improving the usefulness of these solutions. Please feel free to write us care of the Physics Editor, Global Education division, John Wiley & Sons, Inc., 111 River Street, Hoboken, NJ 07030, or contact us at **www.wiley.com/college/cutnell**

CONTENTS

CHAPTER 1 | INTRODUCTION AND MATHEMATICAL CONCEPTS

---

3. **REASONING** We use the facts that 1 mi = 5280 ft, 1 m = 3.281 ft, and 1 yd = 3 ft. With these facts we construct three conversion factors: (5280 ft)/(1 mi) = 1, (1 m)/(3.281 ft) = 1, and (3 ft)/(1 yd) = 1.

SOLUTION By multiplying by the given distance d of the fall by the appropriate conversion factors we find that

$$d = \left(6 \text{ mi}\right)\left(\frac{5280 \text{ ft}}{1 \text{ mi}}\right)\left(\frac{1 \text{ m}}{3.281 \text{ ft}}\right) + \left(551 \text{ yd}\right)\left(\frac{3 \text{ ft}}{1 \text{ yd}}\right)\left(\frac{1 \text{ m}}{3.281 \text{ ft}}\right) = \boxed{10\ 159 \text{ m}}$$

---

7. **REASONING** This problem involves using unit conversions to determine the number of magnums in one jeroboam. The necessary relationships are

$$1.0 \text{ magnum} = 1.5 \text{ liters}$$
$$1.0 \text{ jeroboam} = 0.792 \text{ U. S. gallons}$$
$$1.00 \text{ U. S. gallon} = 3.785 \times 10^{-03} \text{ m}^3 = 3.785 \text{ liters}$$

These relationships may be used to construct the appropriate conversion factors.

SOLUTION By multiplying one jeroboam by the appropriate conversion factors we can determine the number of magnums in a jeroboam as shown below:

$$\left(1.0 \text{ jeroboam}\right)\left(\frac{0.792 \text{ gallons}}{1.0 \text{ jeroboam}}\right)\left(\frac{3.785 \text{ liters}}{1.0 \text{ gallon}}\right)\left(\frac{1.0 \text{ magnum}}{1.5 \text{ liters}}\right) = \boxed{2.0 \text{ magnums}}$$

---

11. **REASONING** The dimension of the spring constant k can be determined by first solving the equation $T = 2\pi\sqrt{m/k}$ for k in terms of the time T and the mass m. Then, the dimensions of T and m can be substituted into this expression to yield the dimension of k.

SOLUTION Algebraically solving the expression above for k gives $k = 4\pi^2 m/T^2$. The term $4\pi^2$ is a numerical factor that does not have a dimension, so it can be ignored in this analysis. Since the dimension for mass is [M] and that for time is [T], the dimension of k is

$$\text{Dimension of } k = \boxed{\frac{[\text{M}]}{[\text{T}]^2}}$$

13. *REASONING* The shortest distance between the two towns is along the line that joins them. This distance, h, is the hypotenuse of a right triangle whose other sides are $h_o = 35.0$ km and $h_a = 72.0$ km, as shown in the figure below.

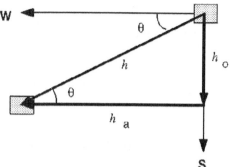

SOLUTION The angle θ is given by $\tan\theta = h_o/h_a$ so that

$$\theta = \tan^{-1}\left(\frac{35.0 \text{ km}}{72.0 \text{ km}}\right) = \boxed{25.9^\circ \text{ S of W}}$$

We can then use the Pythagorean theorem to find h.

$$h = \sqrt{h_o^2 + h_a^2} = \sqrt{(35.0 \text{ km})^2 + (72.0 \text{ km})^2} = \boxed{80.1 \text{ km}}$$

21. *REASONING*   The drawing at the right shows the location of each deer A, B, and C. From the problem statement it follows that

$$b = 62 \text{ m}$$

$$c = 95 \text{ m}$$

$$\gamma = 180^\circ - 51^\circ - 77^\circ = 52^\circ$$

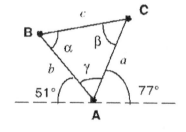

Applying the law of cosines (given in Appendix E) to the geometry in the figure, we have

$$a^2 - 2ab \cos\gamma + (b^2 - c^2) = 0$$

which is an expression that is quadratic in a. It can be simplified to $Aa^2 + Ba + C = 0$, with

$$A = 1$$

$$B = -2b \cos\gamma = -2(62 \text{ m}) \cos 52^\circ = -76 \text{ m}$$

$$C = (b^2 - c^2) = (62 \text{ m})^2 - (95 \text{ m})^2 = -5181 \text{ m}^2$$

This quadratic equation can be solved for the desired quantity a.

SOLUTION   Suppressing units, we obtain from the quadratic formula

$$a = \frac{-(-76) \pm \sqrt{(-76)^2 - 4(1)(-5181)}}{2(1)} = 1.2 \times 10^2 \text{ m} \quad \text{and} \quad -43 \text{ m}$$

Discarding the negative root, which has no physical significance, we conclude that the distance between deer A and C is $\boxed{1.2 \times 10^2 \text{ m}}$.

---

23. **REASONING**

a. Since the two force vectors **A** and **B** have directions due west and due north, they are perpendicular. Therefore, the resultant vector **F** = **A** + **B** has a magnitude given by the Pythagorean theorem: $F^2 = A^2 + B^2$. Knowing the magnitudes of **A** and **B**, we can calculate the magnitude of **F**. The direction of the resultant can be obtained using trigonometry.

b. For the vector **F′** = **A** − **B** we note that the subtraction can be regarded as an addition in the following sense: **F′** = **A** + (−**B**). The vector −**B** points due south, opposite the vector **B**, so the two vectors are once again perpendicular and the magnitude of **F′** again is given by the Pythagorean theorem. The direction again can be obtained using trigonometry.

SOLUTION   a. The drawing shows the two vectors and the resultant vector. According to the Pythagorean theorem, we have

$$F^2 = A^2 + B^2$$

$$F = \sqrt{A^2 + B^2}$$

$$F = \sqrt{(445 \text{ N})^2 + (325 \text{ N})^2}$$

$$= \boxed{551 \text{ N}}$$

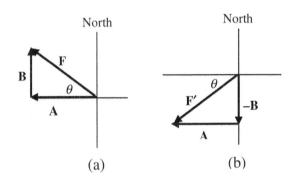

(a)   (b)

Using trigonometry, we can see that the direction of the resultant is

$$\tan\theta = \frac{B}{A} \quad \text{or} \quad \theta = \tan^{-1}\left(\frac{325 \text{ N}}{445 \text{ N}}\right) = \boxed{36.1° \text{ north of west}}$$

b. Referring to the drawing and following the same procedure as in part a, we find

$$F'^2 = A^2 + (-B)^2 \quad \text{or} \quad F' = \sqrt{A^2 + (-B)^2} = \sqrt{(445 \text{ N})^2 + (-325 \text{ N})^2} = \boxed{551 \text{ N}}$$

$$\tan\theta = \frac{B}{A} \quad \text{or} \quad \theta = \tan^{-1}\left(\frac{325 \text{ N}}{445 \text{ N}}\right) = \boxed{36.1° \text{ south of west}}$$

---

25. **REASONING** For convenience, we can assign due east to be the positive direction and due west to be the negative direction. Since all the vectors point along the same east-west line, the vectors can be added just like the usual algebraic addition of positive and negative

scalars. We will carry out the addition for all of the possible choices for the two vectors and identify the resultants with the smallest and largest magnitudes.

SOLUTION There are six possible choices for the two vectors, leading to the following resultant vectors:

$$\mathbf{F_1} + \mathbf{F_2} = 50.0 \text{ newtons} + 10.0 \text{ newtons} = +60.0 \text{ newtons} = 60.0 \text{ newtons, due east}$$

$$\mathbf{F_1} + \mathbf{F_3} = 50.0 \text{ newtons} - 40.0 \text{ newtons} = +10.0 \text{ newtons} = 10.0 \text{ newtons, due east}$$

$$\mathbf{F_1} + \mathbf{F_4} = 50.0 \text{ newtons} - 30.0 \text{ newtons} = +20.0 \text{ newtons} = 20.0 \text{ newtons, due east}$$

$$\mathbf{F_2} + \mathbf{F_3} = 10.0 \text{ newtons} - 40.0 \text{ newtons} = -30.0 \text{ newtons} = 30.0 \text{ newtons, due west}$$

$$\mathbf{F_2} + \mathbf{F_4} = 10.0 \text{ newtons} - 30.0 \text{ newtons} = -20.0 \text{ newtons} = 20.0 \text{ newtons, due west}$$

$$\mathbf{F_3} + \mathbf{F_4} = -40.0 \text{ newtons} - 30.0 \text{ newtons} = -70.0 \text{ newtons} = 70.0 \text{ newtons, due west}$$

The resultant vector with the smallest magnitude is $\boxed{\mathbf{F_1} + \mathbf{F_3} = 10.0 \text{ newtons, due east}}$.

The resultant vector with the largest magnitude is $\boxed{\mathbf{F_3} + \mathbf{F_4} = 70.0 \text{ newtons, due west}}$.

---

31. **REASONING AND SOLUTION** The single force needed to produce the same effect is equal to the resultant of the forces provided by the two ropes. The following figure shows the force vectors drawn to scale and arranged tail to head. The magnitude and direction of the resultant can be found by direct measurement using the scale factor shown in the figure.

a. From the figure, the magnitude of the resultant is $\boxed{5600 \text{ N}}$.

b. The single rope should be directed $\boxed{\text{along the dashed line}}$ in the text drawing.

---

35. **REASONING AND SOLUTION** In order to determine which vector has the largest x and y components, we calculate the magnitude of the x and y components explicitly and compare them. In the calculations, the symbol u denotes the units of the vectors.

$A_x = (100.0 \text{ u}) \cos 90.0° = 0.00 \text{ u}$ $\qquad$ $A_y = (100.0 \text{ u}) \sin 90.0° = 1.00 \times 10^2 \text{ u}$

$B_x = (200.0 \text{ u}) \cos 60.0° = 1.00 \times 10^2 \text{ u}$ $\qquad$ $B_y = (200.0 \text{ u}) \sin 60.0° = 173 \text{ u}$

$C_x = (150.0 \text{ u}) \cos 0.00° = 150.0 \text{ u}$ $\qquad$ $C_y = (150.0 \text{ u}) \sin 0.00° = 0.00 \text{ u}$

a. $\boxed{\mathbf{C} \text{ has the largest } x \text{ component.}}$

b. | **B** has the largest y component. |

---

39. ***REASONING*** The x and y components of **r** are mutually perpendicular; therefore, the magnitude of **r** can be found using the Pythagorean theorem. The direction of **r** can be found using the definition of the tangent function.

SOLUTION According to the Pythagorean theorem, we have

$$r = \sqrt{x^2 + y^2} = \sqrt{(-125 \text{ m})^2 + (-184 \text{ m})^2} = \boxed{222 \text{ m}}$$

The angle θ is

$$\theta = \tan^{-1}\left(\frac{184 \text{ m}}{125 \text{ m}}\right) = \boxed{55.8°}$$

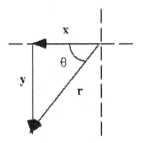

---

43. ***REASONING*** The force **F** and its two components form a right triangle. The hypotenuse is 82.3 newtons, and the side parallel to the $+x$ axis is $F_x = 74.6$ newtons. Therefore, we can use the trigonometric cosine and sine functions to determine the angle of **F** relative to the $+x$ axis and the component F_y of **F** along the $+y$ axis.

SOLUTION
a. The direction of **F** relative to the $+x$ axis is specified by the angle θ as

$$\theta = \cos^{-1}\left(\frac{74.6 \text{ newtons}}{82.3 \text{ newtons}}\right) = \boxed{25.0°} \tag{1.5}$$

b. The component of **F** along the $+y$ axis is

$$F_y = F\sin 25.0° = (82.3 \text{ newtons})\sin 25.0° = \boxed{34.8 \text{ newtons}} \tag{1.4}$$

---

45. ***REASONING*** The individual displacements of the golf ball, **A**, **B**, and **C** are shown in the figure. Their resultant, **R**, is the displacement that would have been needed to "hole the ball" on the very first putt. We will use the component method to find **R**.

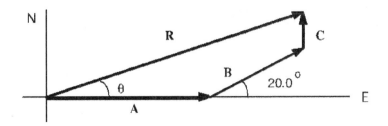

SOLUTION The components of each displacement vector are given in the following table:

Vector	*x* Components	*y* Components
A	(5.0 m) cos 0° = 5.0 m	(5.0 m) sin 0° = 0
B	(2.1 m) cos 20.0° = 2.0 m	(2.1 m) sin 20.0° = 0.72 m
C	(0.50 m) cos 90.0° = 0	(0.50 m) sin 90.0° = 0.50 m
R = A + B + C	7.0 m	1.22 m

The resultant vector **R** has magnitude

$$R = \sqrt{(7.0 \text{ m})^2 + (1.22 \text{ m})^2} = \boxed{7.1 \text{ m}}$$

and the angle θ is

$$\theta = \tan^{-1}\left(\frac{1.22 \text{ m}}{7.0 \text{ m}}\right) = 9.9°$$

Thus, the required direction is $\boxed{9.9° \text{ north of east}}$.

---

53. *REASONING*   Since the finish line is coincident with the starting line, the net displacement of the sailboat is zero. Hence the sum of the components of the displacement vectors of the individual legs must be zero. In the drawing in the text, the directions to the right and upward are taken as positive.

SOLUTION In the horizontal direction $R_h = A_h + B_h + C_h + D_h = 0$

$R_h = (3.20 \text{ km}) \cos 40.0° - (5.10 \text{ km}) \cos 35.0° - (4.80 \text{ km}) \cos 23.0° + D \cos \theta = 0$

$$D \cos \theta = 6.14 \text{ km} \qquad\qquad (1)$$

In the vertical direction $R_v = A_v + B_v + C_v + D_v = 0$.

$R_v = (3.20 \text{ km}) \sin 40.0° + (5.10 \text{ km}) \sin 35.0° - (4.80 \text{ km}) \sin 23.0° - D \sin \theta = 0$.

$$D \sin \theta = 3.11 \text{ km} \qquad\qquad (2)$$

Dividing (2) by (1) gives

$$\tan \theta = (3.11 \text{ km})/(6.14 \text{ km}) \quad \text{ or } \quad \theta = \boxed{26.9°}$$

Solving (1) gives

$$D = (6.14 \text{ km})/\cos 26.9° = \boxed{6.88 \text{ km}}$$

59. **REASONING**   The ostrich's velocity vector **v** and the desired components are shown in the figure at the right. The components of the velocity in the directions due west and due north are v_W and v_N, respectively. The sine and cosine functions can be used to find the components.

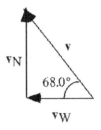

SOLUTION
a. According to the definition of the sine function, we have for the vectors in the figure

$$\sin\ \theta = \frac{v_N}{v} \quad \text{or} \quad v_N = v\sin\ \theta = (17.0\ \text{m/s})\sin 68° = \boxed{15.8\ \text{m/s}}$$

b. Similarly,

$$\cos\ \theta = \frac{v_W}{v} \quad \text{or} \quad v_W = v\cos\ \theta = (17.0\ \text{m/s})\cos 68.0° = \boxed{6.37\ \text{m/s}}$$

63. **REASONING**   The performer walks out on the wire a distance d, and the vertical distance to the net is h. Since these two distances are perpendicular, the magnitude of the displacement is given by the Pythagorean theorem as $s = \sqrt{d^2 + h^2}$. Values for s and h are given, so we can solve this expression for the distance d. The angle that the performer's displacement makes below the horizontal can be found using trigonometry.

SOLUTION
a. Using the Pythagorean theorem, we find that

$$s = \sqrt{d^2 + h^2} \quad \text{or} \quad d = \sqrt{s^2 - h^2} = \sqrt{(26.7\ \text{ft})^2 - (25.0\ \text{ft})^2} = \boxed{9.4\ \text{ft}}$$

b. The angle θ that the performer's displacement makes below the horizontal is given by

$$\tan\theta = \frac{h}{d} \quad \text{or} \quad \theta = \tan^{-1}\left(\frac{h}{d}\right) = \tan^{-1}\left(\frac{25.0\ \text{ft}}{9.4\ \text{ft}}\right) = \boxed{69°}$$

65. **REASONING AND SOLUTION**   We take due north to be the direction of the +y axis. Vectors **A** and **B** are the components of the resultant, **C**. The angle that **C** makes with the x axis is then $\theta = \tan^{-1}(B/A)$. The symbol u denotes the units of the vectors.

a. Solving for B gives

$$B = A \tan \theta = (6.00 \text{ u}) \tan 60.0° = \boxed{10.4 \text{ u}}$$

b. The magnitude of **C** is

$$C = \sqrt{A^2 + B^2} = \sqrt{(6.00 \text{ u})^2 + (10.4 \text{ u})^2} = \boxed{12.0 \text{ u}}$$

---

71. ***CONCEPTS*** **(i)** The magnitude of $\vec{A}$ is given by the Pythagorean theorem in the form $A = \sqrt{A_x^2 + A_y^2}$, since a vector and its components form a right triangle.

(ii) Yes. The vectors $\vec{B}$ and $\vec{R}$ each have a zero value for their y component. This is because these vectors are parallel to the x axis.

(iii) The fact that $\vec{A} + \vec{B} + \vec{C} = \vec{R}$ means that the sum of the x components of $\vec{A}$, $\vec{B}$, and $\vec{C}$ equals the x component of $\vec{R}$: $A_x + B_x + C_x = R_x$. A similar relation holds for the y components: $A_y + B_y + C_y = R_y$.

CALCULATIONS We will use the Pythagorean theorem to relate A to the components of $\vec{A}$: $A = \sqrt{A_x^2 + A_y^2}$.

To obtain the value for A_x we use the fact that the sum of the components of $\vec{A}$, $\vec{B}$, and $\vec{C}$ equals the x component of $\vec{R}$:

$$\underbrace{A_x + \underbrace{10.0 \text{ m}}_{B_x} + \underbrace{(23.0 \text{ m})\cos 50°}_{C_x} = \underbrace{35.0 \text{ m}}_{R_x}} \qquad \text{or } A_x = 10.2 \text{ m}$$

Similarly, for the y components we have

$$A_y + \underbrace{0 \text{ m}}_{B_y} + \underbrace{\left[-(23.0 \text{ m})\sin 50°\right]}_{C_y} = \underbrace{0 \text{ m}}_{R_y} \qquad \text{or} \quad A_y = 17.6 \text{ m}$$

Using these values for A_x and A_y, we find that the magnitude of $\vec{A}$ is

$$A = \sqrt{A_x^2 + A_y^2} = \sqrt{(10.2 \text{ m})^2 + (17.6 \text{ m})^2} = \boxed{20.3 \text{ m}}$$

We also use the components A_x and A_y to find the directional angle θ:

$$\theta = \tan^{-1}\left(\frac{A_y}{A_x}\right) = \tan^{-1}\left(\frac{17.6 \text{ m}}{10.2 \text{ m}}\right) = \boxed{59.9°}$$

CHAPTER 2 | *KINEMATICS IN ONE DIMENSION*

3. **REASONING** The average speed is the distance traveled divided by the elapsed time (Equation 2.1). Since the average speed and distance are known, we can use this relation to find the time.

SOLUTION The time it takes for the continents to drift apart by 1500 m is

$$\text{Elapsed time} = \frac{\text{Distance}}{\text{Average speed}} = \frac{1500\,\text{m}}{\left(3\,\dfrac{\text{cm}}{\text{yr}}\right)\left(\dfrac{1\,\text{m}}{100\,\text{cm}}\right)} = \boxed{5\times10^4\,\text{yr}}$$

11. **REASONING AND SOLUTION**
a. The total displacement traveled by the bicyclist for the entire trip is equal to the sum of the displacements traveled during each part of the trip. The displacement traveled during each part of the trip is given by Equation 2.2: $\Delta x = \bar{v}\Delta t$. Therefore,

$$\Delta x_1 = (7.2\ \text{m/s})(22\ \text{min})\left(\frac{60\ \text{s}}{1\ \text{min}}\right) = 9500\ \text{m}$$

$$\Delta x_2 = (5.1\ \text{m/s})(36\ \text{min})\left(\frac{60\ \text{s}}{1\ \text{min}}\right) = 11\ 000\ \text{m}$$

$$\Delta x_3 = (13\ \text{m/s})(8.0\ \text{min})\left(\frac{60\ \text{s}}{1\ \text{min}}\right) = 6200\ \text{m}$$

The total displacement traveled by the bicyclist during the entire trip is then

$$\Delta x = 9500\ \text{m} + 11\ 000\ \text{m} + 6200\ \text{m} = \boxed{2.67\times10^4\ \text{m}}$$

b. The average velocity can be found from Equation 2.2.

$$\bar{v} = \frac{\Delta x}{\Delta t} = \frac{2.67\times10^4\ \text{m}}{(22\ \text{min} + 36\ \text{min} + 8.0\ \text{min})}\left(\frac{1\ \text{min}}{60\ \text{s}}\right) = \boxed{6.74\ \text{m/s, due north}}$$

25. *REASONING AND SOLUTION*

  a. The magnitude of the acceleration can be found from Equation 2.4 ($v = v_0 + at$) as

$$a = \frac{v - v_0}{t} = \frac{3.0 \text{ m/s} - 0 \text{ m/s}}{2.0 \text{ s}} = \boxed{1.5 \text{ m/s}^2}$$

  b. Similarly the magnitude of the acceleration of the car is

$$a = \frac{v - v_0}{t} = \frac{41.0 \text{ m/s} - 38.0 \text{ m/s}}{2.0 \text{ s}} = \boxed{1.5 \text{ m/s}^2}$$

  c. Assuming that the acceleration is constant, the displacement covered by the car can be found from Equation 2.9 ($v^2 = v_0^2 + 2ax$):

$$x = \frac{v^2 - v_0^2}{2a} = \frac{(41.0 \text{ m/s})^2 - (38.0 \text{ m/s})^2}{2(1.5 \text{ m/s}^2)} = 79 \text{ m}$$

Similarly, the displacement traveled by the jogger is

$$x = \frac{v^2 - v_0^2}{2a} = \frac{(3.0 \text{ m/s})^2 - (0 \text{ m/s})^2}{2(1.5 \text{ m/s}^2)} = 3.0 \text{ m}$$

Therefore, the car travels 79 m – 3.0 m = $\boxed{76 \text{ m}}$ further than the jogger.

___

29. *REASONING AND SOLUTION*   The average acceleration of the plane can be found by solving Equation 2.9 $\left(v^2 = v_0^2 + 2ax \right)$ for a. Taking the direction of motion as positive, we have

$$a = \frac{v^2 - v_0^2}{2x} = \frac{(+6.1 \text{ m/s})^2 - (+69 \text{ m/s})^2}{2(+750 \text{ m})} = \boxed{-3.1 \text{ m/s}^2}$$

The minus sign indicates that the direction of the acceleration is opposite to the direction of motion, and the plane is slowing down.

___

31. **REASONING** The cart has an initial velocity of $v_0 = +5.0$ m/s, so initially it is moving to the right, which is the positive direction. It eventually reaches a point where the displacement is $x = +12.5$ m, and it begins to move to the left. This must mean that the cart comes to a momentary halt at this point (final velocity is $v = 0$ m/s), before beginning to move to the left. In other words, the cart is decelerating, and its acceleration must point opposite to the velocity, or to the left. Thus, the acceleration is negative. Since the initial velocity, the final velocity, and the displacement are known, Equation 2.9 $\left(v^2 = v_0^2 + 2ax\right)$ can be used to determine the acceleration.

SOLUTION Solving Equation 2.9 for the acceleration a shows that

$$a = \frac{v^2 - v_0^2}{2x} = \frac{\left(0 \text{ m/s}\right)^2 - \left(+5.0 \text{ m/s}\right)^2}{2\left(+12.5 \text{ m}\right)} = \boxed{-1.0 \text{ m/s}^2}$$

41. **REASONING**   As the train passes through the crossing, its motion is described by Equations 2.4 ($v = v_0 + at$) and 2.7 $\left[x = \frac{1}{2}(v + v_0)t\right]$, which can be rearranged to give

$$v - v_0 = at \quad \text{and} \quad v + v_0 = \frac{2x}{t}$$

These can be solved simultaneously to obtain the speed v when the train reaches the end of the crossing. Once v is known, Equation 2.4 can be used to find the time required for the train to reach a speed of 32 m/s.

SOLUTION   Adding the above equations and solving for v, we obtain

$$v = \frac{1}{2}\left(at + \frac{2x}{t}\right) = \frac{1}{2}\left[(1.6 \text{ m/s}^2)(2.4 \text{ s}) + \frac{2(20.0 \text{ m})}{2.4 \text{ s}}\right] = 1.0 \times 10^1 \text{ m/s}$$

The motion from the end of the crossing until the locomotive reaches a speed of 32 m/s requires a time

$$t = \frac{v - v_0}{a} = \frac{32 \text{ m/s} - 1.0 \times 10^1 \text{ m/s}}{1.6 \text{ m/s}^2} = \boxed{14 \text{ s}}$$

43. *REASONING AND SOLUTION* When air resistance is neglected, free fall conditions are applicable. The final speed can be found from Equation 2.9;

$$v^2 = v_0^2 + 2ay$$

where v_0 is zero since the stunt man falls from rest. If the origin is chosen at the top of the hotel and the upward direction is positive, then the displacement is $y = -99.4$ m. Solving for v, we have

$$v = -\sqrt{2ay} = -\sqrt{2(-9.80 \text{ m/s}^2)(-99.4 \text{ m})} = -44.1 \text{ m/s}$$

The speed at impact is the magnitude of this result or $\boxed{44.1 \text{ m/s}}$.

---

49. *REASONING* The initial velocity of the compass is $+2.50$ m/s. The initial position of the compass is 3.00 m and its final position is 0 m when it strikes the ground. The displacement of the compass is the final position minus the initial position, or $y = -3.00$ m. As the compass falls to the ground, its acceleration is the acceleration due to gravity, $a = -9.80$ m/s^2. Equation 2.8 $\left(y = v_0 t + \frac{1}{2} a t^2 \right)$ can be used to find how much time elapses before the compass hits the ground.

SOLUTION   Starting with Equation 2.8, we use the quadratic equation to find the elapsed time.

$$t = \frac{-v_0 \pm \sqrt{v_0^2 - 4\left(\frac{1}{2}a\right)(-y)}}{2\left(\frac{1}{2}a\right)} = \frac{-(2.50 \text{ m/s}) \pm \sqrt{(2.50 \text{ m/s})^2 - 4(-4.90 \text{ m/s}^2)\left[-(-3.00 \text{ m})\right]}}{2(-4.90 \text{ m/s}^2)}$$

There are two solutions to this quadratic equation, $t_1 = \boxed{1.08 \text{ s}}$ and $t_2 = -0.568$ s. The second solution, being a negative time, is discarded.

---

53. *REASONING AND SOLUTION*   Since the balloon is released from rest, its initial velocity is zero. The time required to fall through a vertical displacement y can be found from Equation 2.8 $\left(y = v_0 t + \frac{1}{2} a t^2 \right)$ with $v_0 = 0$ m/s.   Assuming upward to be the positive direction, we find

$$t = \sqrt{\frac{2y}{a}} = \sqrt{\frac{2(-6.0 \text{ m})}{-9.80 \text{ m/s}^2}} = \boxed{1.1 \text{ s}}$$

---

59. **REASONING AND SOLUTION**
a. We can use Equation 2.9 to obtain the speed acquired as she falls through the distance H. Taking downward as the positive direction, we find

$$v^2 = v_0^2 + 2ay = (0 \text{ m/s})^2 + 2aH \qquad \text{or} \qquad v = \sqrt{2aH}$$

To acquire a speed of twice this value or $2\sqrt{2aH}$, she must fall an additional distance H'. According to Equation 2.9 $\left(v^2 = v_0^2 + 2ay \right)$, we have

$$\left(2\sqrt{2aH} \right)^2 = \left(\sqrt{2aH} \right)^2 + 2aH' \qquad \text{or} \qquad 4(2aH) = 2aH + 2aH'$$

The acceleration due to gravity a can be eliminated algebraically from this result, giving

$$4H = H + H' \qquad \text{or} \qquad \boxed{H' = 3H}$$

b. In the previous calculation the acceleration due to gravity was eliminated algebraically. Thus, a value other than 9.80 m/s^2 would $\boxed{\text{not have affected the answer to part (a)}}$.

_____

61. **REASONING**   Once the man sees the block, the man must get out of the way in the time it takes for the block to fall through an additional 12.0 m. The velocity of the block at the instant that the man looks up can be determined from Equation 2.9. Once the velocity is known at that instant, Equation 2.8 can be used to find the time required for the block to fall through the additional distance.

SOLUTION   When the man first notices the block, it is 14.0 m above the ground and its displacement from the starting point is $y = 14.0 \text{ m} - 53.0 \text{ m}$. Its velocity is given by Equation 2.9 $\left(v^2 = v_0^2 + 2ay \right)$. Since the block is moving down, its velocity has a negative value,

$$v = -\sqrt{v_0 + 2ay} = -\sqrt{(0 \text{ m/s})^2 + 2(-9.80 \text{ m/s}^2)(14.0 \text{ m} - 53.0 \text{ m})} = -27.7 \text{ m/s}$$

The block then falls the additional 12.0 m to the level of the man's head in a time t which satisfies Equation 2.8:

$$y = v_0 t + \tfrac{1}{2} a t^2$$

where $y = -12.0$ m and $v_0 = -27.7$ m/s. Thus, t is the solution to the quadratic equation

$$4.90 t^2 + 27.7 t - 12.0 = 0$$

where the units have been suppressed for brevity. From the quadratic formula, we obtain

$$t = \frac{-27.7 \pm \sqrt{(27.7)^2 - 4(4.90)(-12.0)}}{2(4.90)} = 0.40 \text{ s} \quad \text{or} \quad -6.1 \text{ s}$$

The negative solution can be rejected as nonphysical, and the time it takes for the block to reach the level of the man is $\boxed{0.40 \text{ s}}$.

---

65. ***REASONING*** The slope of a straight-line segment in a position-versus-time graph is the average velocity. The algebraic sign of the average velocity, therefore, corresponds to the sign of the slope.

SOLUTION
a. The slope, and hence the average velocity, is *positive* for segments A and C, *negative* for segment B, and *zero* for segment D.

b. In the given position-versus-time graph, we find the slopes of the four straight-line segments to be

$$v_A = \frac{1.25 \text{ km} - 0 \text{ km}}{0.20 \text{ h} - 0 \text{ h}} = \boxed{+6.3 \text{ km/h}}$$

$$v_B = \frac{0.50 \text{ km} - 1.25 \text{ km}}{0.40 \text{ h} - 0.20 \text{ h}} = \boxed{-3.8 \text{ km/h}}$$

$$v_C = \frac{0.75 \text{ km} - 0.50 \text{ km}}{0.80 \text{ h} - 0.40 \text{ h}} = \boxed{+0.63 \text{ km/h}}$$

$$v_D = \frac{0.75 \text{ km} - 0.75 \text{ km}}{1.00 \text{ h} - 0.80 \text{ h}} = \boxed{0 \text{ km/h}}$$

---

71. ***REASONING***   The two runners start one hundred meters apart and run toward each other. Each runs ten meters during the first second and, during each second thereafter, each runner runs ninety percent of the distance he ran in the previous second. While the velocity of each runner changes from second to second, it remains constant during any one second.

SOLUTION The following table shows the distance covered during each second for one of the runners, and the position at the end of each second (assuming that he begins at the origin) for the first eight seconds.

Time t (s)	Distance covered (m)	Position x (m)
0.00		0.00
1.00	10.00	10.00
2.00	9.00	19.00
3.00	8.10	27.10
4.00	7.29	34.39
5.00	6.56	40.95
6.00	5.90	46.85
7.00	5.31	52.16
8.00	4.78	56.94

The following graph is the position-time graph constructed from the data in the table above.

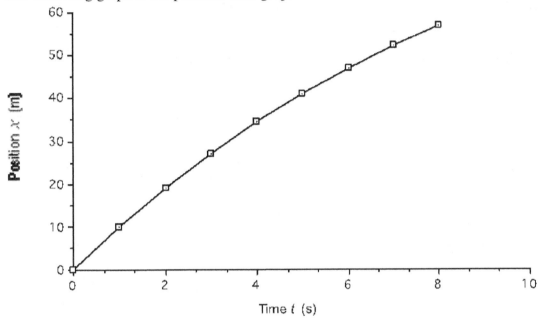

a. Since the two runners are running toward each other in exactly the same way, they will meet halfway between their respective starting points. That is, they will meet at $x = 50.0$ m. According to the graph, therefore, this position corresponds to a time of $\boxed{6.6 \text{ s}}$.

b. Since the runners collide during the seventh second, the speed at the instant of collision can be found by taking the slope of the position-time graph for the seventh second. The speed of either runner in the interval from $t = 6.00$ s to $t = 7.00$ s is

$$v = \frac{\Delta x}{\Delta t} = \frac{52.16 \text{ m} - 46.85 \text{ m}}{7.00 \text{ s} - 6.00 \text{ s}} = 5.3 \text{ m/s}$$

Therefore, at the moment of collision, the speed of either runner is $\boxed{5.3 \text{ m/s}}$.

73. **REASONING AND SOLUTION**

a. Once the pebble has left the slingshot, it is subject only to the acceleration due to gravity. Since the downward direction is negative, the acceleration of the pebble is $\boxed{-9.80 \text{ m/s}^2}$. The pebble is not decelerating. Since its velocity and acceleration both point downward, the magnitude of the pebble's velocity is increasing, not decreasing.

b. The displacement y traveled by the pebble as a function of the time t can be found from Equation 2.8. Using Equation 2.8, we have

$$y = v_0 t + \tfrac{1}{2} a_y t^2 = (-9.0 \text{ m/s})(0.50 \text{ s}) + \tfrac{1}{2}\left[(-9.80 \text{ m/s}^2)(0.50 \text{ s})^2 \right] = -5.7 \text{ m}$$

Thus, after 0.50 s, the pebble is $\boxed{5.7 \text{ m}}$ beneath the cliff-top.

---

75. **REASONING** Since the belt is moving with constant velocity, the displacement ($x_0 = 0$ m) covered by the belt in a time t_{belt} is giving by Equation 2.2 (with x_0 assumed to be zero) as

$$x = v_{\text{belt}} t_{\text{belt}} \tag{1}$$

Since Clifford moves with constant acceleration, the displacement covered by Clifford in a time t_{Cliff} is, from Equation 2.8,

$$x = v_0 t_{\text{Cliff}} + \tfrac{1}{2} a t_{\text{Cliff}}^2 = \tfrac{1}{2} a t_{\text{Cliff}}^2 \tag{2}$$

The speed v_{belt} with which the belt of the ramp is moving can be found by eliminating x between Equations (1) and (2).

SOLUTION Equating the right hand sides of Equations (1) and (2), and noting that $t_{\text{Cliff}} = \tfrac{1}{4} t_{\text{belt}}$, we have

$$v_{\text{belt}} t_{\text{belt}} = \tfrac{1}{2} a \left(\tfrac{1}{4} t_{\text{belt}} \right)^2$$

$$v_{\text{belt}} = \tfrac{1}{32} a t_{\text{belt}} = \tfrac{1}{32}(0.37 \text{ m/s}^2)(64 \text{ s}) = \boxed{0.74 \text{ m/s}}$$

---

81. $\boxed{\text{SSM}}$ **REASONING**   Since the woman runs for a known distance at a known constant speed, we can find the time it takes for her to reach the water from Equation 2.1. We can then use Equation 2.1 to determine the total distance traveled by the dog in this time.

SOLUTION The time required for the woman to reach the water is

$$\text{Elapsed time} = \frac{d_{\text{woman}}}{v_{\text{woman}}} = \left(\frac{4.0 \text{ km}}{2.5 \text{ m/s}}\right)\left(\frac{1000 \text{ m}}{1.0 \text{ km}}\right) = 1600 \text{ s}$$

In 1600 s, the dog travels a total distance of

$$d_{\text{dog}} = v_{\text{dog}}t = (4.5 \text{ m/s})(1600 \text{ s}) = \boxed{7.2 \times 10^3 \text{ m}}$$

---

87. ***REASONING*** Since 1 mile = 1609 m, a quarter-mile race is $L = 402$ m long. If a car crosses the finish line before reaching its maximum speed, then there is only one interval of constant acceleration to consider. We will first determine whether this is true by calculating the car's displacement x_1 while accelerating from rest to top speed from Equation 2.9 $\left(v^2 = v_0^2 + 2ax\right)$, with $v_0 = 0$ m/s and $v = v_{\text{max}}$:

$$v_{\text{max}}^2 = (0 \text{ m/s})^2 + 2ax_1 \quad \text{or} \quad x_1 = \frac{v_{\text{max}}^2}{2a} \qquad (1)$$

If $x_1 > L$, then the car crosses the finish line before reaching top speed, and the total time for its race is found from Equation 2.8 $\left(x = v_0 t + \frac{1}{2}at^2\right)$, with $x = L$ and $v_0 = 0$ m/s:

$$L = (0 \text{ m/s})t + \frac{1}{2}at^2 = \frac{1}{2}at^2 \quad \text{or} \quad t = \sqrt{\frac{2L}{a}} \qquad (2)$$

On the other hand, if a car reaches its maximum speed before crossing the finish line, the race divides into two intervals, each with a different constant acceleration. The displacement x_1 is found as given in Equation (1), but the time t_1 to reach the maximum speed is most easily found from Equation 2.4 $\left(v = v_0 + at\right)$, with $v_0 = 0$ m/s and $v = v_{\text{max}}$:

$$v_{\text{max}} = 0 \text{ m/s} + at_1 \quad \text{or} \quad t_1 = \frac{v_{\text{max}}}{a} \qquad (3)$$

The time t_2 that elapses during the rest of the race is found by solving Equation 2.8 $\left(x = v_0 t + \frac{1}{2}at^2\right)$. Let $x_2 = L - x_1$ represent the displacement for this part of the race. With the aid of Equation (1), this becomes $x_2 = L - \frac{v_{\text{max}}^2}{2a}$. Then, since the car is at its maximum speed, the acceleration is $a = 0$ m/s², and the displacement is

$$x_2 = v_{\text{max}}t_2 + \frac{1}{2}\left(0 \text{ m/s}^2\right)t_2^2 = v_{\text{max}}t_2 \quad \text{or} \quad t_2 = \frac{x_2}{v_{\text{max}}} = \frac{L - \frac{v_{\text{max}}^2}{2a}}{v_{\text{max}}} = \frac{L}{v_{\text{max}}} - \frac{v_{\text{max}}}{2a} \qquad (4)$$

Using this expression for t_2 and Equation (3) for t_1 gives the total time for a two-part race:

$$t = t_1 + t_2 = \frac{v_{max}}{a} + \left(\frac{L}{v_{max}} - \frac{v_{max}}{2a}\right) = \frac{L}{v_{max}} + \frac{v_{max}}{2a} \qquad (5)$$

SOLUTION First, we use Equation (1) to determine whether either car finishes the race while accelerating:

Car A
$$x_1 = \frac{v_{max}^2}{2a} = \frac{(106 \text{ m/s})^2}{2(11.0 \text{ m/s}^2)} = 511 \text{ m}$$

Car B
$$x_1 = \frac{v_{max}^2}{2a} = \frac{(92.4 \text{ m/s})^2}{2(11.6 \text{ m/s}^2)} = 368 \text{ m}$$

Therefore, car A finishes the race before reaching its maximum speed, but car B has 402 m − 368 m = 34 m to travel at its maximum speed. Equation (2) gives the time for car A to reach the finish line as

Car A
$$t = \sqrt{\frac{2L}{a}} = \sqrt{\frac{2(402 \text{ m})}{11.0 \text{ m/s}^2}} = 8.55 \text{ s}$$

Equation (5) gives the time for car B to reach the finish line as

Car B
$$t = \frac{L}{v_{max}} + \frac{v_{max}}{2a} = \frac{402 \text{ m}}{92.4 \text{ m/s}} + \frac{92.4 \text{ m/s}}{2(11.6 \text{ m/s}^2)} = 8.33 \text{ s}$$

Car B wins the race by 8.55 s − 8.33 s = 0.22 s .

---

91. **SSM** **CONCEPTS** (i) Because the dragster has an acceleration of 40.0 m/s², its velocity changes by 40.0 m/s during each second of the travel. Therefore, *since the dragster starts from rest*, the velocity is 40.0 m/s at the end of the first second, 2 × 40.0 m/s at the end of the second second, 3 × 40.0 m/s at the end of the third second, and so on. Thus, when the time doubles, the velocity also doubles. (Be sure to note that this is only true if the initial velocity is equal to zero.)

(ii) The displacement of the dragster is equal to its average velocity multiplied by the elapsed time. The average velocity $\overline{v}$ is just one-half the sum of the initial and final velocities, or $\overline{v} = \frac{1}{2}(v_0 + v)$. Since the initial velocity is zero, $v_0 = 0$ m/s and the average velocity is just one-half the final velocity, or $\overline{v} = \frac{1}{2}v$. However, as we have seen, the final velocity is proportional to the elapsed time, since when the time doubles, the final velocity also doubles.

Therefore, the displacement, being the product of the average velocity and the time, is proportional to the time squared, or t^2. Consequently, as the time doubles, the displacement does not double, but increases by a factor of four. (Again, this is the case since the initial velocity is equal to zero.)

CALCULATIONS (a) According to Equation 2.4, the final velocity v, the initial velocity v_0, the acceleration a, and the elapsed time t are related by $v = v_0 + at$. The final velocities at the two times are:

$$[t = 2.0 \text{ s}] \quad v = v_0 + at = 0 \text{ m/s} + (40.0 \text{ m/s}^2)(2.0 \text{ s}) = \boxed{80 \text{ m/s}}$$

$$[t = 4.0 \text{ s}] \quad v = v_0 + at = 0 \text{ m/s} + (40.0 \text{ m/s}^2)(4.0 \text{ s}) = \boxed{160 \text{ m/s}}$$

We see that the velocity doubles when the time doubles, as expected.

(b) The displacement x is equal to the average velocity multiplied by the time, so

$$x = \underbrace{\tfrac{1}{2}(v_0 + v)}_{\text{Average velocity}} t = \tfrac{1}{2}vt$$

where we have used the fact that $v_0 = 0$ m/s. According to Equation 2.4, the final velocity is related to the acceleration by $v = v_0 + at$, or $v = at$, since $v_0 = 0$ m/s. Therefore, the displacement can be written as $x = \tfrac{1}{2}vt = \tfrac{1}{2}(at)t = \tfrac{1}{2}at^2$. The displacements at the two times are then

$$[t = 2.0 \text{ s}] \quad x = \tfrac{1}{2}at^2 = \tfrac{1}{2}(40.0 \text{ m/s}^2)(2.0 \text{ s})^2 = \boxed{80 \text{ m}}$$

$$[t = 4.0 \text{ s}] \quad x = \tfrac{1}{2}at^2 = \tfrac{1}{2}(40.0 \text{ m/s}^2)(4.0 \text{ s})^2 = \boxed{320 \text{ m}}$$

As predicted, the displacement at $t = 4.0$ s is four times that at $t = 2.0$ s.

CHAPTER 3 │ KINEMATICS IN TWO DIMENSIONS

1. **REASONING** The displacement is a vector drawn from the initial position to the final position. The magnitude of the displacement is the shortest distance between the positions. Note that it is only the initial and final positions that determine the displacement. The fact that the squirrel jumps to an intermediate position before reaching his final position is not important. The trees are perfectly straight and both growing perpendicular to the flat horizontal ground beneath them. Thus, the distance between the trees and the length of the trunk of the second tree below the squirrel's final landing spot form the two perpendicular sides of a right triangle, as the drawing shows. To this triangle, we can apply the Pythagorean theorem and determine the magnitude A of the displacement vector **A**.

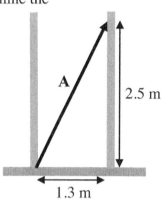

SOLUTION According to the Pythagorean theorem, we have

$$A = \sqrt{(1.3\ \text{m})^2 + (2.5\ \text{m})^2} = \boxed{2.8\ \text{m}}$$

5. **REASONING** The displacement of the elephant seal has two components; 460 m due east and 750 m downward. These components are mutually perpendicular; hence, the Pythagorean theorem can be used to determine their resultant.

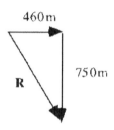

SOLUTION From the Pythagorean theorem,

$$R^2 = (460\ \text{m})^2 + (750\ \text{m})^2$$

Therefore,

$$R = \sqrt{(460\ \text{m})^2 + (750\ \text{m})^2} = \boxed{8.8 \times 10^2\ \text{m}}$$

7. **REASONING AND SOLUTION**

$$x = r \cos\theta = (162\ \text{km}) \cos 62.3° = \boxed{75.3\ \text{km}}$$

$$y = r \sin\theta = (162\ \text{km}) \sin 62.3° = \boxed{143\ \text{km}}$$

9. **REASONING**

a. We designate the direction down and parallel to the ramp as the $+x$ direction, and the table shows the variables that are known. Since three of the five kinematic variables have values, one of the equations of kinematics can be employed to find the acceleration a_x.

x-Direction Data

x	a_x	v_x	v_{0x}	t
+12.0 m	?	+7.70 m/s	0 m/s	

b. The acceleration vector points down and parallel to the ramp, and the angle of the ramp is 25.0° relative to the ground (see the drawing). Therefore, trigonometry can be used to determine the component $a_{parallel}$ of the acceleration that is parallel to the ground.

SOLUTION

a. Equation 3.6a $\left(v_x^2 = v_{0x}^2 + 2a_x x\right)$ can be used to find the acceleration in terms of the three known variables. Solving this equation for a_x gives

$$a_x = \frac{v_x^2 - v_{0x}^2}{2x} = \frac{\left(+7.70 \text{ m/s}\right)^2 - \left(0 \text{ m/s}\right)^2}{2\left(+12.0 \text{ m}\right)} = \boxed{2.47 \text{ m/s}^2}$$

b. The drawing shows that the acceleration vector is oriented 25.0° relative to the ground. The component $a_{parallel}$ of the acceleration that is parallel to the ground is

$$a_{parallel} = a_x \cos 25.0° = \left(2.47 \text{ m/s}^2\right)\cos 25.0° = \boxed{2.24 \text{ m/s}^2}$$

---

13. **REASONING** The vertical component of the ball's velocity $\mathbf{v_0}$ changes as the ball approaches the opposing player. It changes due to the acceleration of gravity. However, the horizontal component does not change, assuming that air resistance can be neglected. Hence, the horizontal component of the ball's velocity when the opposing player fields the ball is the same as it was initially.

SOLUTION Using trigonometry, we find that the horizontal component is

$$v_x = v_0 \cos\theta = \left(15 \text{ m/s}\right)\cos 55° = \boxed{8.6 \text{ m/s}}$$

---

17. **REASONING** Since the spider encounters no appreciable air resistance during its leap, it can be treated as a projectile. The thickness of the magazine is equal to the spider's vertical

displacement y during the leap. The relevant data are as follows (assuming upward to be the $+y$ direction):

y-Direction Data

y	a_y	v_y	v_{0y}	t
?	−9.80 m/s^2		(0.870 m/s) sin 35.0° = +0.499 m/s	0.0770 s

SOLUTION We will calculate the spider's vertical displacement directly from $y = v_{0y}t + \frac{1}{2}a_y t^2$ (Equation 3.5b):

$$y = (0.499 \text{ m/s})(0.0770 \text{ s}) + \frac{1}{2}(-9.80 \text{ m/s}^2)(0.0770 \text{ s})^2 = 0.0094 \text{ m}$$

To express this result in millimeters, we use the fact that 1 m equals 1000 mm:

$$\text{Thickness} = (0.0094 \text{ m})\left(\frac{1000 \text{ mm}}{1 \text{ m}}\right) = \boxed{9.4 \text{ mm}}$$

21. **REASONING** When the skier leaves the ramp, she exhibits projectile motion. Since we know the maximum height attained by the skier, we can find her launch speed v_0 using Equation 3.6b, $v_y^2 = v_{0y}^2 + 2a_y y$, where $v_{0y} = v_0 \sin 63°$.

SOLUTION At the highest point in her trajectory, $v_y = 0$. Solving Equation 3.6b for v_{0y} we obtain, taking upward as the positive direction,

$$v_{0y} = v_0 \sin 63° = \sqrt{-2a_y y} \quad \text{or} \quad v_0 = \frac{\sqrt{-2a_y y}}{\sin 63°} = \frac{\sqrt{-2(-9.80 \text{ m/s}^2)(13 \text{ m})}}{\sin 63°} = \boxed{18 \text{ m/s}}$$

23. **REASONING**   Since the magnitude of the velocity of the fuel tank is given by $v = \sqrt{v_x^2 + v_y^2}$, it is necessary to know the velocity components v_x and v_y just before impact. At the instant of release, the empty fuel tank has the same velocity as that of the plane. Therefore, the magnitudes of the initial velocity components of the fuel tank are given by $v_{0x} = v_0 \cos\theta$ and $v_{0y} = v_0 \sin\theta$, where v_0 is the speed of the plane at the instant of release. Since the x motion has zero acceleration, the x component of the velocity of the plane remains equal to v_{0x} for all later times while the tank is airborne. The y component of the velocity of the tank after it has undergone a vertical displacement y is given by Equation 3.6b.

SOLUTION

a. Taking up as the positive direction, the velocity components of the fuel tank just before it hits the ground are

$$v_x = v_{0x} = v\cos\theta = (135 \text{ m/s}) \cos 15° = 1.30 \times 10^2 \text{ m/s}$$

From Equation 3.6b, we have

$$v_y = -\sqrt{v_{0y}^2 + 2a_y y} = -\sqrt{(v_0 \sin\theta)^2 + 2a_y y}$$

$$= -\sqrt{\left[(135 \text{ m/s}) \sin 15.0°\right]^2 + \left[2(-9.80 \text{ m/s}^2)(-2.00 \times 10^3 \text{ m})\right]} = -201 \text{ m/s}$$

Therefore, the magnitude of the velocity of the fuel tank just before impact is

$$v = \sqrt{v_x^2 + v_y^2} = \sqrt{(1.30 \times 10^2 \text{ m/s})^2 + (201 \text{ m/s})^2} = \boxed{239 \text{ m/s}}$$

The velocity vector just before impact is inclined at an angle ϕ with respect to the horizontal. This angle is

$$\phi = \tan^{-1}\left(\frac{201 \text{ m/s}}{1.30 \times 10^2 \text{ m/s}}\right) = \boxed{57.1°}$$

b. As shown in Conceptual Example 10, once the fuel tank in part a rises and falls to the same altitude at which it was released, its motion is identical to the fuel tank in part b. Therefore, the velocity of the fuel tank in part b just before impact is $\boxed{239 \text{ m/s at an angle of } 57.1° \text{ with respect to the horizontal}}$.

---

27. **REASONING AND SOLUTION** The water exhibits projectile motion. The x component of the motion has zero acceleration while the y component is subject to the acceleration due to gravity. In order to reach the highest possible fire, the displacement of the hose from the building is x, where, according to Equation 3.5a (with $a_x = 0 \text{ m/s}^2$),

$$x = v_{0x}t = (v_0 \cos\theta)t$$

with t equal to the time required for the water the reach its maximum vertical displacement. The time t can be found by considering the vertical motion. From Equation 3.3b,

$$v_y = v_{0y} + a_y t$$

When the water has reached its maximum vertical displacement, $v_y = 0 \text{ m/s}$. Taking up and to the right as the positive directions, we find that

$$t = \frac{-v_{0y}}{a_y} = \frac{-v_0 \sin\theta}{a_y}$$

and

$$x = (v_0 \cos\theta)\left(\frac{-v_0 \sin\theta}{a_y}\right)$$

Therefore, we have

$$x = -\frac{v_0^2 \cos\theta \sin\theta}{a_y} = -\frac{(25.0 \text{ m/s})^2 \cos 35.0° \sin 35.0°}{-9.80 \text{ m/s}^2} = \boxed{30.0 \text{ m}}$$

---

31. ***REASONING*** The speed of the fish at any time t is given by $v = \sqrt{v_x^2 + v_y^2}$, where v_x and v_y are the x and y components of the velocity at that instant. Since the horizontal motion of the fish has zero acceleration, $v_x = v_{0x}$ for all times t. Since the fish is dropped by the eagle, v_{0x} is equal to the horizontal speed of the eagle and $v_{0y} = 0$. The y component of the velocity of the fish for any time t is given by Equation 3.3b with $v_{0y} = 0$. Thus, the speed at any time t is given by $v = \sqrt{v_{0x}^2 + (a_y t)^2}$.

SOLUTION

a. The initial speed of the fish is $v_0 = \sqrt{v_{0x}^2 + v_{0y}^2} = \sqrt{v_{0x}^2 + 0^2} = v_{0x}$. When the fish's speed doubles, $v = 2v_{0x}$. Therefore,

$$2v_{0x} = \sqrt{v_{0x}^2 + (a_y t)^2} \qquad \text{or} \qquad 4v_{0x}^2 = v_{0x}^2 + (a_y t)^2$$

Assuming that downward is positive and solving for t, we have

$$t = \sqrt{3}\frac{v_{0x}}{a_y} = \sqrt{3}\left(\frac{6.0 \text{ m/s}}{9.80 \text{ m/s}^2}\right) = \boxed{1.1 \text{ s}}$$

b. When the fish's speed doubles again, $v = 4v_{0x}$. Therefore,

$$4v_{0x} = \sqrt{v_{0x}^2 + (a_y t)^2} \qquad \text{or} \qquad 16v_{0x}^2 = v_{0x}^2 + (a_y t)^2$$

Solving for t, we have

$$t = \sqrt{15}\frac{v_{0x}}{a_y} = \sqrt{15}\left(\frac{6.0 \text{ m/s}}{9.80 \text{ m/s}^2}\right) = 2.37 \text{ s}$$

Therefore, the additional time for the speed to double again is $(2.4 \text{ s}) - (1.1 \text{ s}) = \boxed{1.3 \text{ s}}$.

37. ***REASONING***

a. The drawing shows the initial velocity v_0 of the package when it is released. The initial speed of the package is 97.5 m/s. The component of its displacement along the ground is labeled as x. The data for the x direction are indicated in the data table below.

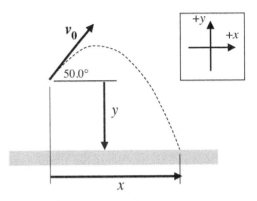

x-Direction Data

x	a_x	v_x	v_{0x}	t
?	0 m/s^2		+(97.5 m/s) cos 50.0° = +62.7 m/s	

Since only two variables are known, it is not possible to determine x from the data in this table. A value for a third variable is needed. We know that the time of flight t is the same for both the x and y motions, so let's now look at the data in the y direction.

y-Direction Data

y	a_y	v_y	v_{0y}	t
−732 m	−9.80 m/s^2		+(97.5 m/s) sin 50.0° = +74.7 m/s	?

Note that the displacement y of the package points from its initial position toward the ground, so its value is negative, i.e., $y = -732$ m. The data in this table, along with the appropriate equation of kinematics, can be used to find the time of flight t. This value for t can, in turn, be used in conjunction with the x-direction data to determine x.

b. The drawing at the right shows the velocity of the package just before impact. The angle that the velocity makes with respect to the ground can be found from the inverse tangent function as $\theta = \tan^{-1}\left(v_y / v_x\right)$. Once the time has been found in part (a), the values of v_y and v_x can be determined from the data in the tables and the appropriate equations of kinematics.

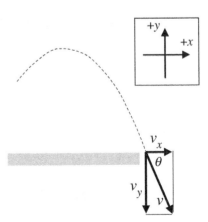

SOLUTION

a. To determine the time that the package is in the air, we

will use Equation 3.5b $\left(y = v_{0y}t + \frac{1}{2}a_y t^2\right)$ and the data in the y-direction data table. Solving this quadratic equation for the time yields

$$t = \frac{-v_{0y} \pm \sqrt{v_{0y}^2 - 4\left(\frac{1}{2}a_y\right)(-y)}}{2\left(\frac{1}{2}a_y\right)}$$

$$t = \frac{-(74.7 \text{ m/s}) \pm \sqrt{(74.7 \text{ m/s})^2 - 4\left(\frac{1}{2}\right)(-9.80 \text{ m/s}^2)(732 \text{ m})}}{2\left(\frac{1}{2}\right)(-9.80 \text{ m/s}^2)} = -6.78 \text{ s} \quad \text{and} \quad 22.0 \text{ s}$$

We discard the first solution, since it is a negative value and, hence, unrealistic. The displacement x can be found using $t = 22.0$ s, the data in the x-direction data table, and Equation 3.5a:

$$x = v_{0x}t + \frac{1}{2}a_x t^2 = (+62.7 \text{ m/s})(22.0 \text{ s}) + \underbrace{\frac{1}{2}(0 \text{ m/s}^2)(22.0 \text{ s})^2}_{= 0} = \boxed{+1380 \text{ m}}$$

b. The angle θ that the velocity of the package makes with respect to the ground is given by $\theta = \tan^{-1}(v_y / v_x)$. Since there is no acceleration in the x direction ($a_x = 0$ m/s²), v_x is the same as v_{0x}, so that $v_x = v_{0x} = +62.7$ m/s. Equation 3.3b can be employed with the y-direction data to find v_y:

$$v_y = v_{0y} + a_y t = +74.7 \text{ m/s} + \left(-9.80 \text{ m/s}^2\right)(22.0 \text{ s}) = -141 \text{ m/s}$$

Therefore,

$$\theta = \tan^{-1}\left(\frac{v_y}{v_x}\right) = \tan^{-1}\left(\frac{-141 \text{ m/s}}{+62.7 \text{ m/s}}\right) = -66.0°$$

where the minus sign indicates that the angle is $\boxed{66.0° \text{ below the horizontal}}$.

---

43. **REASONING** The angle θ can be found from

$$\theta = \tan^{-1}\left(\frac{2400 \text{ m}}{x}\right) \tag{1}$$

where x is the horizontal displacement of the flare. Since $a_x = 0 \text{m/s}^2$, it follows that $x = (v_0 \cos 30.0°)t$. The flight time t is determined by the vertical motion. In particular, the time t can be found from Equation 3.5b. Once the time is known, x can be calculated.

SOLUTION From Equation 3.5b, assuming upward is the positive direction, we have

$$y = -(v_0 \sin 30.0°)t + \frac{1}{2}a_y t^2$$

which can be rearranged to give the following equation that is quadratic in t:

$$\tfrac{1}{2}a_y t^2 - (v_0 \sin 30.0^\circ)t - y = 0$$

Using $y = -2400$ m and $a_y = -9.80$ m/s^2 and suppressing the units, we obtain the quadratic equation

$$4.9t^2 + 120t - 2400 = 0$$

Using the quadratic formula, we obtain $t = 13$ s. Therefore, we find that

$$x = (v_0 \cos 30.0^\circ)t = (240 \text{ m/s})(\cos 30.0^\circ)(13 \text{ s}) = 2700 \text{ m}$$

Equation (1) then gives

$$\theta = \tan^{-1}\left(\frac{2400 \text{ m}}{2700 \text{ m}}\right) = \boxed{42^\circ}$$

---

47. ***REASONING AND SOLUTION*** In the absence of air resistance, the bullet exhibits projectile motion. The x component of the motion has zero acceleration while the y component of the motion is subject to the acceleration due to gravity. The horizontal distance traveled by the bullet is given by Equation 3.5a (with $a_x = 0$ m/s^2):

$$x = v_{0x}t = (v_0 \cos \theta)t$$

with t equal to the time required for the bullet to reach the target. The time t can be found by considering the vertical motion. From Equation 3.3b,

$$v_y = v_{0y} + a_y t$$

When the bullet reaches the target, $v_y = -v_{0y}$. Assuming that up and to the right are the positive directions, we have

$$t = \frac{-2v_{0y}}{a_y} = \frac{-2v_0 \sin \theta}{a_y} \qquad \text{and} \qquad x = (v_0 \cos \theta)\left(\frac{-2v_0 \sin \theta}{a_y}\right)$$

Using the fact that $2\sin \theta \cos \theta = \sin 2\theta$, we have

$$x = -\frac{2v_0^2 \cos \theta \sin \theta}{a_y} = -\frac{v_0^2 \sin 2\theta}{a_y}$$

Thus, we find that

$$\sin 2\theta = -\frac{x\,a_y}{v_0^2} = -\frac{(91.4 \text{ m})\,(-9.80 \text{ m/s}^2)}{(427 \text{ m/s})^2} = 4.91 \times 10^{-3}$$

and

$$2\theta = 0.281° \quad \text{or} \quad 2\theta = 180.000° - 0.281° = 179.719°$$

Therefore,

$$\theta = \boxed{0.141° \text{ and } 89.860°}$$

---

49. **REASONING**   Since the horizontal motion is not accelerated, we know that the x component of the velocity remains constant at 340 m/s. Thus, we can use Equation 3.5a (with $a_x = 0\text{m/s}^2$) to determine the time that the bullet spends in the building before it is embedded in the wall. Since we know the vertical displacement of the bullet after it enters the building, we can use the flight time in the building and Equation 3.5b to find the y component of the velocity of the bullet as it enters the window. Then, Equation 3.6b can be used (with $v_{0y} = 0\text{m/s}$) to determine the vertical displacement y of the bullet as it passes between the buildings. We can determine the distance H by adding the magnitude of y to the vertical distance of 0.50 m within the building.

Once we know the vertical displacement of the bullet as it passes between the buildings, we can determine the time t_1 required for the bullet to reach the window using Equation 3.4b. Since the motion in the x direction is not accelerated, the distance D can then be found from $D = v_{0x}t_1$.

SOLUTION   Assuming that the direction to the right is positive, we find that the time that the bullet spends in the building is (according to Equation 3.5a)

$$t = \frac{x}{v_{0x}} = \frac{6.9 \text{ m}}{340 \text{ m/s}} = 0.0203 \text{ s}$$

The vertical displacement of the bullet after it enters the building is, taking down as the negative direction, equal to –0.50 m. Therefore, the vertical component of the velocity of the bullet as it passes through the window is, from Equation 3.5b,

$$v_{0y(\text{window})} = \frac{y - \frac{1}{2}a_y t^2}{t} = \frac{y}{t} - \frac{1}{2}a_y t = \frac{-0.50 \text{ m}}{0.0203 \text{ s}} - \frac{1}{2}(-9.80 \text{ m/s}^2)(0.0203 \text{ s}) = -24.5 \text{ m/s}$$

The vertical displacement of the bullet as it travels between the buildings is (according to Equation 3.6b with $v_{0y} = 0\text{m/s}$)

$$y = \frac{v_y^2}{2a_y} = \frac{(-24.5 \text{ m/s})^2}{2(-9.80 \text{ m/s}^2)} = -30.6 \text{ m}$$

Therefore, the distance H is

$$H = 30.6 \text{ m} + 0.50 \text{ m} = \boxed{31 \text{ m}}$$

The time for the bullet to reach the window, according to Equation 3.4b, is

$$t_1 = \frac{2y}{v_{0y} + v_y} = \frac{2y}{v_y} = \frac{2(-30.6 \text{ m})}{(-24.5 \text{ m/s})} = 2.50 \text{ s}$$

Hence, the distance D is given by

$$D = v_{0x} t_1 = (340 \text{ m/s})(2.50 \text{ s}) = \boxed{850 \text{ m}}$$

---

53. **REASONING** The velocity $\mathbf{v}_{SG}$ of the swimmer relative to the ground is the vector sum of the velocity $\mathbf{v}_{SW}$ of the swimmer relative to the water and the velocity $\mathbf{v}_{WG}$ of the water relative to the ground as shown at the right: $\mathbf{v}_{SG} = \mathbf{v}_{SW} + \mathbf{v}_{WG}$.

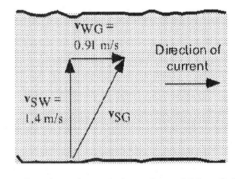

The component of $\mathbf{v}_{SG}$ that is parallel to the width of the river determines how fast the swimmer is moving across the river; this parallel component is v_{SW}. The time for the swimmer to cross the river is equal to the width of the river divided by the magnitude of this velocity component.

The component of $\mathbf{v}_{SG}$ that is parallel to the direction of the current determines how far the swimmer is carried down stream; this component is v_{WG}. Since the motion occurs with constant velocity, the distance that the swimmer is carried downstream while crossing the river is equal to the magnitude of v_{WG} multiplied by the time it takes for the swimmer to cross the river.

SOLUTION
a. The time t for the swimmer to cross the river is

$$t = \frac{\text{width}}{v_{SW}} = \frac{2.8 \times 10^3 \text{ m}}{1.4 \text{ m/s}} = \boxed{2.0 \times 10^3 \text{ s}}$$

b. The distance x that the swimmer is carried downstream while crossing the river is

$$x = v_{WG}t = (0.91 \text{ m/s})(2.0 \times 10^3 \text{ s}) = \boxed{1.8 \times 10^3 \text{ m}}$$

---

59. **REASONING** The velocity v_{AB} of train **A** relative to train **B** is the vector sum of the velocity v_{AG} of train **A** relative to the ground and the velocity v_{GB} of the ground relative to train **B**, as indicated by Equation 3.7: $v_{AB} = v_{AG} + v_{GB}$. The values of v_{AG} and v_{BG} are given in the statement of the problem. We must also make use of the fact that $v_{GB} = -v_{BG}$.

SOLUTION
a. Taking east as the positive direction, the velocity of **A** relative to **B** is, according to Equation 3.7,

$$v_{AB} = v_{AG} + v_{GB} = v_{AG} - v_{BG} = (+13 \text{ m/s}) - (-28 \text{ m/s}) = \boxed{+41 \text{ m/s}}$$

The positive sign indicates that the direction of v_{AB} is $\boxed{\text{due east}}$.

b. Similarly, the velocity of **B** relative to **A** is

$$v_{BA} = v_{BG} + v_{GA} = v_{BG} - v_{AG} = (-28 \text{ m/s}) - (+13 \text{ m/s}) = \boxed{-41 \text{ m/s}}$$

The negative sign indicates that the direction of v_{BA} is $\boxed{\text{due west}}$.

---

63. **REASONING** The velocity v_{PM} of the puck relative to Mario is the vector sum of the velocity v_{PI} of the puck relative to the ice and the velocity v_{IM} of the ice relative to Mario as indicated by Equation 3.7: $v_{PM} = v_{PI} + v_{IM}$. The values of v_{MI} and v_{PI} are given in the statement of the problem. In order to use the data, we must make use of the fact that $v_{IM} = -v_{MI}$, with the result that $v_{PM} = v_{PI} - v_{MI}$.

SOLUTION The first two rows of the following table give the east/west and north/south components of the vectors v_{PI} and $-v_{MI}$. The third row gives the components of their resultant $v_{PM} = v_{PI} - v_{MI}$. Due east and due north have been taken as positive.

Vector	East/West Component	North/South Component
$\mathbf{v}_{PI}$	$-(11.0 \text{ m/s}) \sin 22° = -4.1 \text{ m/s}$	$-(11.0 \text{ m/s}) \cos 22° = -10.2 \text{ m/s}$
$-\mathbf{v}_{MI}$	0	$+7.0 \text{ m/s}$
$\mathbf{v}_{PM} = \mathbf{v}_{PI} - \mathbf{v}_{MI}$	-4.1 m/s	-3.2 m/s

Now that the components of $\mathbf{v}_{PM}$ are known, the Pythagorean theorem can be used to find the magnitude.

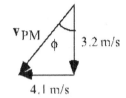

$$v_{PM} = \sqrt{(-4.1 \text{ m/s})^2 + (-3.2 \text{ m/s})^2} = \boxed{5.2 \text{ m/s}}$$

The direction of $\mathbf{v}_{PM}$ is found from

$$\phi = \tan^{-1}\left(\frac{4.1 \text{ m/s}}{3.2 \text{ m/s}}\right) = \boxed{52° \text{ west of south}}$$

65. **REASONING** The relative velocities in this problem are:

$\mathbf{v}_{PS}$ = velocity of the **P**assenger relative to the **S**hore
$\mathbf{v}_{P2}$ = velocity of the **P**assenger relative to Boat **2** (1.20 m/s, due east)
$\mathbf{v}_{2S}$ = velocity of Boat **2** relative to the **S**hore
$\mathbf{v}_{21}$ = velocity of Boat **2** relative to Boat **1** (1.60 m/s, at 30.0° north of east)
$\mathbf{v}_{1S}$ = velocity of Boat **1** relative to the **S**hore (3.00 m/s, due north)

The velocity $\mathbf{v}_{PS}$ of the passenger relative to the shore is related to $\mathbf{v}_{P2}$ and $\mathbf{v}_{2S}$ by (see the method of subscripting discussed in Section 3.4):

$$\mathbf{v}_{PS} = \mathbf{v}_{P2} + \mathbf{v}_{2S}$$

But $\mathbf{v}_{2S}$, the velocity of Boat 2 relative to the shore, is related to $\mathbf{v}_{21}$ and $\mathbf{v}_{1S}$ by

$$\mathbf{v}_{2S} = \mathbf{v}_{21} + \mathbf{v}_{1S}$$

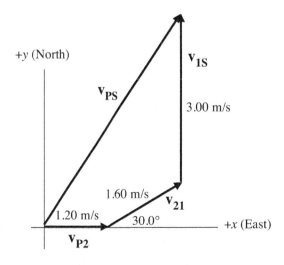

Substituting this expression for $\mathbf{v}_{2S}$ into the first equation yields

$$\mathbf{v}_{PS} = \mathbf{v}_{P2} + \mathbf{v}_{21} + \mathbf{v}_{1S}$$

This vector sum is shown in the diagram. We will determine the magnitude of $\mathbf{v}_{PS}$ from the equation above by using the method of scalar components.

SOLUTION The following table lists the scalar components of the four vectors in the drawing.

Vector	x Component	y Component
$\mathbf{v}_{P2}$	+1.20 m/s	0 m/s
$\mathbf{v}_{21}$	+(1.60 m/s) cos 30.0° = +1.39 m/s	+(1.60 m/s) sin 30.0° = +0.80 m/s
$\mathbf{v}_{1S}$	0 m/s	+3.00 m/s
$\mathbf{v}_{PS} = \mathbf{v}_{P2} + \mathbf{v}_{21} + \mathbf{v}_{1S}$	+1.20 m/s + 1.39 m/s = +2.59 m/s	+0.80 m/s +3.00 m/s = +3.80 m/s

The magnitude of $\mathbf{v}_{PS}$ can be found by applying the Pythagorean theorem to its x and y components:

$$v_{PS} = \sqrt{(2.59 \text{ m/s})^2 + (3.80 \text{ m/s})^2} = \boxed{4.60 \text{ m/s}}$$

---

71. **REASONING**   Once the diver is airborne, he moves in the x direction with constant velocity while his motion in the y direction is accelerated (at the acceleration due to gravity). Therefore, the magnitude of the x component of his velocity remains constant at 1.20 m/s for all times t. The magnitude of the y component of the diver's velocity after he has fallen through a vertical displacement y can be determined from Equation 3.6b: $v_y^2 = v_{0y}^2 + 2a_y y$. Since the diver runs off the platform horizontally, $v_{0y} = 0$ m/s. Once the x and y components of the velocity are known for a particular vertical displacement y, the speed of the diver can be obtained from $v = \sqrt{v_x^2 + v_y^2}$.

SOLUTION   For convenience, we will take downward as the positive y direction. After the diver has fallen 10.0 m, the y component of his velocity is, from Equation 3.6b,

$$v_y = \sqrt{v_{0y}^2 + 2a_y y} = \sqrt{0^2 + 2(9.80 \text{ m/s}^2)(10.0 \text{ m})} = 14.0 \text{ m/s}$$

Therefore,

$$v = \sqrt{v_x^2 + v_y^2} = \sqrt{(1.20 \text{ m/s})^2 + (14.0 \text{ m/s})^2} = \boxed{14.1 \text{ m/s}}$$

75. **REASONING** The horizontal distance covered by stone **1** is equal to the distance covered by stone **2** after it passes point **P** in the following diagram. Thus, the distance Δx between the points where the stones strike the ground is equal to x_2, the horizontal distance covered by stone **2** when it reaches **P**. In the diagram, we assume up and to the right are positive.

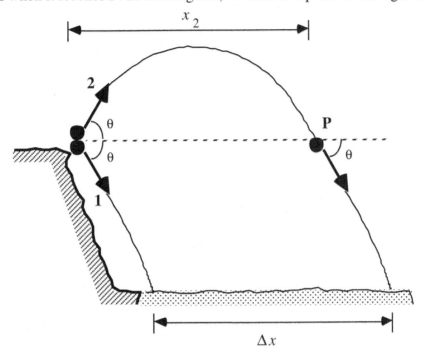

SOLUTION If t_P is the time required for stone **2** to reach **P**, then

$$x_2 = v_{0x}t_P = (v_0 \cos\theta)t_P$$

For the vertical motion of stone **2**, $v_y = v_0 \sin\theta + a_y t$. Solving for t gives

$$t = \frac{v_y - v_0 \sin\theta}{a_y}$$

When stone **2** reaches **P**, $v_y = -v_0 \sin\theta$, so the time required to reach **P** is

$$t_P = \frac{-2v_0 \sin\theta}{a_y}$$

Then,

$$x_2 = v_{0x}t_P = (v_0 \cos\theta)\left(\frac{-2v_0 \sin\theta}{a_y}\right)$$

$$x_2 = \frac{-2v_0^2 \sin\theta \, \cos\theta}{a_y} = \frac{-2(13.0 \text{ m/s})^2 \sin\ 30.0° \cos 30.0°}{-9.80 \text{ m/s}^2} = \boxed{14.9 \text{ m}}$$

81.   **SSM** *CONCEPTS* **(i)** The y components of their velocities are equal: $v_{0y} = 10.0$ m/s Note that the initial velocity of the blue ball has no x component. **(ii)** The x components of their velocities are equal: $v_{0x} = 4.6$ m/s. Note that the initial velocity of the yellow ball has no y component.

CALCULATIONS Based on the concept questions and answers, we know that the x and y components of the red ball are $v_{0x} = 4.6$ m/s and $v_{0y} = 10.0$ m/s respectively.

The initial speed (launch speed) v_0 is the magnitude of the initial velocity, and it, along with the directional angle θ, can be determined from the components:

$$v_0 = \sqrt{\left(v_{0x}\right)^2 + \left(v_{0y}\right)^2} = \sqrt{\left(4.6 \text{ m/s}\right)^2 + \left(10.0 \text{ m/s}\right)^2} = \boxed{11 \text{ m/s}}$$

$$\theta = \tan^{-1}\left(\frac{v_{0y}}{v_{0x}}\right) = \tan^{-1}\left(\frac{10.0 \text{ m/s}}{4.6 \text{ m/s}}\right) = \boxed{65^\circ}$$

CHAPTER 4 | FORCES AND NEWTON'S LAWS OF MOTION

PROBLEMS

---

5. **REASONING** The magnitude ΣF of the net force acting on the kayak is given by Newton's second law as $\Sigma F = ma$ (Equation 4.1), where m is the combined mass of the person and kayak, and a is their acceleration. Since the initial and final velocities, v_0 and v, and the displacement x are known, we can employ one of the equations of kinematics from Chapter 2 to find the acceleration.

SOLUTION Solving Equation 2.9 $\left(v^2 = v_0^{\ 2} + 2ax\right)$ from the equations of kinematics for the acceleration, we have

$$a = \frac{v^2 - v_0^{\ 2}}{2x}$$

Substituting this result into Newton's second law gives

$$\Sigma F = ma = m\left(\frac{v^2 - v_0^{\ 2}}{2x}\right) = \left(73 \text{ kg}\right)\left[\frac{\left(0.60 \text{ m/s}\right)^2 - \left(0 \text{ m/s}\right)^2}{2\left(0.41 \text{ m}\right)}\right] = \boxed{32 \text{ N}}$$

---

7. **REASONING AND SOLUTION** The acceleration required is

$$a = \frac{v^2 - v_0^2}{2x} = \frac{-\left(15.0 \text{ m/s}\right)^2}{2\left(50.0 \text{ m}\right)} = -2.25 \text{ m/s}^2$$

Newton's second law then gives the magnitude of the net force as

$$F = ma = (1580 \text{ kg})(2.25 \text{ m/s}^2) = \boxed{3560 \text{ N}}$$

---

9. **REASONING** Let due east be chosen as the positive direction. Then, when both forces point due east, Newton's second law gives

$$\underbrace{F_A + F_B}_{\Sigma F} = ma_1 \tag{1}$$

where $a_1 = 0.50 \text{ m/s}^2$. When F_A points due east and F_B points due west, Newton's second law gives

$$\underbrace{F_A - F_B}_{\Sigma F} = ma_2 \tag{2}$$

where $a_2 = 0.40 \text{ m/s}^2$. These two equations can be used to find the magnitude of each force.

SOLUTION
a. Adding Equations 1 and 2 gives

$$F_A = \frac{m(a_1 + a_2)}{2} = \frac{(8.0 \text{ kg})(0.50 \text{ m/s}^2 + 0.40 \text{ m/s}^2)}{2} = \boxed{3.6 \text{ N}}$$

b. Subtracting Equation 2 from Equation 1 gives

$$F_B = \frac{m(a_1 - a_2)}{2} = \frac{(8.0 \text{ kg})(0.50 \text{ m/s}^2 - 0.40 \text{ m/s}^2)}{2} = \boxed{0.40 \text{ N}}$$

---

13. **REASONING** To determine the acceleration we will use Newton's second law $\Sigma\mathbf{F} = m\mathbf{a}$. Two forces act on the rocket, the thrust T and the rocket's weight W, which is $mg = (4.50 \times 10^5 \text{ kg})(9.80 \text{ m/s}^2) = 4.41 \times 10^6 \text{ N}$. Both of these forces must be considered when determining the net force $\Sigma\mathbf{F}$. The direction of the acceleration is the same as the direction of the net force.

SOLUTION In constructing the free-body diagram for the rocket we choose upward and to the right as the positive directions. The free-body diagram is as follows:
The x component of the net force is

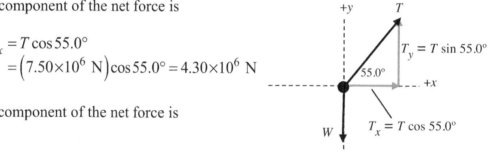

$$\Sigma F_x = T\cos 55.0°$$
$$= (7.50 \times 10^6 \text{ N})\cos 55.0° = 4.30 \times 10^6 \text{ N}$$

The y component of the net force is

$$\Sigma F_y = T\sin 55.0° - W = (7.50 \times 10^6 \text{ N})\sin 55.0° - 4.41 \times 10^6 \text{ N} = 1.73 \times 10^6 \text{ N}$$

The magnitudes of the net force and of the acceleration are

$$\Sigma F = \sqrt{(\Sigma F_x)^2 + (\Sigma F_y)^2}$$

$$a = \frac{\sqrt{(\Sigma F_x)^2 + (\Sigma F_y)^2}}{m} = \frac{\sqrt{(4.30 \times 10^6 \text{ N})^2 + (1.73 \times 10^6 \text{ N})^2}}{4.50 \times 10^5 \text{ kg}} = \boxed{10.3 \text{ m/s}^2}$$

The direction of the acceleration is the same as the direction of the net force. Thus, it is directed above the horizontal at an angle of

$$\theta = \tan^{-1}\left(\frac{\Sigma F_y}{\Sigma F_x}\right) = \tan^{-1}\left(\frac{1.73\times10^6 \text{ N}}{4.30\times10^6 \text{ N}}\right) = \boxed{21.9^\circ}$$

19. **REASONING** We first determine the acceleration of the boat. Then, using Newton's second law, we can find the net force $\Sigma\mathbf{F}$ that acts on the boat. Since two of the three forces are known, we can solve for the unknown force $\mathbf{F_W}$ once the net force $\Sigma\mathbf{F}$ is known.

SOLUTION Let the direction due east be the positive x direction and the direction due north be the positive y direction. The x and y components of the initial velocity of the boat are then

$$v_{0x} = (2.00 \text{ m/s}) \cos 15.0^\circ = 1.93 \text{ m/s}$$

$$v_{0y} = (2.00 \text{ m/s}) \sin 15.0^\circ = 0.518 \text{ m/s}$$

Thirty seconds later, the x and y velocity components of the boat are

$$v_x = (4.00 \text{ m/s}) \cos 35.0^\circ = 3.28 \text{ m/s}$$

$$v_y = (4.00 \text{ m/s}) \sin 35.0^\circ = 2.29 \text{ m/s}$$

Therefore, according to Equations 3.3a and 3.3b, the x and y components of the acceleration of the boat are

$$a_x = \frac{v_x - v_{0x}}{t} = \frac{3.28 \text{ m/s} - 1.93 \text{ m/s}}{30.0 \text{ s}} = 4.50\times10^{-2} \text{ m/s}^2$$

$$a_y = \frac{v_y - v_{0y}}{t} = \frac{2.29 \text{ m/s} - 0.518 \text{ m/s}}{30.0 \text{ s}} = 5.91\times10^{-2} \text{ m/s}^2$$

Thus, the x and y components of the net force that act on the boat are

$$\Sigma F_x = ma_x = (325 \text{ kg})(4.50\times10^{-2} \text{ m/s}^2) = 14.6 \text{ N}$$

$$\Sigma F_y = ma_y = (325 \text{ kg})(5.91\times10^{-2} \text{ m/s}^2) = 19.2 \text{ N}$$

The following table gives the x and y components of the net force $\Sigma\mathbf{F}$ and the two known forces that act on the boat. The fourth row of that table gives the components of the unknown force $\mathbf{F_W}$.

Force	x-Component	y-Component
$\sum \mathbf{F}$	14.6 N	19.2 N
$\mathbf{F}_1$	$(31.0\ \text{N})\cos 15.0° = 29.9\ \text{N}$	$(31.0\ \text{N})\sin 15.0° = 8.02\ \text{N}$
$\mathbf{F}_2$	$-(23.0\ \text{N})\cos 15.0° = -22.2\ \text{N}$	$-(23.0\ \text{N})\sin 15.0° = -5.95\ \text{N}$
$\mathbf{F}_\text{W} = \sum \mathbf{F} - \mathbf{F}_1 - \mathbf{F}_2$	$14.6\ \text{N} - 29.9\ \text{N} + 22.2\ \text{N} = 6.9\ \text{N}$	$19.2\ \text{N} - 8.02\ \text{N} + 5.95\ \text{N} = 17.1\ \text{N}$

The magnitude of $\mathbf{F}_\text{W}$ is given by the Pythagorean theorem as

$$F_\text{W} = \sqrt{(6.9\ \text{N})^2 + (17.1\ \text{N})^2} = \boxed{18.4\ \text{N}}$$

The angle θ that $\mathbf{F}_\text{W}$ makes with the x axis is

$$\theta = \tan^{-1}\left(\frac{17.1\ \text{N}}{6.9\ \text{N}}\right) = 68°$$

Therefore, the direction of $\mathbf{F}_\text{W}$ is $\boxed{68°, \text{north of east}}$.

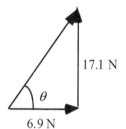

17.1 N

θ

6.9 N

---

25. **REASONING** Newton's law of universal gravitation indicates that the gravitational force that each uniform sphere exerts on the other has a magnitude that is inversely proportional to the square of the distance between the centers of the spheres. Therefore, the maximum gravitational force between two uniform spheres occurs when the centers of the spheres are as close together as possible, that is, when the spherical surfaces touch. Then the distance between the centers of the spheres is the sum of the two radii.

SOLUTION When the bowling ball and the billiard ball are touching, the distance between their centers is $r = r_\text{Bowling} + r_\text{Billiard}$. Using this expression in Newton's law of universal gravitation gives

$$F = \frac{Gm_\text{Bowling}m_\text{Billiard}}{r^2} = \frac{Gm_\text{Bowling}m_\text{Billiard}}{\left(r_\text{Bowling} + r_\text{Billiard}\right)^2}$$

$$= \frac{\left(6.67\times10^{-11}\ \text{N}\cdot\text{m}^2/\text{kg}^2\right)(7.2\ \text{kg})(0.38\ \text{kg})}{\left(0.11\ \text{m} + 0.028\ \text{m}\right)^2} = \boxed{9.6\times10^{-9}\ \text{N}}$$

27. **REASONING AND SOLUTION**   According to Equations 4.4 and 4.5, the weight of an object of mass m at a distance r from the *center* of the earth is

$$mg = \frac{GM_E m}{r^2}$$

In a circular orbit that is 3.59×10^7 m above the surface of the earth (radius = 6.38×10^6 m, mass = 5.98×10^{24} kg), the total distance from the center of the earth is $r = 3.59 \times 10^7$ m + 6.38×10^6 m. Thus the acceleration g due to gravity is

$$g = \frac{GM_E}{r^2} = \frac{(6.67 \times 10^{-11} \text{N} \cdot \text{m}^2/\text{kg}^2)(5.98 \times 10^{24}\text{kg})}{(3.59 \times 10^7\text{m} + 6.38 \times 10^6\text{m})^2} = \boxed{0.223 \text{ m/s}^2}$$

---

31. **REASONING**   According to Equation 4.4, the weights of an object of mass m on the surfaces of planet A (mass = M_A, radius = R) and planet B (mass = M_B, radius = R) are

$$W_A = \frac{GM_A m}{R^2} \quad \text{and} \quad W_B = \frac{GM_B m}{R^2}$$

The difference between these weights is given in the problem.

SOLUTION The difference in weights is

$$W_A - W_B = \frac{GM_A m}{R^2} - \frac{GM_B m}{R^2} = \frac{Gm}{R^2}(M_A - M_B)$$

Rearranging this result, we find

$$M_A - M_B = \frac{(W_A - W_B)R^2}{Gm} = \frac{(3620 \text{ N})(1.33 \times 10^7 \text{ m})^2}{(6.67 \times 10^{-11}\text{N} \cdot \text{m}^2/\text{kg}^2)(5450 \text{ kg})} = \boxed{1.76 \times 10^{24} \text{ kg}}$$

---

35. **REASONING** The gravitational force that the sun exerts on a person standing on the earth is given by Equation 4.3 as $F_{sun} = GM_{sun}m/r^2_{sun\text{-}earth}$, where M_{sun} is the mass of the sun, m is the mass of the person, and $r_{sun\text{-}earth}$ is the distance from the sun to the earth. Likewise, the gravitational force that the moon exerts on a person standing on the earth is given by $F_{moon} = GM_{moon}m/r^2_{moon\text{-}earth}$, where M_{moon} is the mass of the moon and $r_{moon\text{-}earth}$ is the distance from the moon to the earth. These relations will allow us to determine whether the sun or the moon exerts the greater gravitational force on the person.

SOLUTION Taking the ratio of F_{sun} to F_{moon}, and using the mass and distance data from the inside of the text's front cover, we find

$$\frac{F_{sun}}{F_{moon}} = \frac{\dfrac{GM_{sun}m}{r_{sun\text{-}earth}^2}}{\dfrac{GM_{moon}m}{r_{moon\text{-}earth}^2}} = \left(\frac{M_{sun}}{M_{moon}}\right)\left(\frac{r_{moon\text{-}earth}}{r_{sun\text{-}earth}}\right)^2$$

$$= \left(\frac{1.99\times10^{30}\text{ kg}}{7.35\times10^{22}\text{ kg}}\right)\left(\frac{3.85\times10^8\text{ m}}{1.50\times10^{11}\text{ m}}\right)^2 = \boxed{178}$$

Therefore, the sun exerts the greater gravitational force.

___

39. **REASONING**   In order to start the crate moving, an external agent must supply a force that is at least as large as the maximum value $f_s^{MAX} = \mu_s F_N$, where μ_s is the coefficient of static friction (see Equation 4.7). Once the crate is moving, the magnitude of the frictional force is very nearly constant at the value $f_k = \mu_k F_N$, where μ_k is the coefficient of kinetic friction (see Equation 4.8). In both cases described in the problem statement, there are only two vertical forces that act on the crate; they are the upward normal force F_N, and the downward pull of gravity (the weight) mg. Furthermore, the crate has no vertical acceleration in either case. Therefore, if we take upward as the positive direction, Newton's second law in the vertical direction gives $F_N - mg = 0$, and we see that, in both cases, the magnitude of the normal force is $F_N = mg$.

SOLUTION
a. Therefore, the applied force needed to start the crate moving is

$$f_s^{MAX} = \mu_s mg = (0.760)(60.0\text{ kg})(9.80\text{ m/s}^2) = \boxed{447\text{ N}}$$

b. When the crate moves in a straight line at constant speed, its velocity does not change, and it has zero acceleration. Thus, Newton's second law in the horizontal direction becomes $P - f_k = 0$, where P is the required pushing force. Thus, the applied force required to keep the crate sliding across the dock at a constant speed is

$$P = f_k = \mu_k mg = (0.410)(60.0\text{ kg})(9.80\text{ m/s}^2) = \boxed{241\text{ N}}$$

___

45. **REASONING**   In each of the three cases under consideration the kinetic frictional force is given by $f_k = \mu_k F_N$. However, the normal force F_N varies from case to case. To determine the normal force, we use Equation 4.6 ($F_N = mg + ma$) and thereby take into account the acceleration of the elevator. The normal force is greatest when the elevator accelerates upward (a positive) and smallest when the elevator accelerates downward (a negative).

SOLUTION

a. When the elevator is stationary, its acceleration is $a = 0$ m/s^2. Using Equation 4.6, we can express the kinetic frictional force as

$$f_k = \mu_k F_N = \mu_k (mg + ma) = \mu_k m(g + a)$$

$$= (0.360)(6.00 \text{ kg})\left[\left(9.80 \text{ m/s}^2\right) + \left(0 \text{ m/s}^2\right)\right] = \boxed{21.2 \text{ N}}$$

b. When the elevator accelerates upward, $a = +1.20$ m/s^2. Then,

$$f_k = \mu_k F_N = \mu_k (mg + ma) = \mu_k m(g + a)$$

$$= (0.360)(6.00 \text{ kg})\left[\left(9.80 \text{ m/s}^2\right) + \left(1.20 \text{ m/s}^2\right)\right] = \boxed{23.8 \text{ N}}$$

c. When the elevator accelerates downward, $a = -1.20$ m/s^2. Then,

$$f_k = \mu_k F_N = \mu_k (mg + ma) = \mu_k m(g + a)$$

$$= (0.360)(6.00 \text{ kg})\left[\left(9.80 \text{ m/s}^2\right) + \left(-1.20 \text{ m/s}^2\right)\right] = \boxed{18.6 \text{ N}}$$

---

49. *REASONING AND SOLUTION* The free-body diagram is shown at the right. The forces that act on the picture are the pressing force P, the normal force $\mathbf{F_N}$ exerted on the picture by the wall, the weight mg of the picture, and the force of static friction $\mathbf{f}_s^{MAX}$. The maximum magnitude for the frictional force is given by Equation 4.7: $f_s^{MAX} = \mu_s F_N$. The picture is in equilibrium, and, if we take the directions to the right and up as positive, we have in the x direction

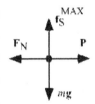

$$\Sigma F_x = P - F_N = 0 \qquad \text{or} \qquad P = F_N$$

and in the y direction

$$\Sigma F_y = f_s^{MAX} - mg = 0 \qquad \text{or} \qquad f_s^{MAX} = mg$$

Therefore,

$$f_s^{MAX} = \mu_s F_N = mg$$

But since $F_N = P$, we have

$$\mu_s P = mg$$

Solving for P, we have

$$P = \frac{mg}{\mu_s} = \frac{(1.10 \text{ kg})(9.80 \text{ m/s}^2)}{0.660} = \boxed{16.3 \text{ N}}$$

53. **REASONING** In order for the object to move with constant velocity, the net force on the object must be zero. Therefore, the north/south component of the third force must be equal in magnitude and opposite in direction to the 80.0 N force, while the east/west component of the third force must be equal in magnitude and opposite in direction to the 60.0 N force. Therefore, the third force has components: 80.0 N due south and 60.0 N due east. We can use the Pythagorean theorem and trigonometry to find the magnitude and direction of this third force.

SOLUTION The magnitude of the third force is

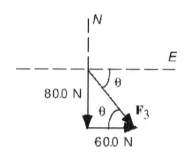

$$F_3 = \sqrt{(80.0 \text{ N})^2 + (60.0 \text{ N})^2} = \boxed{1.00 \times 10^2 \text{ N}}$$

The direction of F_3 is specified by the angle θ where

$$\theta = \tan^{-1}\left(\frac{80.0 \text{ N}}{60.0 \text{ N}}\right) = \boxed{53.1°, \text{ south of east}}$$

57. **REASONING** The worker is standing still. Therefore, he is in equilibrium, and the net force acting on him is zero. The static frictional force $\mathbf{f}_s$ that prevents him from slipping points upward along the roof, an angle of θ degrees above the horizontal; we choose this as the $-x$ direction (see the free-body diagram). The normal force $\mathbf{F}_N$ is perpendicular to the roof and thus has no x component. But the gravitational force $m\mathbf{g}$ of the earth on the worker points straight down, and thus has a component parallel to the roof. We will use this free-body diagram and find the worker's mass m by applying Newton's second law with the acceleration equal to zero.

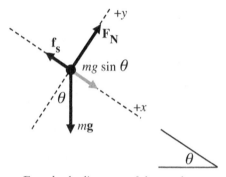

Free-body diagram of the worker

SOLUTION The static frictional force $\mathbf{f}_s$ points in the $-x$ direction, and the x component of the worker's weight $m\mathbf{g}$ points in the $+x$ direction. Because there are no other forces with x components, and the worker's acceleration is zero, these two forces must balance each other. The x component of the worker's weight is $mg \sin \theta$, therefore $f_s = mg \sin \theta$. Solving this relation for the worker's mass, we obtain

$$m = \frac{f_s}{g \sin \theta} = \frac{390 \text{ N}}{(9.80 \text{ m/s}^2)(\sin 36°)} = \boxed{68 \text{ kg}}$$

63. **REASONING** There are four forces that act on the chandelier; they are the forces of tension T in each of the three wires, and the downward force of gravity mg. Under the influence of these forces, the chandelier is at rest and, therefore, in equilibrium. Consequently, the sum of the x components as well as the sum of the y components of the forces must each be zero. The figure below shows a quasi-free-body diagram for the chandelier and the force components for a suitable system of x, y axes. Note that the diagram only shows one of the forces of tension; the second and third tension forces are not shown in the interest of clarity. The triangle at the right shows the geometry of one of the cords, where l is the length of the cord, and d is the distance from the ceiling.

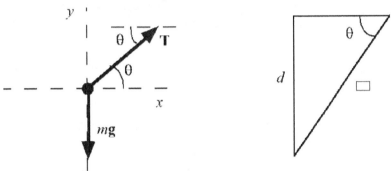

We can use the forces in the y direction to find the magnitude T of the tension in any one wire.

SOLUTION Remembering that there are three tension forces, we see from the diagram that

$$3T \sin \theta = mg \qquad \text{or} \qquad T = \frac{mg}{3\sin\theta} = \frac{mg}{3(d/l)} = \frac{mgl}{3d}$$

Therefore, the magnitude of the tension in any one of the cords is

$$T = \frac{(44 \text{ kg})(9.80 \text{ m/s}^2)(2.0 \text{ m})}{3(1.5 \text{ m})} = \boxed{1.9\times10^2\,\text{N}}$$

67. **REASONING** When the bicycle is coasting straight down the hill, the forces that act on it are the normal force F_N exerted by the surface of the hill, the force of gravity mg, and the force of air resistance R. When the bicycle climbs the hill, there is one additional force; it is the applied force that is required for the bicyclist to climb the hill at constant speed. We can use our knowledge of the motion of the bicycle down the hill to find R. Once R is known, we can analyze the motion of the bicycle as it climbs the hill.

SOLUTION The following figure (on the left) shows the free-body diagram for the forces during the downhill motion. The hill is inclined at an angle θ above the horizontal. The figure (on the right) also shows these forces resolved into components parallel to and perpendicular to the line of motion.

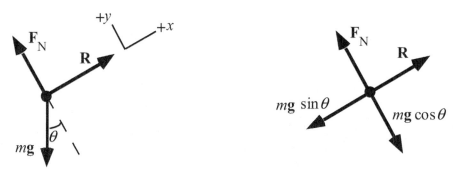

Since the bicyclist is traveling at a constant velocity, his acceleration is zero. Therefore, according to Newton's second law, we have $\sum F_x = 0$ and $\sum F_y = 0$. Taking the direction up the hill as positive, we have $\sum F_x = R - mg \sin \theta = 0$, or

$$R = mg \sin \theta = (80.0 \text{ kg})(9.80 \text{ m/s}^2) \sin 15.0° = 203 \text{ N}$$

When the bicyclist climbs the same hill at constant speed, an applied force P must push the system up the hill. Since the speed is the same, the magnitude of the force of air resistance will remain 203 N. However, the air resistance will oppose the motion by pointing down the hill. The figure at the right shows the resolved forces that act on the system during the uphill motion.

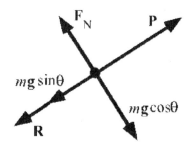

Using the same sign convention as above, we have $\sum F_x = P - mg \sin \theta - R = 0$, or

$$P = R + mg \sin \theta = 203 \text{ N} + 203 \text{ N} = \boxed{406 \text{ N}}$$

---

71. **REASONING** We can use the appropriate equation of kinematics to find the acceleration of the bullet. Then Newton's second law can be used to find the average net force on the bullet.

SOLUTION According to Equation 2.4, the acceleration of the bullet is

$$a = \frac{v - v_0}{t} = \frac{715 \text{ m/s} - 0 \text{ m/s}}{2.50 \times 10^{-3} \text{ s}} = 2.86 \times 10^5 \text{ m/s}^2$$

Therefore, the net average force on the bullet is

$$\sum F = ma = (15 \times 10^{-3} \text{ kg})(2.86 \times 10^5 \text{ m/s}^2) = \boxed{4290 \text{ N}}$$

73. **REASONING** According to Newton's second law, the acceleration has the same direction as the net force and a magnitude given by $a = \Sigma F/m$.

SOLUTION Since the two forces are perpendicular, the magnitude of the net force is given by the Pythagorean theorem as $\Sigma F = \sqrt{(40.0 \text{ N})^2 + (60.0 \text{ N})^2}$. Thus, according to Newton's second law, the magnitude of the acceleration is

$$a = \frac{\Sigma F}{m} = \frac{\sqrt{(40.0 \text{ N})^2 + (60.0 \text{ N})^2}}{4.00 \text{ kg}} = \boxed{18.0 \text{ m/s}^2}$$

The direction of the acceleration vector is given by

$$\theta = \tan^{-1}\left(\frac{60.0 \text{ N}}{40.0 \text{ N}}\right) = \boxed{56.3° \text{ above the } +x \text{ axis}}$$

---

79. **REASONING** The speed of the skateboarder at the bottom of the ramp can be found by solving Equation 2.9 ($v^2 = v_0^2 + 2ax$, where x is the distance that the skater moves down the ramp) for v. The figure at the right shows the free-body diagram for the skateboarder. The net force ΣF, which accelerates the skateboarder down the ramp, is the component of

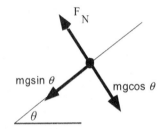

the weight that is parallel to the incline: $\Sigma F = mg \sin\theta$. Therefore, we know from Newton's second law that the acceleration of the skateboarder down the ramp is

$$a = \frac{\Sigma F}{m} = \frac{mg \sin\theta}{m} = g \sin\theta$$

SOLUTION Thus, the speed of the skateboarder at the bottom of the ramp is

$$v = \sqrt{v_0^2 + 2ax} = \sqrt{v_0^2 + 2gx \sin\theta} = \sqrt{(2.6 \text{ m/s})^2 + 2(9.80 \text{ m/s}^2)(6.0 \text{ m}) \sin 18°} = \boxed{6.6 \text{ m/s}}$$

---

83. **REASONING AND SOLUTION** The system is shown in the drawing. We will let $m_1 = 21.0 \text{ kg}$, and $m_2 = 45.0 \text{ kg}$. Then, m_1 will move upward, and m_2 will move downward. There are two forces that act on each object; they are the tension T in the cord and the weight mg of the object. The forces are shown in the free-body diagrams at the far right.

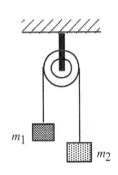

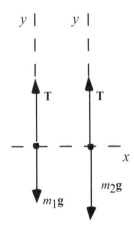

We will take upward as the positive direction. If the acceleration of m_1 is a, then the acceleration of m_2 must be $-a$. From Newton's second law, we have for m_1 that

$$\sum F_y = T - m_1 g = m_1 a \tag{1}$$

and for m_2

$$\sum F_y = T - m_2 g = -m_2 a \tag{2}$$

a. Eliminating T between these two equations, we obtain

$$a = \frac{m_2 - m_1}{m_2 + m_1} g = \left(\frac{45.0 \text{ kg} - 21.0 \text{ kg}}{45.0 \text{ kg} + 21.0 \text{ kg}} \right) (9.80 \text{ m/s}^2) = \boxed{3.56 \text{ m/s}^2}$$

b. Eliminating a between Equations (1) and (2), we find

$$T = \frac{2m_1 m_2}{m_1 + m_2} g = \left[\frac{2(21.0 \text{ kg})(45.0 \text{ kg})}{21.0 \text{ kg} + 45.0 \text{ kg}} \right] (9.80 \text{ m/s}^2) = \boxed{281 \text{ N}}$$

---

89. **REASONING**   The shortest time to pull the person from the cave corresponds to the maximum acceleration, a_y, that the rope can withstand. We first determine this acceleration and then use kinematic Equation 3.5b ($y = v_{0y} t + \frac{1}{2} a_y t^2$) to find the time t.

SOLUTION   As the person is being pulled from the cave, there are two forces that act on him; they are the tension T in the rope that points vertically upward, and the weight of the person mg that points vertically downward. Thus, if we take upward as the positive direction, Newton's second law gives $\sum F_y = T - mg = ma_y$. Solving for a_y, we have

$$a_y = \frac{T}{m} - g = \frac{T}{W/g} - g = \frac{569 \text{ N}}{(5.20 \times 10^2 \text{ N})/(9.80 \text{ m/s}^2)} - 9.80 \text{ m/s}^2 = 0.92 \text{ m/s}^2$$

Therefore, from Equation 3.5b with $v_{0y} = 0$ m/s, we have $y = \frac{1}{2} a_y t^2$. Solving for t, we find

$$t = \sqrt{\frac{2y}{a_y}} = \sqrt{\frac{2(35.1 \text{ m})}{0.92 \text{ m/s}^2}} = \boxed{8.7 \text{ s}}$$

---

93. **REASONING AND SOLUTION**   The penguin comes to a halt on the horizontal surface because the kinetic frictional force opposes the motion and causes it to slow down. The time required for the penguin to slide to a halt ($v = 0$ m/s) after entering the horizontal patch of ice is, according to Equation 2.4,

$$t = \frac{v - v_0}{a_x} = \frac{-v_0}{a_x}$$

We must, therefore, determine the acceleration of the penguin as it slides along the horizontal patch (see the following drawing).

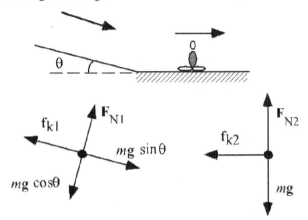

Free-body diagram A   **Free-body diagram B**

For the penguin sliding on the horizontal patch of ice, we find from free-body diagram B and Newton's second law in the x direction (motion to the right is taken as positive) that

$$\Sigma F_x = -f_{k2} = ma_x \qquad \text{or} \qquad a_x = \frac{-f_{k2}}{m} = \frac{-\mu_k F_{N2}}{m}$$

In the y direction in free-body diagram B, we have $\Sigma F_y = F_{N2} - mg = 0$, or $F_{N2} = mg$. Therefore, the acceleration of the penguin is

$$a_x = \frac{-\mu_k mg}{m} = -\mu_k g \qquad\qquad (1)$$

Equation (1) indicates that, in order to find the acceleration a_x, we must find the coefficient of kinetic friction.

We are told in the problem statement that the coefficient of kinetic friction between the penguin and the ice is the same for the incline as for the horizontal patch. Therefore, we can use the motion of the penguin on the incline to determine the coefficient of friction and use it in Equation (1).

For the penguin sliding down the incline, we find from free-body diagram A (see the previous drawing) and Newton's second law (taking the direction of motion as positive) that

$$\Sigma F_x = mg \sin\theta - f_{k1} = ma_x = 0 \qquad \text{or} \qquad f_{k1} = mg \sin\theta \qquad (2)$$

Here, we have used the fact that the penguin slides down the incline with a constant velocity, so that it has zero acceleration. From Equation 4.8, we know that $f_{k1} = \mu_k F_{N1}$. Applying Newton's second law in the direction perpendicular to the incline, we have

$$\Sigma F_y = F_{N1} - mg\cos\theta = 0 \qquad \text{or} \qquad F_{N1} = mg\cos\theta$$

Therefore, $f_{k1} = \mu_k mg\cos\theta$, so that according to Equation (2), we find

$$f_{k1} = \mu_k mg\cos\theta = mg\sin\theta$$

Solving for the coefficient of kinetic friction, we have

$$\mu_k = \frac{\sin\theta}{\cos\theta} = \tan\theta$$

Finally, the time required for the penguin to slide to a halt after entering the horizontal patch of ice is

$$t = \frac{-v_0}{a_x} = \frac{-v_0}{-\mu_k g} = \frac{v_0}{g\tan\theta} = \frac{1.4 \text{ m/s}}{(9.80 \text{ m/s}^2)\tan 6.9°} = \boxed{1.2 \text{ s}}$$

---

95. **REASONING** With air resistance neglected, only two forces act on the bungee jumper at this instant (see the free-body diagram): the bungee cord pulls up on her with a force B, and the earth pulls down on her with a gravitational force mg. Because we know the jumper's mass and acceleration, we can apply Newton's second law to this free-body diagram and solve for B.

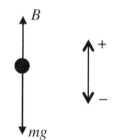

Free-body diagram of jumper

SOLUTION We will take the direction of the jumper's acceleration (downward) as negative. Then, the net force acting on the bungee jumper is $\Sigma F = B - mg$. With Newton's second law ($\Sigma F = ma$), this becomes $ma = B - mg$. We now solve for B:

$$B = ma + mg = m(a + g) = (55 \text{ kg})(-7.6 \text{ m/s}^2 + 9.80 \text{ m/s}^2) = \boxed{120 \text{ N}}$$

As indicated in the free-body diagram, the direction of the force applied by the bungee cord is $\boxed{\text{upward}}$.

---

97. **REASONING AND SOLUTION**
    a. The apparent weight of the person is given by Equation 4.6 as

$$F_N = mg + ma$$
$$= (95.0 \text{ kg})(9.80 \text{ m/s}^2 + 1.80 \text{ m/s}^2) = \boxed{1.10 \times 10^3 \text{ N}}$$

   b. $$F_N = (95.0 \text{ kg})(9.80 \text{ m/s}^2) = \boxed{931 \text{ N}}$$

   c. $$F_N = (95.0 \text{ kg})(9.80 \text{ m/s}^2 - 1.30 \text{ m/s}^2) = \boxed{808 \text{ N}}$$

99. **REASONING** The book is kept from falling as long as the total static frictional force balances the weight of the book. The forces that act on the book are shown in the following free-body diagram, where P is the magnitude of the pressing force applied by each hand.

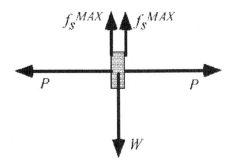

In this diagram, note that there are two pressing forces, one from each hand. Each hand also applies a static frictional force, and, therefore, two static frictional forces are shown. The maximum static frictional force is related in the usual way to a normal force F_N, but in this problem the normal force is provided by the pressing force, so that $F_N = P$.

SOLUTION Since the frictional forces balance the weight, we have

$$2f_s^{MAX} = 2(\mu_s F_N) = 2(\mu_s P) = W$$

Solving for P, we find that

$$P = \frac{W}{2\mu_s} = \frac{31 \text{ N}}{2(0.40)} = \boxed{39 \text{ N}}$$

103. **REASONING** If we assume that the acceleration is constant, we can use Equation 2.4 ($v = v_0 + at$) to find the acceleration of the car. Once the acceleration is known, Newton's second law ($\sum F = ma$) can be used to find the magnitude and direction of the net force that produces the deceleration of the car.

SOLUTION We choose due east to be the positive direction. Given this choice, the average acceleration of the car is, according to Equation 2.4,

$$a = \frac{v - v_0}{t} = \frac{17.0 \text{ m/s} - 27.0 \text{ m/s}}{8.00 \text{ s}} = -1.25 \text{ m/s}^2$$

where the minus sign indicates that the acceleration points due west. According to Newton's Second law, the net force on the car is

$$\sum F = ma = (1380 \text{ kg})(-1.25 \text{ m/s}^2) = -1730 \text{ N}$$

The magnitude of the net force is $\boxed{1730 \text{ N}}$. From Newton's second law, we know that the direction of the force is the same as the direction of the acceleration, so the force also points $\boxed{\text{due west}}$.

107. **REASONING**   The free-body diagrams for Robin (mass = m) and for the chandelier (mass = M) are given at the right. The tension T in the rope applies an upward force to both. Robin accelerates upward, while the chandelier accelerates downward, each acceleration having the same magnitude. Our solution is based on separate applications of Newton's second law to Robin and the chandelier.

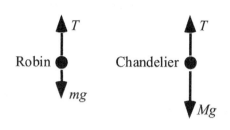

SOLUTION Applying Newton's second law, we find

$$\underbrace{T - mg = ma}_{\text{Robin Hood}} \qquad \text{and} \qquad \underbrace{T - Mg = -Ma}_{\text{Chandelier}}$$

In these applications we have taken upward as the positive direction, so that Robin's acceleration is a, while the chandelier's acceleration is $-a$. Solving the Robin-Hood equation for T gives

$$T = mg + ma$$

Substituting this expression for T into the Chandelier equation gives

$$mg + ma - Mg = -Ma \qquad \text{or} \qquad a = \left(\frac{M-m}{M+m} \right) g$$

a. Robin's acceleration is

$$a = \left(\frac{M-m}{M+m} \right) g = \left[\frac{(195 \text{ kg}) - (77.0 \text{ kg})}{(195 \text{ kg}) + (77.0 \text{ kg})} \right] (9.80 \text{ m/s}^2) = \boxed{4.25 \text{ m/s}^2}$$

b. Substituting the value of a into the expression for T gives

$$T = mg + ma = (77.0 \text{ kg})(9.80 \text{ m/s}^2 + 4.25 \text{ m/s}^2) = \boxed{1080 \text{ N}}$$

---

111. **REASONING** The tension in each coupling bar is responsible for accelerating the objects behind it. The masses of the cars are m_1, m_2, and m_3. We can use Newton's second law to express the tension in each coupling bar, since friction is negligible:

$$\underbrace{T_A = \left(m_1 + m_2 + m_3 \right) a}_{\text{Coupling bar A}} \qquad \underbrace{T_B = \left(m_2 + m_3 \right) a}_{\text{Coupling bar B}} \qquad \underbrace{T_C = m_3 a}_{\text{Coupling bar C}}$$

In these expressions $a = 0.12 \text{ m/s}^2$ remains constant. Consequently, the tension in a given bar will change only if the total mass of the objects accelerated by that bar changes as a result of the luggage transfer. Using Δ (Greek capital delta) to denote a change in the usual fashion, we can express the changes in the above tensions as follows:

$$\underbrace{\Delta T_A = \left[\Delta\left(m_1 + m_2 + m_3\right)\right]a}_{\text{Coupling bar A}} \qquad \underbrace{\Delta T_B = \left[\Delta\left(m_2 + m_3\right)\right]a}_{\text{Coupling bar B}} \qquad \underbrace{\Delta T_C = \left(\Delta m_3\right)a}_{\text{Coupling bar C}}$$

SOLUTION

a. Moving luggage from car 2 to car 1 does not change the total mass $m_1 + m_2 + m_3$, so $\Delta(m_1 + m_2 + m_3) = 0$ kg and $\boxed{\Delta T_A = 0 \text{ N}}$.

The transfer from car 2 to car 1 causes the total mass $m_2 + m_3$ to decrease by 39 kg, so $\Delta(m_2 + m_3) = -39$ kg and

$$\Delta T_B = \left[\Delta\left(m_2 + m_3\right)\right]a = \left(-39 \text{ kg}\right)\left(0.12 \text{ m/s}^2\right) = \boxed{-4.7 \text{ N}}$$

The transfer from car 2 to car 1 does not change the mass m_3, so $\Delta m_3 = 0$ kg and $\boxed{\Delta T_C = 0 \text{ N}}$.

b. Moving luggage from car 2 to car 3 does not change the total mass $m_1 + m_2 + m_3$, so $\Delta(m_1 + m_2 + m_3) = 0$ kg and $\boxed{\Delta T_A = 0 \text{ N}}$.

The transfer from car 2 to car 3 does not change the total mass $m_2 + m_3$, so $\Delta(m_2 + m_3) = 0$ kg and $\boxed{\Delta T_B = 0 \text{ N}}$.

The transfer from car 2 to car 3 causes the mass m_3 to increase by 39 kg, so $\Delta m_3 = +39$ kg and

$$\Delta T_C = \left(\Delta m_3\right)a = \left(+39 \text{ kg}\right)\left(0.12 \text{ m/s}^2\right) = \boxed{+4.7 \text{ N}}$$

---

115. **REASONING** Let us assume that the skater is moving horizontally along the $+x$ axis. The time t it takes for the skater to reduce her velocity to $v_x = +2.8$ m/s from $v_{0x} = +6.3$ m/s can be obtained from one of the equations of kinematics:

$$v_x = v_{0x} + a_x t \tag{3.3a}$$

The initial and final velocities are known, but the acceleration is not. We can obtain the acceleration from Newton's second law $\left(\Sigma F_x = ma_x, \text{ Equation 4.2a}\right)$ in the following manner. The kinetic frictional force is the only horizontal force that acts on the skater, and, since it is a resistive force, it acts opposite to the direction of the motion. Thus, the net force in the x direction is $\Sigma F_x = -f_k$, where f_k is the magnitude of the kinetic frictional force. Therefore, the acceleration of the skater is $a_x = \Sigma F_x/m = -f_k/m$.

The magnitude of the frictional force is $f_k = \mu_k F_N$ (Equation 4.8), where μ_k is the coefficient of kinetic friction between the ice and the skate blades and F_N is the magnitude of the normal force. There are two vertical forces acting on the skater: the upward-acting normal force $\mathbf{F_N}$ and the downward pull of gravity (her weight) $m\mathbf{g}$. Since the skater has no vertical acceleration, Newton's second law in the vertical direction gives (taking upward as the positive direction) $\Sigma F_y = F_N - mg = 0$. Therefore, the magnitude of the normal force is $F_N = mg$ and the magnitude of the acceleration is

$$a_x = \frac{-f_k}{m} = \frac{-\mu_k F_N}{m} = \frac{-\mu_k \cancel{m} g}{\cancel{m}} = -\mu_k g$$

SOLUTION
Solving the equation $v_x = v_{0x} + a_x t$ for the time and substituting the expression above for the acceleration yields

$$t = \frac{v_x - v_{0x}}{a_x} = \frac{v_x - v_{0x}}{-\mu_k g} = \frac{2.8 \text{ m/s} - 6.3 \text{ m/s}}{-(0.081)(9.80 \text{ m/s}^2)} = \boxed{4.4 \text{ s}}$$

---

117. *REASONING AND SOLUTION*

a. The left mass (mass 1) has a tension T_1 pulling it up. Newton's second law gives

$$T_1 - m_1 g = m_1 a \qquad (1)$$

The right mass (mass 3) has a different tension, T_3, trying to pull it up. Newton's second for it is

$$T_3 - m_3 g = -m_3 a \qquad (2)$$

The middle mass (mass 2) has both tensions acting on it along with friction. Newton's second law for its horizontal motion is

$$T_3 - T_1 - \mu_k m_2 g = m_2 a \qquad (3)$$

Solving Equation (1) and Equation (2) for T_1 and T_3, respectively, and substituting into Equation (3) gives

$$a = \frac{(m_3 - m_1 - \mu_k m_2) g}{m_1 + m_2 + m_3}$$

Hence,

$$a = \frac{\left[25.0 \text{ kg} - 10.0 \text{ kg} - (0.100)(80.0 \text{ kg})\right](9.80 \text{ m/s}^2)}{10.0 \text{ kg} + 80.0 \text{ kg} + 25.0 \text{ kg}} = \boxed{0.60 \text{ m/s}^2}$$

b. From part a:

$$T_1 = m_1(g + a) = (10.0 \text{ kg})(9.80 \text{ m/s}^2 + 0.60 \text{ m/s}^2) = \boxed{104 \text{ N}}$$

$$T_3 = m_3(g - a) = (25.0 \text{ kg})(9.80 \text{ m/s}^2 - 0.60 \text{ m/s}^2) = \boxed{230 \text{ N}}$$

---

121. $\boxed{\text{SSM}}$ *CONCEPTS* (i) If the spacecraft were stationary, its acceleration would be zero. According to Newton's second law, the acceleration of an object is proportional to the net force acting on it. Thus, the net force must also be zero. But the net force is the vector sum of the three forces in this case. Therefore, the force $\vec{\mathbf{F}}_3$ must have a direction such that it balances to zero the forces $\vec{\mathbf{F}}_1$ and $\vec{\mathbf{F}}_2$. Since $\vec{\mathbf{F}}_1$ and $\vec{\mathbf{F}}_2$ point along the +x axis, $\vec{\mathbf{F}}_3$ must then point along the −x axis.

(ii) Since the velocity is constant, the acceleration is still zero. As a result, everything stated for the stationary case (from part (i)) applies again here. The net force is zero, and the force $\vec{\mathbf{F}}_3$ must point along the −x axis.

CALCULATIONS Since the velocity is constant, the acceleration is zero. The net force must also be zero, so that

$$\sum F_x = F_1 + F_2 + F_3 = 0$$

Solving for F_3 yields

$$F_3 = -(F_1 + F_2) = -(3000 \text{ N} + 5000 \text{ N}) = \boxed{-8000 \text{ N}}$$

The negative sign in the answer means that $\vec{\mathbf{F}}_3$ points opposite to the sum of $\vec{\mathbf{F}}_1$ and $\vec{\mathbf{F}}_2$, or along the −x axis. The force $\vec{\mathbf{F}}_3$ has a magnitude of 8000 N, which is the magnitude of the sum of the forces $\vec{\mathbf{F}}_1$ and $\vec{\mathbf{F}}_2$. The answer is independent of the velocity of the spacecraft, as long as that velocity remains constant.

CHAPTER 5 | *DYNAMICS OF UNIFORM CIRCULAR MOTION*

1. ***REASONING*** The magnitude a_c of the car's centripetal acceleration is given by Equation 5.2 as $a_c = v^2/r$, where v is the speed of the car and r is the radius of the track. The radius is $r = 2.6 \times 10^3$ m. The speed can be obtained from Equation 5.1 as the circumference $(2\pi r)$ of the track divided by the period T of the motion. The period is the time for the car to go once around the track ($T = 360$ s).

SOLUTION Since $a_c = v^2/r$ and $v = (2\pi r)/T$, the magnitude of the car's centripetal acceleration is

$$a_c = \frac{v^2}{r} = \frac{\left(\dfrac{2\pi r}{T}\right)^2}{r} = \frac{4\pi^2 r}{T^2} = \frac{4\pi^2 \left(2.6 \times 10^3 \text{ m}\right)}{\left(360 \text{ s}\right)^2} = \boxed{0.79 \text{ m/s}^2}$$

5. ***REASONING*** The speed of the plane is given by Equation 5.1: $v = 2\pi r/T$, where T is the period or the time required for the plane to complete one revolution.

SOLUTION Solving Equation 5.1 for T we have

$$T = \frac{2\pi r}{v} = \frac{2\pi (2850 \text{ m})}{110 \text{ m/s}} = \boxed{160 \text{ s}}$$

9. ***REASONING AND SOLUTION*** Since the magnitude of the centripetal acceleration is given by Equation 5.2, $a_C = v^2/r$, we can solve for r and find that

$$r = \frac{v^2}{a_C} = \frac{(98.8 \text{ m/s})^2}{3.00(9.80 \text{ m/s}^2)} = \boxed{332 \text{ m}}$$

13. ***REASONING*** In Example 3, it was shown that the magnitudes of the centripetal acceleration for the two cases are

Radius = 33 m	$a_C = 35$ m/s^2
Radius = 24 m	$a_C = 48$ m/s^2

According to Newton's second law, the centripetal force is $F_C = ma_C$ (see Equation 5.3).

SOLUTION a. Therefore, when the sled undergoes the turn of radius 33 m,

$$F_C = ma_C = (350 \text{ kg})(35 \text{ m/s}^2) = \boxed{1.2 \times 10^4 \text{ N}}$$

b. Similarly, when the radius of the turn is 24 m,

$$F_C = ma_C = (350 \text{ kg})(48 \text{ m/s}^2) = \boxed{1.7 \times 10^4 \text{ N}}$$

---

21. **REASONING** Let v_0 be the initial speed of the ball as it begins its projectile motion. Then, the centripetal force is given by Equation 5.3: $F_C = mv_0^2/r$. We are given the values for m and r; however, we must determine the value of v_0 from the details of the projectile motion after the ball is released.

In the absence of air resistance, the x component of the projectile motion has zero acceleration, while the y component of the motion is subject to the acceleration due to gravity. The horizontal distance traveled by the ball is given by Equation 3.5a (with $a_x = 0 \text{m/s}^2$):

$$x = v_{0x}t = (v_0 \cos\theta)t$$

with t equal to the flight time of the ball while it exhibits projectile motion. The time t can be found by considering the vertical motion. From Equation 3.3b,

$$v_y = v_{0y} + a_y t$$

After a time t, $v_y = -v_{0y}$. Assuming that up and to the right are the positive directions, we have

$$t = \frac{-2v_{0y}}{a_y} = \frac{-2v_0 \sin\theta}{a_y}$$

and

$$x = (v_0 \cos\theta)\left(\frac{-2v_0 \sin\theta}{a_y}\right)$$

Using the fact that $2\sin\theta \cos\theta = \sin 2\theta$, we have

$$x = -\frac{2v_0^2 \cos\theta \sin\theta}{a_y} = -\frac{v_0^2 \sin 2\theta}{a_y} \qquad (1)$$

Equation (1) (with upward and to the right chosen as the positive directions) can be used to determine the speed v_0 with which the ball begins its projectile motion. Then Equation 5.3 can be used to find the centripetal force.

SOLUTION Solving equation (1) for v_0, we have

$$v_0 = \sqrt{\frac{-x\,a_y}{\sin 2\theta}} = \sqrt{\frac{-(86.75 \text{ m})(-9.80 \text{ m/s}^2)}{\sin 2(41°)}} = 29.3 \text{ m/s}$$

Then, from Equation 5.3,

$$F_C = \frac{mv_0^2}{r} = \frac{(7.3 \text{ kg})(29.3 \text{ m/s})^2}{1.8 \text{ m}} = \boxed{3500 \text{ N}}$$

---

33. **REASONING** Equation 5.5 gives the orbital speed for a satellite in a circular orbit around the earth. It can be modified to determine the orbital speed around any planet **P** by replacing the mass of the earth M_E by the mass of the planet M_P: $v = \sqrt{GM_P/r}$.

SOLUTION The ratio of the orbital speeds is, therefore,

$$\frac{v_2}{v_1} = \frac{\sqrt{GM_P/r_2}}{\sqrt{GM_P/r_1}} = \sqrt{\frac{r_1}{r_2}}$$

Solving for v_2 gives

$$v_2 = v_1\sqrt{\frac{r_1}{r_2}} = (1.70 \times 10^4 \text{ m/s})\sqrt{\frac{5.25 \times 10^6 \text{ m}}{8.60 \times 10^6 \text{ m}}} = \boxed{1.33 \times 10^4 \text{ m/s}}$$

---

37. **REASONING** Equation 5.2 for the centripetal acceleration applies to both the plane and the satellite, and the centripetal acceleration is the same for each. Thus, we have

$$a_c = \frac{v_{plane}^2}{r_{plane}} = \frac{v_{satellite}^2}{r_{satellite}} \quad \text{or} \quad v_{plane} = \left(\sqrt{\frac{r_{plane}}{r_{satellite}}}\right)v_{satellite}$$

The speed of the satellite can be obtained directly from Equation 5.5.

SOLUTION Using Equation 5.5, we can express the speed of the satellite as

$$v_{satellite} = \sqrt{\frac{Gm_E}{r_{satellite}}}$$

Substituting this expression into the expression obtained in the reasoning for the speed of the plane gives

$$v_{plane} = \left(\sqrt{\frac{r_{plane}}{r_{satellite}}}\right) v_{satellite} = \left(\sqrt{\frac{r_{plane}}{r_{satellite}}}\right)\sqrt{\frac{Gm_E}{r_{satellite}}} = \frac{\sqrt{r_{plane}Gm_E}}{r_{satellite}}$$

$$v_{plane} = \frac{\sqrt{(15 \text{ m})(6.67\times10^{-11} \text{ N}\cdot\text{m}^2/\text{kg}^2)(5.98\times10^{24} \text{ kg})}}{6.7\times10^6 \text{ m}} = \boxed{12 \text{ m/s}}$$

41. **REASONING**   As the motorcycle passes over the top of the hill, it will experience a centripetal force, the magnitude of which is given by Equation 5.3: $F_C = mv^2/r$. The centripetal force is provided by the net force on the cycle + driver system. At that instant, the net force on the system is composed of the normal force, which points upward, and the weight, which points downward. Taking the direction toward the center of the circle (downward) as the positive direction, we have $F_C = mg - F_N$. This expression can be solved for F_N, the normal force.

SOLUTION
a. The magnitude of the centripetal force is

$$F_C = \frac{mv^2}{r} = \frac{(342 \text{ kg})(25.0 \text{ m/s})^2}{126 \text{ m}} = \boxed{1.70\times10^3 \text{ N}}$$

b. The magnitude of the normal force is

$$F_N = mg - F_C = (342 \text{ kg})(9.80 \text{ m/s}^2) - 1.70\times10^3 \text{ N} = \boxed{1.66\times10^3 \text{ N}}$$

43. **REASONING**   The centripetal force is the name given to the net force pointing toward the center of the circular path. At point 3 at the top the net force pointing toward the center of the circle consists of the normal force and the weight, both pointing toward the center. At point 1 at the bottom the net force consists of the normal force pointing upward toward the center and the weight pointing downward or away from the center. In either case the centripetal force is given by Equation 5.3 as $F_c = mv^2/r$.

SOLUTION   At point 3 we have

$$F_c = F_N + mg = \frac{mv_3^2}{r}$$

At point 1 we have

$$F_c = F_N - mg = \frac{mv_1^2}{r}$$

Subtracting the second equation from the first gives

$$2mg = \frac{mv_3^2}{r} - \frac{mv_1^2}{r}$$

Rearranging gives

$$v_3^2 = 2gr + v_1^2$$

Thus, we find that

$$v_3 = \sqrt{2(9.80 \text{ m/s}^2)(3.0 \text{ m}) + (15 \text{ m/s})^2} = \boxed{17 \text{ m/s}}$$

---

49. **REASONING** The magnitude F_c of the centripetal force that acts on the skater is given by Equation 5.3 as $F_c = mv^2/r$, where m and v are the mass and speed of the skater, and r is the distance of the skater from the pivot. Since all of these variables are known, we can find the magnitude of the centripetal force.

SOLUTION The magnitude of the centripetal force is

$$F_c = \frac{mv^2}{r} = \frac{(80.0 \text{ kg})(6.80 \text{ m/s})^2}{6.10 \text{ m}} = \boxed{606 \text{ N}}$$

---

55. **REASONING** According to Equation 5.3, the magnitude F_c of the centripetal force that acts on each passenger is $F_c = mv^2/r$, where m and v are the mass and speed of a passenger and r is the radius of the turn. From this relation we see that the speed is given by $v = \sqrt{F_c r/m}$. The centripetal force is the net force required to keep each passenger moving on the circular path and points toward the center of the circle. With the aid of a free-body diagram, we will evaluate the net force and, hence, determine the speed.

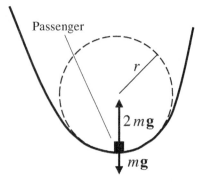

SOLUTION The free-body diagram shows a passenger at the bottom of the circular dip. There are two forces acting: her downward-acting weight $m\mathbf{g}$ and the upward-acting force $2m\mathbf{g}$ that the seat exerts on her. The net force is $+2mg - mg = +mg$, where we have taken "up" as the positive direction. Thus, $F_c = mg$. The speed of the passenger can be found by using this result in the equation above.

Substituting $F_c = mg$ into the relation $v = \sqrt{F_c r/m}$ yields

$$v = \sqrt{\frac{F_c r}{m}} = \sqrt{\frac{(mg)r}{m}} = \sqrt{gr} = \sqrt{(9.80 \text{ m/s}^2)(20.0 \text{ m})} = \boxed{14.0 \text{ m/s}}$$

57. ***REASONING AND SOLUTION*** The centripetal acceleration for any point on the blade a distance r from center of the circle, according to Equation 5.2, is $a_c = v^2/r$. From Equation 5.1, we know that $v = 2\pi r/T$ where T is the period of the motion. Combining these two equations, we obtain

$$a_c = \frac{(2\pi r/T)^2}{r} = \frac{4\pi^2 r}{T^2}$$

a. Since the turbine blades rotate at 617 rev/s, all points on the blades rotate with a period of $T = (1/617)$ s $= 1.62 \times 10^{08}$ s. Therefore, for a point with $r = 0.020$ m, the magnitude of the centripetal acceleration is

$$a_c = \frac{4\pi^2 (0.020 \text{ m})}{(1.62 \times 10^{-3} \text{ s})^2} = \boxed{3.0 \times 10^5 \text{ m/s}^2}$$

b. Expressed as a multiple of g, this centripetal acceleration is

$$a_c = \left(3.0 \times 10^5 \text{ m/s}^2\right) \left(\frac{1.00 \text{ } g}{9.80 \text{ m/s}^2}\right) = \boxed{3.1 \times 10^4 \text{ } g}$$

61. ***REASONING*** If the effects of gravity are not ignored in Example 5, the plane will make an angle θ with the vertical as shown in figure **A** below. The figure **B** shows the forces that act on the plane, and figure **C** shows the horizontal and vertical components of these forces.

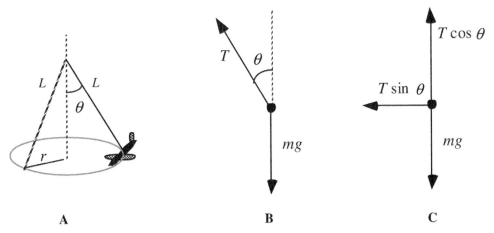

A	**B**	**C**

From figure **C** we see that the resultant force in the horizontal direction is the horizontal component of the tension in the guideline and provides the centripetal force. Therefore,

$$T \sin \theta = \frac{mv^2}{r}$$

From figure **A**, the radius r is related to the length L of the guideline by $r = L \sin\theta$; therefore,

$$T \sin \theta = \frac{mv^2}{L \sin \theta} \tag{1}$$

The resultant force in the vertical direction is zero: $T\cos\theta - mg = 0$, so that

$$T\cos\theta = mg \tag{2}$$

From equation (2) we have

$$T = \frac{mg}{\cos \theta} \tag{3}$$

Equation (3) contains two unknown, T and θ. First we will solve equations (1) and (3) simultaneously to determine the value(s) of the angle θ. Once θ is known, we can calculate the tension using equation (3).

SOLUTION   Substituting equation (3) into equation (1):

$$\left(\frac{mg}{\cos \theta}\right) \sin \theta = \frac{mv^2}{L \sin \theta}$$

Thus,

$$\frac{\sin^2 \theta}{\cos \theta} = \frac{v^2}{gL} \tag{4}$$

Using the fact that $\cos^2 \theta + \sin^2 \theta = 1$, equation (4) can be written

$$\frac{1 - \cos^2 \theta}{\cos \theta} = \frac{v^2}{gL}$$

or

$$\frac{1}{\cos \theta} - \cos \theta = \frac{v^2}{gL}$$

This can be put in the form of an equation that is quadratic in $\cos \theta$. Multiplying both sides by $\cos \theta$ and rearranging yields:

$$\cos^2 \theta + \frac{v^2}{gL}\cos \theta - 1 = 0 \tag{5}$$

Equation (5) is of the form

$$ax^2 + bx + c = 0 \tag{6}$$

with $x = \cos \theta$, $a = 1$, $b = v^2/(gL)$, and $c = -1$. The solution to equation (6) is found from the quadratic formula:

$$x = \frac{-b \pm \sqrt{b^2 - 4ac}}{2a}$$

When $v = 19.0$ m/s, $b = 2.17$. The positive root from the quadratic formula gives $x = \cos\theta = 0.391$. Substitution into equation (3) yields

$$T = \frac{mg}{\cos\theta} = \frac{(0.900 \text{ kg})(9.80 \text{ m/s}^2)}{0.391} = \boxed{23 \text{ N}}$$

When $v = 38.0$ m/s, $b = 8.67$. The positive root from the quadratic formula gives $x = \cos\theta = 0.114$. Substitution into equation (3) yields

$$T = \frac{mg}{\cos\theta} = \frac{(0.900 \text{ kg})(9.80 \text{ m/s}^2)}{0.114} = \boxed{77 \text{ N}}$$

---

63. **SSM** *CONCEPTS* **(i)** Acceleration is a vector and, for it to be constant, both its magnitude and direction must be constant. Automobile A has a constant acceleration because the magnitude of its accelerations is constant (3.5 m/s^2), and its direction always points forward along the straight road. Automobile B has an acceleration with a constant magnitude (again, 3.5 m/s^2), but the acceleration does not have a constant direction. Automobile B is in uniform circular motion, so it has a centripetal acceleration which points towards the center of the circle at every instant. Therefore, for automobile B, the direction of the acceleration vector continually changes, and the vector is not constant.

(ii) The equations of kinematics apply for automobile A because it has a constant acceleration, which must be the case when you apply these equations. They do not apply to automobile B because it does not have a constant a constant acceleration.

CALCULATIONS To determine the speed of automobile A at $t = 2.0$ s we use Equation 2.4 from the equations of kinematics:

$$v = v_0 + at = (18 \text{ m/s}) + (3.5 \text{ m/s}^2)(2.0 \text{ s}) = \boxed{25 \text{ m/s}}$$

To determine the speed of automobile B, we note that this car is in uniform circular motion and, therefore, has a constant speed as it goes around the turn. At a time of $t = 2.0$ s its speed is the same as at $t = 0$ s, or

$$v = \boxed{18 \text{ m/s}}$$

CHAPTER 6 | *WORK AND ENERGY*

1. **REASONING AND SOLUTION** We will assume that the tug-of-war rope remains parallel to the ground, so that the force that moves team B is in the same direction as the displacement. According to Equation 6.1, the work done by team A is

$$W = (F\cos\theta)s = (1100 \text{ N})(\cos 0°)(2.0 \text{ m}) = \boxed{2.2 \times 10^3 \text{ J}}$$

5. **REASONING AND SOLUTION** Solving Equation 6.1 for the angle θ, we obtain

$$\theta = \cos^{-1}\left(\frac{W}{Fs}\right) = \cos^{-1}\left[\frac{1.10 \times 10^3 \text{ J}}{(30.0 \text{ N})(50.0 \text{ m})}\right] = \boxed{42.8°}$$

7. **REASONING AND SOLUTION**
    a. In both cases, the lift force **L** is perpendicular to the displacement of the plane, and, therefore, does no work. As shown in the drawing in the text, when the plane is in the dive, there is a component of the weight **W** that points in the direction of the displacement of the plane. When the plane is climbing, there is a component of the weight that points opposite to the displacement of the plane. Thus, since the thrust **T** is the same for both cases, the net force in the direction of the displacement is greater for the case where the plane is diving. Since the displacement **s** is the same in both cases, $\boxed{\text{more net work is done during the dive}}$.

    b. The work done during the dive is $W_{\text{dive}} = (T + W\cos 75°)\,s$, while the work done during the climb is $W_{\text{climb}} = (T + W\cos 115°)\,s$. Therefore, the difference between the net work done during the dive and the climb is

$$W_{\text{dive}} - W_{\text{climb}} = (T + W\cos 75°)\,s - (T + W\cos 115°)\,s = Ws\,(\cos 75° - \cos 115°)$$

$$= (5.9 \times 10^4 \text{ N})(1.7 \times 10^3 \text{ m})(\cos 75° - \cos 115°) = \boxed{6.8 \times 10^7 \text{ J}}$$

15. **REASONING** The car's kinetic energy depends upon its mass and speed via $\text{KE} = \frac{1}{2}mv^2$ (Equation 6.2). The total amount of work done on the car is equal to the difference between its final and initial kinetic energies: $W = \text{KE}_f - \text{KE}_0$ (Equation 6.3). We will use these two relationships to determine the car's mass.

SOLUTION Combining $\mathrm{KE} = \frac{1}{2}mv^2$ (Equation 6.2) and $W = \mathrm{KE_f} - \mathrm{KE_0}$ (Equation 6.3), we obtain

$$W = \tfrac{1}{2}mv_{\mathrm{f}}^2 - \tfrac{1}{2}mv_0^2 \qquad \text{or} \qquad \tfrac{1}{2}m\left(v_{\mathrm{f}}^2 - v_0^2\right) = W$$

Solving this expression for the car's mass m, and noting that 185 kJ = 185 000 J, we find that

$$m = \frac{2W}{v_{\mathrm{f}}^2 - v_0^2} = \frac{2(185\,000\text{ J})}{(28.0\text{ m/s})^2 - (23.0\text{ m/s})^2} = \boxed{1450\text{ kg}}$$

---

19. ***REASONING*** The work done to launch either object can be found from Equation 6.3, the work-energy theorem, $W = \mathrm{KE_f} - \mathrm{KE_0} = \frac{1}{2}mv_{\mathrm{f}}^2 - \frac{1}{2}mv_0^2$.

SOLUTION
a. The work required to launch the hammer is

$$W = \tfrac{1}{2}mv_{\mathrm{f}}^2 - \tfrac{1}{2}mv_0^2 = \tfrac{1}{2}m\left(v_{\mathrm{f}}^2 - v_0^2\right) = \tfrac{1}{2}(7.3\text{ kg})\left[(29\text{ m/s})^2 - (0\text{ m/s})^2\right] = \boxed{3.1 \times 10^3\text{ J}}$$

b. Similarly, the work required to launch the bullet is

$$W = \tfrac{1}{2}m\left(v_{\mathrm{f}}^2 - v_0^2\right) = \tfrac{1}{2}(0.0026\text{ kg})\left[(410\text{ m/s})^2 - (0\text{ m/s})^2\right] = \boxed{2.2 \times 10^2\text{ J}}$$

---

23. ***REASONING*** When the satellite goes from the first to the second orbit, its kinetic energy changes. The net work that the external force must do to change the orbit can be found from the work-energy theorem: $W = \mathrm{KE_f} - \mathrm{KE_0} = \frac{1}{2}mv_{\mathrm{f}}^2 - \frac{1}{2}mv_0^2$. The speeds v_{f} and v_0 can be obtained from Equation 5.5 for the speed of a satellite in a circular orbit of radius r. Given the speeds, the work energy theorem can be used to obtain the work.

SOLUTION According to Equation 5.5, $v = \sqrt{GM_{\mathrm{E}}/r}$. Substituting into the work-energy theorem, we have

$$W = \tfrac{1}{2}mv_{\mathrm{f}}^2 - \tfrac{1}{2}mv_0^2 = \tfrac{1}{2}m\left(v_{\mathrm{f}}^2 - v_0^2\right) = \tfrac{1}{2}m\left[\left(\sqrt{\frac{GM_{\mathrm{E}}}{r_{\mathrm{f}}}}\right)^2 - \left(\sqrt{\frac{GM_{\mathrm{E}}}{r_0}}\right)^2\right] = \frac{GM_{\mathrm{E}}m}{2}\left(\frac{1}{r_{\mathrm{f}}} - \frac{1}{r_0}\right)$$

Therefore,

$$W = \frac{(6.67 \times 10^{-11}\text{ N}\cdot\text{m}^2/\text{kg}^2)(5.98 \times 10^{24}\text{ kg})(6200\text{ kg})}{2}$$

$$\times\left(\frac{1}{7.0 \times 10^6\text{ m}} - \frac{1}{3.3 \times 10^7\text{ m}}\right) = \boxed{1.4 \times 10^{11}\text{ J}}$$

25. **_REASONING_** According to the work-energy theorem, the kinetic energy of the sled increases in each case because work is done on the sled. The work-energy theorem is given by Equation 6.3: $W = KE_f - KE_0 = \frac{1}{2}mv_f^2 - \frac{1}{2}mv_0^2$. The work done on the sled is given by Equation 6.1: $W = (F\cos\theta)s$. The work done in each case can, therefore, be expressed as

$$W_1 = (F\cos 0°)s = \frac{1}{2}mv_f^2 - \frac{1}{2}mv_0^2 = \Delta KE_1$$

and

$$W_2 = (F\cos 62°)s = \frac{1}{2}mv_f^2 - \frac{1}{2}mv_0^2 = \Delta KE_2$$

The fractional increase in the kinetic energy of the sled when $\theta = 0°$ is

$$\frac{\Delta KE_1}{KE_0} = \frac{(F\cos 0°)s}{KE_0} = 0.38$$

Therefore,

$$Fs = (0.38)\,KE_0 \tag{1}$$

The fractional increase in the kinetic energy of the sled when $\theta = 62°$ is

$$\frac{\Delta KE_2}{KE_0} = \frac{(F\cos 62°)s}{KE_0} = \frac{Fs}{KE_0}(\cos 62°) \tag{2}$$

Equation (1) can be used to substitute for Fs in Equation (2).

SOLUTION Combining Equations (1) and (2), we have

$$\frac{\Delta KE_2}{KE_0} = \frac{Fs}{KE_0}(\cos 62°) = \frac{(0.38)\,KE_0}{KE_0}(\cos 62°) = (0.38)(\cos 62°) = 0.18$$

Thus, the sled's kinetic energy would increase by $\boxed{18\,\%}$.

---

33. **_REASONING_** During each portion of the trip, the work done by the resistive force is given by Equation 6.1, $W = (F\cos\theta)s$. Since the resistive force always points opposite to the displacement of the bicyclist, $\theta = 180°$; hence, on each part of the trip, $W = (F\cos 180°)s = -Fs$. The work done by the resistive force during the round trip is the algebraic sum of the work done during each portion of the trip.

SOLUTION
a. The work done during the round trip is, therefore,

$$W_{total} = W_1 + W_2 = -F_1 s_1 - F_2 s_2$$

$$= -(3.0\text{ N})(5.0\times10^3\text{ m}) - (3.0\text{ N})(5.0\times10^3\text{ m}) = \boxed{-3.0\times10^4\text{ J}}$$

b. Since the work done by the resistive force over the closed path is *not* zero, we can conclude that the resistive force is *not* a conservative force.

---

35. **REASONING AND SOLUTION**

a. The work done by non-conservative forces is given by Equation 6.7b as

$$W_{nc} = \Delta KE + \Delta PE \qquad \text{so} \qquad \Delta PE = W_{nc} - \Delta KE$$

Now

$$\Delta KE = \tfrac{1}{2}mv_f^2 - \tfrac{1}{2}mv_0^2 = \tfrac{1}{2}(55.0 \text{ kg})[(6.00 \text{ m/s})^2 - (1.80 \text{ m/s})^2] = 901 \text{ J}$$

and

$$\Delta PE = 80.0 \text{ J} - 265 \text{ J} - 901 \text{ J} = \boxed{-1086 \text{ J}}$$

b. $\Delta PE = mg(h - h_0)$ so

$$h - h_0 = \frac{\Delta PE}{mg} = \frac{-1086 \text{ J}}{(55.0 \text{ kg})(9.80 \text{ m/s}^2)} = -2.01 \text{ m}$$

This answer is negative, so that the skater's vertical position has changed by $\boxed{2.01 \text{ m}}$ and the skater is $\boxed{\text{below the starting point}}$.

---

37. **REASONING** The only two forces that act on the gymnast are his weight and the force exerted on his hands by the high bar. The latter is the (non-conservative) reaction force to the force exerted on the bar by the gymnast, as predicted by Newton's third law. This force, however, does no work because it points perpendicular to the circular path of motion. Thus, $W_{nc} = 0$ J, and we can apply the principle of conservation of mechanical energy.

SOLUTION The conservation principle gives

$$\underbrace{\tfrac{1}{2}mv_f^2 + mgh_f}_{E_f} = \underbrace{\tfrac{1}{2}mv_0^2 + mgh_0}_{E_0}$$

Since the gymnast's speed is momentarily zero at the top of the swing, $v_0 = 0$ m/s. If we take $h_f = 0$ m at the bottom of the swing, then $h_0 = 2r$, where r is the radius of the circular path followed by the gymnast's waist. Making these substitutions in the above expression and solving for v_f, we obtain

$$v_f = \sqrt{2gh_0} = \sqrt{2g(2r)} = \sqrt{2(9.80 \text{ m/s}^2)(2 \times 1.1 \text{ m})} = \boxed{6.6 \text{ m/s}}$$

43. ***REASONING*** To find the maximum height H above the end of the track we will analyze the projectile motion of the skateboarder after she leaves the track. For this analysis we will use the principle of conservation of mechanical energy, which applies because friction and air resistance are being ignored. In applying this principle to the projectile motion, however, we will need to know the speed of the skateboarder when she leaves the track. Therefore, we will begin by determining this speed, also using the conservation principle in the process. Our approach, then, uses the conservation principle twice.

SOLUTION Applying the conservation of mechanical energy in the form of Equation 6.9b, we have

$$\underbrace{\tfrac{1}{2}mv_f^2 + mgh_f}_{\substack{\text{Final mechanical energy}\\\text{at end of track}}} = \underbrace{\tfrac{1}{2}mv_0^2 + mgh_0}_{\substack{\text{Initial mechanical energy}\\\text{on flat part of track}}}$$

We designate the flat portion of the track as having a height $h_0 = 0$ m and note from the drawing that its end is at a height of $h_f = 0.40$ m above the ground. Solving for the final speed at the end of the track gives

$$v_f = \sqrt{v_0^2 + 2g(h_0 - h_f)} = \sqrt{(5.4 \text{ m/s})^2 + 2(9.80 \text{ m/s}^2)[(0 \text{ m}) - (0.40 \text{ m})]} = 4.6 \text{ m/s}$$

This speed now becomes the initial speed $v_0 = 4.6$ m/s for the next application of the conservation principle. At the maximum height of her trajectory she is traveling horizontally with a speed v_f that equals the horizontal component of her launch velocity. Thus, for the next application of the conservation principle $v_f = (4.6 \text{ m/s}) \cos 48°$. Applying the conservation of mechanical energy again, we have

$$\underbrace{\tfrac{1}{2}mv_f^2 + mgh_f}_{\substack{\text{Final mechanical energy}\\\text{at maximum height of trajectory}}} = \underbrace{\tfrac{1}{2}mv_0^2 + mgh_0}_{\substack{\text{Initial mechanical energy}\\\text{upon leaving the track}}}$$

Recognizing that $h_0 = 0.40$ m and $h_f = 0.40$ m $+ H$ and solving for H give

$$\tfrac{1}{2}mv_f^2 + mg[(0.40 \text{ m}) + H] = \tfrac{1}{2}mv_0^2 + mg(0.40 \text{ m})$$

$$H = \frac{v_0^2 - v_f^2}{2g} = \frac{(4.6 \text{ m/s})^2 - [(4.6 \text{ m/s})\cos 48°]^2}{2(9.80 \text{ m/s}^2)} = \boxed{0.60 \text{ m}}$$

45. ***REASONING*** Gravity is the only force acting on the vaulters, since friction and air resistance are being ignored. Therefore, the net work done by the nonconservative forces is zero, and the principle of conservation of mechanical energy holds.

SOLUTION Let E_{2f} represent the total mechanical energy of the second vaulter at the ground, and E_{20} represent the total mechanical energy of the second vaulter at the bar. Then, the principle of mechanical energy is written for the second vaulter as

$$\underbrace{\tfrac{1}{2}mv_{2f}^2 + mgh_{2f}}_{E_{2f}} = \underbrace{\tfrac{1}{2}mv_{20}^2 + mgh_{20}}_{E_{20}}$$

Since the mass m of the vaulter appears in every term of the equation, m can be eliminated algebraically. The quantity h_{20} is $h_{20} = h$, where h is the height of the bar. Furthermore, when the vaulter is at ground level, $h_{2f} = 0$ m. Solving for v_{20} we have

$$v_{20} = \sqrt{v_{2f}^2 - 2gh} \qquad (1)$$

In order to use Equation (1), we must first determine the height h of the bar. The height h can be determined by applying the principle of conservation of mechanical energy to the first vaulter on the ground and at the bar. Using notation similar to that above, we have

$$\underbrace{\tfrac{1}{2}mv_{1f}^2 + mgh_{1f}}_{E_{1f}} = \underbrace{\tfrac{1}{2}mv_{10}^2 + mgh_{10}}_{E_{10}}$$

Where E_{10} corresponds to the total mechanical energy of the first vaulter at the bar. The height of the bar is, therefore,

$$h = h_{10} = \frac{v_{1f}^2 - v_{10}^2}{2g} = \frac{(8.90 \text{ m/s})^2 - (1.00 \text{ m/s})^2}{2(9.80 \text{ m/s}^2)} = 3.99 \text{ m}$$

The speed at which the second vaulter clears the bar is, from Equation (1),

$$v_{20} = \sqrt{v_{2f}^2 - 2gh} = \sqrt{(9.00 \text{ m/s})^2 - 2(9.80 \text{ m/s}^2)(3.99 \text{ m})} = \boxed{1.7 \text{ m/s}}$$

---

51. **REASONING**

a. Since there is no air friction, the only force that acts on the projectile is the conservative gravitational force (its weight). The initial and final speeds of the ball are known, so the conservation of mechanical energy can be used to find the maximum height that the projectile attains.

b. When air resistance, a nonconservative force, is present, it does negative work on the projectile and slows it down. Consequently, the projectile does not rise as high as when there is no air resistance. The work-energy theorem, in the form of Equation 6.6, may be used to find the work done by air friction. Then, using the definition of work, Equation 6.1, the average force due to air resistance can be found.

SOLUTION

a. The conservation of mechanical energy, as expressed by Equation 6.9b, states that

$$\underbrace{\tfrac{1}{2}mv_{\mathrm f}^2 + mgh_{\mathrm f}}_{E_{\mathrm f}} = \underbrace{\tfrac{1}{2}mv_0^2 + mgh_0}_{E_0}$$

The mass m can be eliminated algebraically from this equation since it appears as a factor in every term. Solving for the final height $h_{\mathrm f}$ gives

$$h_{\mathrm f} = \frac{\tfrac{1}{2}\left(v_0^2 - v_{\mathrm f}^2\right)}{g} + h_0$$

Setting $h_0 = 0$ m and $v_{\mathrm f} = 0$ m/s, the final height, in the absence of air resistance, is

$$h_{\mathrm f} = \frac{v_0^2 - v_{\mathrm f}^2}{2g} = \frac{(18.0 \text{ m/s})^2 - (0 \text{ m/s})^2}{2(9.80 \text{ m/s}^2)} = \boxed{16.5 \text{ m}}$$

b. The work-energy theorem is

$$W_{\mathrm{nc}} = \left(\tfrac{1}{2}mv_{\mathrm f}^2 - \tfrac{1}{2}mv_0^2\right) + \left(mgh_{\mathrm f} - mgh_0\right) \tag{6.6}$$

where W_{nc} is the nonconservative work done by air resistance. According to Equation 6.1, the work can be written as $W_{\mathrm{nc}} = \left(\overline{F}_{\mathrm R} \cos 180°\right)s$, where $\overline{F}_{\mathrm R}$ is the average force of air resistance. As the projectile moves upward, the force of air resistance is directed downward, so the angle between the two vectors is $\theta = 180°$ and $\cos\theta = -1$. The magnitude s of the displacement is the difference between the final and initial heights, $s = h_{\mathrm f} - h_0 = 11.8$ m. With these substitutions, the work-energy theorem becomes

$$-\overline{F}_{\mathrm R}\, s = \tfrac{1}{2}m\left(v_{\mathrm f}^2 - v_0^2\right) + mg\left(h_{\mathrm f} - h_0\right)$$

Solving for $\overline{F}_{\mathrm R}$ gives

$$\overline{F}_{\mathrm R} = \frac{\tfrac{1}{2}m\left(v_{\mathrm f}^2 - v_0^2\right) + mg\left(h_{\mathrm f} - h_0\right)}{-s}$$

$$= \frac{\tfrac{1}{2}(0.750 \text{ kg})\left[(0 \text{ m/s})^2 - (18.0 \text{ m/s})^2\right] + (0.750 \text{ kg})(9.80 \text{ m/s}^2)(11.8 \text{ m})}{-(11.8 \text{ m})} = \boxed{2.9 \text{ N}}$$

---

55. ***REASONING*** The work-energy theorem can be used to determine the net work done on the car by the nonconservative forces of friction and air resistance. The work-energy theorem is, according to Equation 6.8,

$$W_{\mathrm{nc}} = \left(\tfrac{1}{2}mv_{\mathrm f}^2 + mgh_{\mathrm f}\right) - \left(\tfrac{1}{2}mv_0^2 + mgh_0\right)$$

The nonconservative forces are the force of friction, the force of air resistance, and the force provided by the engine. Thus, we can rewrite Equation 6.8 as

$$W_{\text{friction}} + W_{\text{air}} + W_{\text{engine}} = \left(\tfrac{1}{2}mv_f^2 + mgh_f\right) - \left(\tfrac{1}{2}mv_0^2 + mgh_0\right)$$

This expression can be solved for $W_{\text{friction}} + W_{\text{air}}$.

SOLUTION We will measure the heights from sea level, where $h_0 = 0$ m. Since the car starts from rest, $v_0 = 0$ m/s. Thus, we have

$$W_{\text{friction}} + W_{\text{air}} = m\left(\tfrac{1}{2}v_f^2 + gh_f\right) - W_{\text{engine}}$$

$$= (1.50\times10^3 \text{ kg})\left[\tfrac{1}{2}(27.0 \text{ m/s})^2 + (9.80 \text{ m/s}^2)(2.00\times10^2 \text{ m})\right] - (4.70\times10^6 \text{ J})$$

$$= \boxed{-1.21\times10^6 \text{ J}}$$

---

61. **REASONING** After the wheels lock, the only nonconservative force acting on the truck is friction. The work done by this conservative force can be determined from the work-energy theorem. According to Equation 6.8,

$$W_{\text{nc}} = E_f - E_0 = \left(\tfrac{1}{2}mv_f^2 + mgh_f\right) - \left(\tfrac{1}{2}mv_0^2 + mgh_0\right) \tag{1}$$

where W_{nc} is the work done by friction. According to Equation 6.1, $W = (F\cos\theta)s$; since the force of kinetic friction points opposite to the displacement, $\theta = 180°$. According to Equation 4.8, the kinetic frictional force has a magnitude of $f_k = \mu_k F_N$, where μ_k is the coefficient of kinetic friction and F_N is the magnitude of the normal force. Thus,

$$W_{\text{nc}} = (f_k\cos\theta)s = \mu_k F_N\,(\cos 180°)\,s = -\mu_k F_N s \tag{2}$$

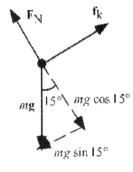

Since the truck is sliding down an incline, we refer to the free-body diagram in order to determine the magnitude of the normal force F_N. The free-body diagram at the right shows the three forces that act on the truck. They are the normal force $\mathbf{F_N}$, the force of kinetic friction $\mathbf{f_k}$, and the weight $m\mathbf{g}$. The weight has been resolved into its components, and these vectors are shown as dashed arrows. From the free body diagram and the fact that the truck does not accelerate perpendicular to the incline, we can see that the magnitude of the normal force is given by

$$F_N = mg\cos 15.0°$$

Therefore, Equation (2) becomes

$$W_{nc} = -\mu_k \left(mg \cos 15.0° \right) s \tag{3}$$

Combining Equations (1), (2), and (3), we have

$$-\mu_k \left(mg \cos 15.0° \right) s = \left(\tfrac{1}{2} mv_f^2 + mgh_f \right) - \left(\tfrac{1}{2} mv_0^2 + mgh_0 \right) \tag{4}$$

This expression may be solved for s.

SOLUTION When the car comes to a stop, $v_f = 0$ m/s. If we take $h_f = 0$ m at ground level, then, from the drawing at the right, we see that $h_0 = s \sin 15.0°$. Equation (4) then gives

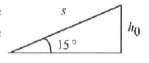

$$s = \frac{v_0^2}{2g \left(\mu_k \cos 15.0° - \sin 15.0° \right)} = \frac{\left(11.1 \text{ m/s} \right)^2}{2 \left(9.80 \text{ m/s}^2 \right) \left[\left(0.750 \right) \cos 15.0° - \sin 15.0° \right]} = \boxed{13.5 \text{ m}}$$

---

63. **REASONING** The work W done is equal to the average power $\bar{P}$ multiplied by the time t, or

$$W = \bar{P}t \tag{6.10a}$$

The average power is the average power generated per kilogram of body mass multiplied by Armstrong's mass. The time of the race is the distance s traveled divided by the average speed $\bar{v}$, or $t = s / \bar{v}$ (see Equation 2.1).

SOLUTION
a. Substituting $t = s / \bar{v}$ into Equation 6.10a gives

$$W = \bar{P}t = \bar{P} \left(\frac{s}{\bar{v}} \right) = \underbrace{\left(6.50 \ \frac{\text{W}}{\text{kg}} \right) \left(75.0 \text{ kg} \right)}_{\bar{P}} \left(\frac{135 \times 10^3 \text{ m}}{12.0 \text{ m/s}} \right) = \boxed{5.48 \times 10^6 \text{ J}}$$

b. Since 1 joule = 2.389×10^{-4} nutritional calories, the work done is

$$W = 5.48 \times 10^6 \text{ joules} = \left(5.48 \times 10^6 \text{ joules} \right) \left(\frac{2.389 \times 10^{-4} \text{ nutritional calories}}{1 \text{ joule}} \right)$$

$$= \boxed{1.31 \times 10^3 \text{ nutritional calories}}$$

69. **REASONING AND SOLUTION** In the following drawings, the positive direction is to the right. When the boat is not pulling a skier, the engine must provide a force F_1 to overcome the resistive force of the water F_R. Since the boat travels with constant speed, these forces must be equal in magnitude and opposite in direction.

$$-F_R + F_1 = ma = 0$$

or

$$F_R = F_1 \qquad (1)$$

When the boat is pulling a skier, the engine must provide a force F_2 to balance the resistive force of the water F_R and the tension in the tow rope T.

$$-F_R + F_2 - T = ma = 0$$

or

$$F_2 = F_R + T \qquad (2)$$

The average power is given by $\overline{P} = \overline{F}\overline{v}$, according to Equation 6.11 in the text. Since the boat moves with the same speed in both cases, $v_1 = v_2$, and we have

$$\frac{\overline{P_1}}{F_1} = \frac{\overline{P_2}}{F_2} \qquad \text{or} \qquad F_2 = F_1 \frac{\overline{P_2}}{\overline{P_1}}$$

Using Equations (1) and (2), this becomes

$$F_1 + T = F_1 \frac{\overline{P_2}}{\overline{P_1}}$$

Solving for T gives

$$T = F_1 \left(\frac{\overline{P_2}}{\overline{P_1}} - 1 \right)$$

The force F_1 can be determined from $\overline{P_1} = F_1 v_1$, thereby giving

$$T = \frac{\overline{P_1}}{v_1} \left(\frac{\overline{P_2}}{\overline{P_1}} - 1 \right) = \frac{7.50 \times 10^4 \text{ W}}{12 \text{ m/s}} \left(\frac{8.30 \times 10^4 \text{ W}}{7.50 \times 10^4 \text{ W}} - 1 \right) = \boxed{6.7 \times 10^2 \text{ N}}$$

___

71. **REASONING** The area under the force-versus-displacement graph over any displacement interval is equal to the work done over that displacement interval.

SOLUTION
a. Since the area under the force-versus-displacement graph from $s = 0$ to 0.50 m is greater for bow 1, it follows that $\boxed{\text{bow 1 requires more work to draw the bow fully}}$.

b. The work required to draw bow 1 is equal to the area of the triangular region under the force-versus-displacement curve for bow 1. Since the area of a triangle is equal one-half times the base of the triangle times the height of the triangle, we have

$$W_1 = \frac{1}{2}(0.50\text{m})(350\,\text{N}) = 88\,\text{J}$$

For bow 2, we note that each small square under the force-versus-displacement graph has an area of

$$(0.050\text{ m})(40.0\text{ N}) = 2.0\text{ J}$$

We estimate that there are approximately 31.3 squares under the force-versus-displacement graph for bow 2; therefore, the total work done is

$$(31.3\text{ squares})\left(\frac{2.0\text{ J}}{\text{square}}\right) = 63\text{ J}$$

Therefore, the additional work required to stretch bow 1 as compared to bow 2 is

$$88\text{ J} - 63\text{ J} = \boxed{25\text{ J}}$$

---

77. **REASONING** No forces other than gravity act on the rock since air resistance is being ignored. Thus, the net work done by nonconservative forces is zero, $W_{nc} = 0$ J. Consequently, the principle of conservation of mechanical energy holds, so the total mechanical energy remains constant as the rock falls.

If we take $h = 0$ m at ground level, the gravitational potential energy at any height h is, according to Equation 6.5, $\text{PE} = mgh$. The kinetic energy of the rock is given by Equation 6.2: $\text{KE} = \frac{1}{2}mv^2$. In order to use Equation 6.2, we must have a value for v^2 at each desired height h. The quantity v^2 can be found from Equation 2.9 with $v_0 = 0$ m/s, since the rock is released from rest. At a height h, the rock has fallen through a distance $(20.0\text{ m}) - h$, and according to Equation 2.9, $v^2 = 2ay = 2a[(20.0\text{ m}) - h]$. Therefore, the kinetic energy at any height h is given by $\text{KE} = ma[(20.0\text{ m}) - h]$. The total energy at any height h is the sum of the kinetic energy and potential energy at the particular height.

SOLUTION The calculations are performed below for $h = 10.0$ m. The table that follows also shows the results for $h = 20.0$ m and $h = 0$ m.

$$\text{PE} = mgh = (2.00\text{ kg})(9.80\text{ m/s}^2)(10.0\text{ m}) = \boxed{196\text{ J}}$$

$$\text{KE} = ma[(20.0\text{ m}) - h] = (2.00\text{ kg})(9.80\text{ m/s}^2)(20.0\text{ m} - 10.0\text{ m}) = \boxed{196\text{ J}}$$

$$E = \text{KE} + \text{PE} = 196\text{ J} + 196\text{ J} = \boxed{392\text{ J}}$$

h (m)	KE (J)	PE (J)	E (J)
20.0	0	392	392
10.0	196	196	392
0	392	0	392

Note that in each case, the value of E is the same, because mechanical energy is conserved.

---

79. **REASONING** The work W done by the tension in the tow rope is given by Equation 6.1 as $W = (F\cos\theta)s$, where F is the magnitude of the tension, s is the magnitude of the skier's displacement, and θ is the angle between the tension and the displacement vectors. The magnitude of the displacement (or the distance) is the speed v of the skier multiplied by the time t (see Equation 2.1), or $s = vt$.

SOLUTION Substituting $s = vt$ into the expression for the work, $W = (F\cos\theta)s$, we have $W = (F\cos\theta)vt$. Since the skier moves parallel to the boat and since the tow rope is parallel to the water, the angle between the tension and the skier's displacement is $\theta = 37.0°$. Thus, the work done by the tension is

$$W = (F\cos\theta)vt = [(135\text{ N})\cos 37.0°](9.30\text{ m/s})(12.0\text{ s}) = \boxed{1.20\times10^4\text{ J}}$$

---

83. **REASONING AND SOLUTION**
a. The lost mechanical energy is

$$E_{lost} = E_0 - E_f$$

The ball is dropped from rest, so its initial energy is purely potential. The ball is momentarily at rest at the highest point in its rebound, so its final energy is also purely potential. Then

$$E_{lost} = mgh_0 - mgh_f = (0.60\text{ kg})(9.80\text{ m/s}^2)[(1.05\text{ m})-(0.57\text{ m})] = \boxed{2.8\text{ J}}$$

b. The work done by the player must compensate for this loss of energy.

$$E_{lost} = \underbrace{(F\cos\theta)s}_{\text{work done by player}} \Rightarrow F = \frac{E_{lost}}{(\cos\theta)s} = \frac{2.8\text{ J}}{(\cos 0°)(0.080\text{ m})} = \boxed{35\text{ N}}$$

---

89. **CONCEPTS**

The work done by a force depends on the component of the force that points along the direction of the displacement. The work is positive if that component points in the same direction as the displacement. The work is negative if that component points opposite to the displacement. The work done by the weight $\vec{W}$ is positive because the weight has a component that points in the same direction as the displacement. The top drawing in part (*b*) of the figure shows this weight component.

The work done by the <u>frictional force</u> $\vec{f}$ is negative because it points opposite to the direction of the displacement ($\theta = 180°$ in the middle drawing).

The work done by the <u>normal force</u> $\vec{F}_N$ is zero, because the normal force is perpendicular to the displacement ($\theta = 90°$ in the bottom drawing) and does not have a component along the displacement.

CALCULATIONS The work W done by a force is given by Equation 6.1 as $W = (F\cos\theta)s$ where F is the magnitude of the force, s is the magnitude of the displacement, and θ is the angle between the force and the displacement. The figure shows this angle for each of the three forces. The work done by each force is computed in the following table. The net work is the algebraic sum of the three values.

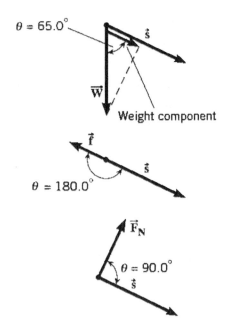

Force	Magnitude F of the Force	Angle θ	Magnitude s of the displacement	Work done by the Force $W = (F\cos\theta)s$
$\vec{W}$	675 N	65.0°	9.2 m	$W = (675\ \text{N})(\cos 65.0°)$ $\times (9.2\ \text{m}) = +2620\ \text{J}$
$\vec{f}$	125 N	180.0°	9.2 m	$W = (125\ \text{N})(\cos 180.0°)$ $\times (9.2\ \text{m}) = -1150\ \text{J}$
$\vec{F}_N$	612 N	90.0°	9.2 m	$W = (612\ \text{N})(\cos 90.0°)$ $\times (9.2\ \text{m}) = 0\ \text{J}$

The net work done by the three forces is: $(+2620 \text{ J}) + (-1150 \text{ J}) + (0 \text{ J}) = \boxed{+1470 \text{ J}}$

CHAPTER 7 | *IMPULSE AND MOMENTUM*

1. ***REASONING*** The impulse that the wall exerts on the skater can be found from the impulse-momentum theorem, Equation 7.4. The average force $\overline{F}$ exerted on the skater by the wall is the only force exerted on her in the horizontal direction, so it is the net force; $\Sigma\overline{F} = \overline{F}$.

   SOLUTION From Equation 7.4, the average force exerted on the skater by the wall is

   $$\overline{F} = \frac{m\mathbf{v}_f - m\mathbf{v}_0}{\Delta t} = \frac{(46 \text{ kg})(-1.2 \text{ m/s}) - (46 \text{ kg})(0 \text{ m/s})}{0.80 \text{ s}} = -69 \text{ N}$$

   From Newton's third law, the average force exerted on the wall by the skater is equal in magnitude and opposite in direction to this force. Therefore,

   $$\textbf{Force exerted on wall} = \boxed{+69 \text{ N}}$$

   The plus sign indicates that this force points $\boxed{\text{opposite to the velocity}}$ of the skater.

---

5. ***REASONING*** The impulse that the volleyball player applies to the ball can be found from the impulse-momentum theorem, Equation 7.4. Two forces act on the volleyball while it's being spiked: an average force $\overline{F}$ exerted by the player, and the weight of the ball. As in Example 1, we will assume that $\overline{F}$ is much greater than the weight of the ball, so the weight can be neglected. Thus, the net average force $(\Sigma\overline{F})$ is equal to $\overline{F}$.

   SOLUTION From Equation 7.4, the impulse that the player applies to the volleyball is

   $$\underbrace{\overline{F}\,\Delta t}_{\text{Impulse}} = \underbrace{m\mathbf{v}_f}_{\substack{\text{Final} \\ \text{momentum}}} - \underbrace{m\mathbf{v}_0}_{\substack{\text{Initial} \\ \text{momentum}}}$$

   $$= m(\mathbf{v}_f - \mathbf{v}_0) = (0.35 \text{ kg})\left[(-21 \text{ m/s}) - (+4.0 \text{ m/s})\right] = \boxed{-8.7 \text{ kg}\cdot\text{m/s}}$$

   The minus sign indicates that the direction of the impulse is the same as that of the final velocity of the ball.

---

19. ***REASONING*** The system consists of the lumberjack and the log. For this system, the sum of the external forces is zero. This is because the weight of the system is balanced by the corresponding normal force (provided by the buoyant force of the water) and the water is assumed to be frictionless. The lumberjack and the log, then, constitute an isolated system, and the principle of conservation of linear momentum holds.

SOLUTION

a.   The total linear momentum of the system before the lumberjack begins to move is zero, since all parts of the system are at rest. Momentum conservation requires that the total momentum remains zero during the motion of the lumberjack.

$$\underbrace{m_1 v_{f1} + m_2 v_{f2}}_{\substack{\text{Total momentum} \\ \text{just before the jump}}} = \underbrace{0}_{\text{Initial momentum}}$$

Here the subscripts "1" and "2" refer to the first log and lumberjack, respectively. Let the direction of motion of the lumberjack be the positive direction. Then, solving for v_{1f} gives

$$v_{f1} = -\frac{m_2 v_{f2}}{m_1} = -\frac{(98 \text{ kg})(+3.6 \text{ m/s})}{230 \text{ kg}} = \boxed{-1.5 \text{ m/s}}$$

The minus sign indicates that the first log recoils as the lumberjack jumps off.

b.   Now the system is composed of the lumberjack, just before he lands on the second log, and the second log. Gravity acts on the system, but for the short time under consideration while the lumberjack lands, the effects of gravity in changing the linear momentum of the system are negligible. Therefore, to a very good approximation, we can say that the linear momentum of the system is very nearly conserved. In this case, the initial momentum is not zero as it was in part (a); rather the initial momentum of the system is the momentum of the lumberjack just before he lands on the second log. Therefore,

$$\underbrace{m_1 v_{f1} + m_2 v_{f2}}_{\substack{\text{Total momentum} \\ \text{just after lumberjack lands}}} = \underbrace{m_1 v_{01} + m_2 v_{02}}_{\text{Initial momentum}}$$

In this expression, the subscripts "1" and "2" now represent the second log and lumberjack, respectively. Since the second log is initially at rest, $v_{01} = 0$. Furthermore, since the lumberjack and the second log move with a common velocity, $v_{f1} = v_{f2} = v_f$. The statement of momentum conservation then becomes

$$m_1 v_f + m_2 v_f = m_2 v_{02}$$

Solving for v_f, we have

$$v_f = \frac{m_2 v_{02}}{m_1 + m_2} = \frac{(98 \text{ kg})(+3.6 \text{ m/s})}{230 \text{ kg} + 98 \text{ kg}} = \boxed{+1.1 \text{ m/s}}$$

The positive sign indicates that the system moves in the same direction as the original direction of the lumberjack's motion.

---

21. *REASONING* The two-stage rocket constitutes the system. The forces that act to cause the separation during the explosion are, therefore, forces that are internal to the system. Since no external forces act on this system, it is isolated and the principle of conservation of linear momentum applies:

$$\underbrace{m_1 v_{f1} + m_2 v_{f2}}_{\substack{\text{Total momentum} \\ \text{after separation}}} = \underbrace{(m_1 + m_2) v_0}_{\substack{\text{Total momentum} \\ \text{before separation}}}$$

where the subscripts "1" and "2" refer to the lower and upper stages, respectively. This expression can be solved for v_{f1}.

SOLUTION   Solving for v_{f1} gives

$$v_{f1} = \frac{(m_1 + m_2) v_0 - m_2 v_{f2}}{m_1}$$

$$= \frac{\left[2400 \text{ kg} + 1200 \text{ kg}\right](4900 \text{ m/s}) - (1200 \text{ kg})(5700 \text{ m/s})}{2400 \text{ kg}} = \boxed{+4500 \text{ m/s}}$$

Since v_{f1} is positive, $\boxed{\text{its direction is the same as the rocket before the explosion}}$.

---

25. **REASONING** No net external force acts on the plate parallel to the floor; therefore, the component of the momentum of the plate that is parallel to the floor is conserved as the plate breaks and flies apart. Initially, the total momentum parallel to the floor is zero. After the collision with the floor, the component of the total momentum parallel to the floor must remain zero. The drawing in the text shows the pieces in the plane parallel to the floor just after the collision. Clearly, the linear momentum in the plane parallel to the floor has two components; therefore the linear momentum of the plate must be conserved in each of these two mutually perpendicular directions. Using the drawing in the text, with the positive directions taken to be up and to the right, we have

$$\boxed{x \text{ direction}} \qquad -m_1 v_1 (\sin 25.0°) + m_2 v_2 (\cos 45.0°) = 0 \qquad (1)$$

$$\boxed{y \text{ direction}} \qquad m_1 v_1 (\cos 25.0°) + m_2 v_2 (\sin 45.0°) - m_3 v_3 = 0 \qquad (2)$$

These equations can be solved simultaneously for the masses m_1 and m_2.

SOLUTION Using the values given in the drawing for the velocities after the plate breaks, we have,

$$-m_1 (3.00 \text{ m/s}) \sin 25.0° + m_2 (1.79 \text{ m/s}) \cos 45.0° = 0 \qquad (1)$$

$$m_1 (3.00 \text{ m/s}) \cos 25.0° + m_2 (1.79 \text{ m/s}) \sin 45.0° - (1.30 \text{ kg})(3.07 \text{ m/s}) = 0 \qquad (2)$$

Subtracting (2) from (1), and noting that $\cos 45.0° = \sin 45.0°$, gives $\boxed{m_1 = 1.00 \text{ kg}}$.

Substituting this value into either (1) or (2) then yields $\boxed{m_2 = 1.00 \text{ kg}}$.

---

27. **REASONING** The cannon and the shell constitute the system. Since no external force hinders the motion of the system after the cannon is unbolted, conservation of linear momentum applies in that case. If we assume that the burning gun powder imparts the same kinetic energy to the system in each case, we have sufficient information to develop a mathematical description of this situation, and solve it for the velocity of the shell fired by the loose cannon.

SOLUTION For the case where the cannon is unbolted, momentum conservation gives

$$\underbrace{m_1 v_{f1} + m_2 v_{f2}}_{\substack{\text{Total momentum} \\ \text{after shell is fired}}} = \underbrace{0}_{\substack{\text{Initial momentum} \\ \text{of system}}} \tag{1}$$

where the subscripts "1" and "2" refer to the cannon and shell, respectively. In both cases, the burning gun power imparts the same kinetic energy to the system. When the cannon is bolted to the ground, only the shell moves and the kinetic energy imparted to the system is

$$\text{KE} = \tfrac{1}{2} m_{\text{shell}} v_{\text{shell}}^2 = \tfrac{1}{2}(85.0 \text{ kg})(551 \text{ m/s})^2 = 1.29 \times 10^7 \text{ J}$$

The kinetic energy imparted to the system when the cannon is unbolted has the same value and can be written using the same notation as in equation (1):

$$\text{KE} = \tfrac{1}{2} m_1 v_{f1}^2 + \tfrac{1}{2} m_2 v_{f2}^2 \tag{2}$$

Solving equation (1) for v_{f1}, the velocity of the cannon after the shell is fired, and substituting the resulting expression into Equation (2) gives

$$\text{KE} = \frac{m_2^2 v_{f2}^2}{2m_1} + \tfrac{1}{2} m_2 v_{f2}^2 \tag{3}$$

Solving equation (3) for v_{f2} gives

$$v_{f2} = \sqrt{\frac{2\text{KE}}{m_2\left(\dfrac{m_2}{m_1}+1\right)}} = \sqrt{\frac{2(1.29 \times 10^7 \text{ J})}{(85.0 \text{ kg})\left(\dfrac{85.0 \text{ kg}}{5.80 \times 10^3 \text{ kg}}+1\right)}} = \boxed{+547 \text{ m/s}}$$

---

31. **REASONING** Batman and the boat with the criminal constitute the system. Gravity acts on this system as an external force; however, gravity acts vertically, and we are concerned only with the horizontal motion of the system. If we neglect air resistance and friction, there are no external forces that act horizontally; therefore, the total linear momentum in the horizontal direction is conserved. When Batman collides with the boat, the horizontal component of his velocity is zero, so the statement of conservation of linear momentum in the horizontal direction can be written as

$$\underbrace{(m_1 + m_2)v_f}_{\substack{\text{Total horizontal momentum} \\ \text{after collision}}} = \underbrace{m_1 v_{01} + 0}_{\substack{\text{Total horizontal momentum} \\ \text{before collision}}}$$

Here, m_1 is the mass of the boat, and m_2 is the mass of Batman. This expression can be solved for v_f, the velocity of the boat after Batman lands in it.

SOLUTION Solving for v_f gives

$$v_f = \frac{m_1 v_{01}}{m_1 + m_2} = \frac{(510 \text{ kg})(+11 \text{ m/s})}{510 \text{ kg} + 91 \text{ kg}} = \boxed{+9.3 \text{ m/s}}$$

The plus sign indicates that the boat continues to move in its initial direction of motion.

---

33. **REASONING** The system consists of the two balls. The total linear momentum of the two-ball system is conserved because the net external force acting on it is zero. The principle of conservation of linear momentum applies whether or not the collision is elastic.

$$\underbrace{m_1 v_{f1} + m_2 v_{f2}}_{\substack{\text{Total momentum} \\ \text{after collision}}} = \underbrace{m_1 v_{01} + 0}_{\substack{\text{Total momentum} \\ \text{before collision}}}$$

When the collision is elastic, the kinetic energy is also conserved during the collision

$$\underbrace{\tfrac{1}{2}m_1 v_{f1}^2 + \tfrac{1}{2}m_2 v_{f2}^2}_{\substack{\text{Total kinetic energy} \\ \text{after collision}}} = \underbrace{\tfrac{1}{2}m_1 v_{01}^2 + 0}_{\substack{\text{Total kinetic energy} \\ \text{before collision}}}$$

SOLUTION
a. The final velocities for an elastic collision are determined by simultaneously solving the above equations for the final velocities. The procedure is discussed in Example 7 in the text, and leads to Equations 7.8a and 7.8b. According to Equation 7.8:

$$v_{f1} = \left(\frac{m_1 - m_2}{m_1 + m_2} \right) v_{01} \qquad \text{and} \qquad v_{f2} = \left(\frac{2m_1}{m_1 + m_2} \right) v_{01}$$

Let the initial direction of motion of the 5.00-kg ball define the positive direction. Substituting the values given in the text, these equations give

[5.00-kg ball] $\qquad v_{f1} = \left(\dfrac{5.00 \text{ kg} - 7.50 \text{ kg}}{5.00 \text{ kg} + 7.50 \text{ kg}} \right)(2.00 \text{ m/s}) = \boxed{-0.400 \text{ m/s}}$

[7.50-kg ball] $\qquad v_{f2} = \left(\dfrac{2(5.00 \text{ kg})}{5.00 \text{ kg} + 7.50 \text{ kg}} \right)(2.00 \text{ m/s}) = \boxed{+1.60 \text{ m/s}}$

The signs indicate that, after the collision, the 5.00-kg ball reverses its direction of motion, while the 7.50-kg ball moves in the direction in which the 5.00-kg ball was initially moving.

b. When the collision is completely inelastic, the balls stick together, giving a composite body of mass $m_1 + m_2$ that moves with a velocity v_f. The statement of conservation of linear momentum then becomes

$$\underbrace{(m_1 + m_2)v_f}_{\substack{\text{Total momentum} \\ \text{after collision}}} = \underbrace{m_1 v_{01} + 0}_{\substack{\text{Total momentum} \\ \text{before collision}}}$$

The final velocity of the two balls after the collision is, therefore,

$$v_f = \frac{m_1 v_{01}}{m_1 + m_2} = \frac{(5.00 \text{ kg})(2.00 \text{ m/s})}{5.00 \text{ kg} + 7.50 \text{ kg}} = \boxed{+0.800 \text{ m/s}}$$

---

35. **REASONING** We obtain the desired percentage in the usual way, as the kinetic energy of the target (with the projectile in it) divided by the projectile's incident kinetic energy, multiplied by a factor of 100. Each kinetic energy is given by Equation 6.2 as $\frac{1}{2}mv^2$, where m and v are mass and speed, respectively. Data for the masses are given, but the speeds are not provided. However, information about the speeds can be obtained by using the principle of conservation of linear momentum.

SOLUTION We define the following quantities:

$\qquad$ KE_{TP} = kinetic energy of the target with the projectile in it
$\qquad$ KE_{0P} = kinetic energy of the incident projectile
$\qquad$ m_P = mass of incident projectile = 0.20 kg
$\qquad$ m_T = mass of target = 2.50 kg
$\qquad$ v_f = speed at which the target with the projectile in it flies off after being struck
$\qquad$ v_{0P} = speed of incident projectile

The desired percentage is

$$\text{Percentage} = \frac{KE_{TP}}{KE_{OP}} \times 100 = \frac{\frac{1}{2}\left(m_T + m_P\right)v_f^2}{\frac{1}{2}m_P v_{OP}^2} \times 100\% \tag{1}$$

According to the momentum-conservation principle, we have

$$\underbrace{\left(m_T + m_P\right)v_f}_{\substack{\text{Total momentum of target and} \\ \text{projectile after target is struck}}} = \underbrace{0 + m_P v_{OP}}_{\substack{\text{Total momentum of target and} \\ \text{projectile before target is struck}}}$$

Note that the target is stationary before being struck and, hence, has zero initial momentum. Solving for the ratio v_f/v_{OP}, we find that

$$\frac{v_f}{v_{OP}} = \frac{m_P}{m_T + m_P}$$

Substituting this result into Equation (1) gives

$$\text{Percentage} = \frac{\frac{1}{2}\left(m_T + m_P\right)v_f^2}{\frac{1}{2}m_P v_{OP}^2} \times 100\% = \frac{\frac{1}{2}\left(m_T + m_P\right)}{\frac{1}{2}m_P}\left(\frac{m_P}{m_T + m_P}\right)^2 \times 100\%$$

$$= \frac{m_P}{m_T + m_P} \times 100\% = \frac{0.20 \text{ kg}}{2.50 \text{ kg} + 0.20 \text{ kg}} \times 100\% = \boxed{7.4\%}$$

---

41. **REASONING** The two skaters constitute the system. Since the net external force acting on

$$(m_1 + m_2)v_f \cos\theta = m_1 v_{01}$$

Note that since the skaters hold onto each other, they move away with a common velocity v_f. In the y direction, $P_{f_y} = P_{0_y}$, or

$$(m_1 + m_2)v_f \sin\theta = m_2 v_{02}$$

These equations can be solved simultaneously to obtain both the angle θ and the velocity v_f.

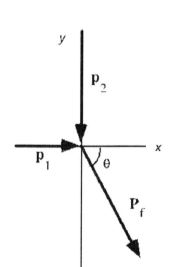

SOLUTION
a. Division of the equations above gives

$$\theta = \tan^{-1}\left(\frac{m_2 v_{02}}{m_1 v_{01}}\right) = \tan^{-1}\left[\frac{(70.0 \text{ kg})(7.00 \text{ m/s})}{(50.0 \text{ kg})(3.00 \text{ m/s})}\right] = \boxed{73.0°}$$

the system is zero, the total linear momentum of the system is conserved. In the x direction (the east/west direction), conservation of linear momentum gives $P_{fx} = P_{0x}$, or

b. Solution of the first of the momentum equations gives

$$v_f = \frac{m_1 v_{01}}{(m_1 + m_2)\cos\theta} = \frac{(50.0 \text{ kg})(3.00 \text{ m/s})}{(50.0 \text{ kg} + 70.0 \text{ kg})(\cos 73.0°)} = \boxed{4.28 \text{ m/s}}$$

---

45. **REASONING** The two balls constitute the system. The tension in the wire is the only nonconservative force that acts on the ball. The tension does no work since it is perpendicular to the displacement of the ball. Since $W_{nc} = 0$ J, the principle of conservation of mechanical energy holds and can be used to find the speed of the 1.50-kg ball just before the collision. Momentum is conserved during the collision, so the principle of conservation of momentum can be used to find the velocities of both balls just after the collision. Once the collision has occurred, energy conservation can be used to determine how high each ball rises.

SOLUTION
a. Applying the principle of energy conservation to the 1.50-kg ball, we have

$$\underbrace{\tfrac{1}{2}mv_f^2 + mgh_f}_{E_f} = \underbrace{\tfrac{1}{2}mv_0^2 + mgh_0}_{E_0}$$

If we measure the heights from the lowest point in the swing, $h_f = 0$ m, and the expression above simplifies to

$$\tfrac{1}{2}mv_f^2 = \tfrac{1}{2}mv_0^2 + mgh_0$$

Solving for v_f, we have

$$v_f = \sqrt{v_0^2 + 2gh_0} = \sqrt{(5.00 \text{ m/s})^2 + 2(9.80 \text{ m/s}^2)(0.300 \text{ m})} = \boxed{5.56 \text{ m/s}}$$

b. If we assume that the collision is elastic, then the velocities of both balls just after the collision can be obtained from Equations 7.8a and 7.8b:

$$v_{f1} = \left(\frac{m_1 - m_2}{m_1 + m_2}\right)v_{01} \qquad \text{and} \qquad v_{f2} = \left(\frac{2m_1}{m_1 + m_2}\right)v_{01}$$

Since v_{01} corresponds to the speed of the 1.50-kg ball just before the collision, it is equal to the quantity v_f calculated in part (a). With the given values of $m_1 = 1.50$ kg and $m_2 = 4.60$ kg, and the value of $v_{01} = 5.56$ m/s obtained in part (a), Equations 7.8a and 7.8b yield the following values:

$$\boxed{v_{f1} = -2.83 \text{ m/s}} \qquad \text{and} \qquad \boxed{v_{f2} = +2.73 \text{ m/s}}$$

The minus sign in v_{f1} indicates that the first ball reverses its direction as a result of the collision.

c. If we apply the conservation of mechanical energy to either ball after the collision we have

$$\underbrace{\tfrac{1}{2}mv_f^2 + mgh_f}_{E_f} = \underbrace{\tfrac{1}{2}mv_0^2 + mgh_0}_{E_0}$$

where v_0 is the speed of the ball just after the collision, and h_f is the final height to which the ball rises. For either ball, $h_0 = 0$ m, and when either ball has reached its maximum height, $v_f = 0$ m/s. Therefore, the expression of energy conservation reduces to

$$gh_f = \tfrac{1}{2}v_0^2 \qquad \text{or} \qquad h_f = \frac{v_0^2}{2g}$$

Thus, the heights to which each ball rises after the collision are

$\boxed{1.50\text{-kg ball}}$ $\qquad h_f = \dfrac{v_0^2}{2g} = \dfrac{(2.83 \text{ m/s})^2}{2(9.80 \text{ m/s}^2)} = \boxed{0.409 \text{ m}}$

$\boxed{4.60\text{-kg ball}}$ $\qquad h_f = \dfrac{v_0^2}{2g} = \dfrac{(2.73 \text{ m/s})^2}{2(9.80 \text{ m/s}^2)} = \boxed{0.380 \text{ m}}$

---

49. ***REASONING AND SOLUTION***   The velocity of the center of mass of a system is given by Equation 7.11. Using the data and the results obtained in Example 5, we obtain the following:

a. The velocity of the center of mass of the two-car system before the collision is

$$\left(v_{cm}\right)_{before} = \frac{m_1 v_{01} + m_2 v_{02}}{m_1 + m_2}$$

$$= \frac{(65 \times 10^3 \text{ kg})(+0.80 \text{ m/s}) + (92 \times 10^3 \text{ kg})(+1.2 \text{ m/s})}{65 \times 10^3 \text{ kg} + 92 \times 10^3 \text{ kg}} = \boxed{+1.0 \text{ m/s}}$$

b. The velocity of the center of mass of the two-car system after the collision is

$$\left(v_{cm}\right)_{after} = \frac{m_1 v_f + m_2 v_f}{m_1 + m_2} = v_f = \boxed{+1.0 \text{ m/s}}$$

c. The answer in part (b) should be the same as the common velocity v_f. Since the cars are coupled together, every point of the two-car system, including the center of mass, must move with the same velocity.

---

53. **REASONING** Let m be Al's mass, which means that Jo's mass is $168 \text{ kg} - m$. Since friction is negligible and since the downward-acting weight of each person is balanced by the upward-acting normal force from the ice, the net external force acting on the two-person system is zero. Therefore, the system is isolated, and the conservation of linear momentum applies. The initial total momentum must be equal to the final total momentum.

SOLUTION Applying the principle of conservation of linear momentum and assuming that the direction in which Al moves is the positive direction, we find

$$\underbrace{m(0 \text{ m/s}) + (168 \text{ kg} - m)(0 \text{ m/s})}_{\text{Initial total momentum}} = \underbrace{m(0.90 \text{ m/s}) + (168 \text{ kg} - m)(-1.2 \text{ m/s})}_{\text{Final total momentum}}$$

Solving this equation for m, we find that

$$0 = m(0.90 \text{ m/s}) - (168 \text{ kg})(1.2 \text{ m/s}) + m(1.2 \text{ m/s})$$

$$m = \frac{(168 \text{ kg})(1.2 \text{ m/s})}{0.90 \text{ m/s} + 1.2 \text{ m/s}} = \boxed{96 \text{ kg}}$$

---

57. **REASONING** According to Equation 7.1, the impulse $\mathbf{J}$ produced by an average force $\overline{\mathbf{F}}$ is $\mathbf{J} = \overline{\mathbf{F}}\Delta t$, where Δt is the time interval during which the force acts. We will apply this definition for each of the forces and then set the two impulses equal to one another. The fact that one average force has a magnitude that is three times as large as that of the other average force will then be used to obtain the desired time interval.

SOLUTION Applying Equation 7.1, we write the impulse of each average force as follows:

$$\mathbf{J_1} = \overline{\mathbf{F}}_1 \Delta t_1 \quad \text{and} \quad \mathbf{J_2} = \overline{\mathbf{F}}_2 \Delta t_2$$

But the impulses $\mathbf{J_1}$ and $\mathbf{J_2}$ are the same, so we have that $\overline{\mathbf{F}}_1 \Delta t_1 = \overline{\mathbf{F}}_2 \Delta t_2$. Writing this result in terms of the magnitudes of the forces gives

$$\overline{F}_1 \Delta t_1 = \overline{F}_2 \Delta t_2 \quad \text{or} \quad \Delta t_2 = \left(\frac{\overline{F}_1}{\overline{F}_2}\right) \Delta t_1$$

The ratio of the force magnitudes is given as $\overline{F}_1 / \overline{F}_2 = 3$, so we find that

$$\Delta t_2 = \left(\frac{\overline{F}_1}{\overline{F}_2} \right) \Delta t_1 = 3(3.2 \text{ ms}) = \boxed{9.6 \text{ ms}}$$

---

65. ***REASONING***   We will define the system to be the platform, the two people and the ball. Since the ball travels nearly horizontally, the effects of gravity are negligible. Momentum is conserved. Since the initial momentum of the system is zero, it must remain zero as the ball is thrown and caught. While the ball is in motion, the platform will recoil in such a way that the total momentum of the system remains zero. As the ball is caught, the system must come to rest so that the total momentum remains zero. The distance that the platform moves before coming to rest again can be determined by using the expressions for momentum conservation and the kinematic description for this situation.

SOLUTION   While the ball is in motion, we have

$$MV + mv = 0 \tag{1}$$

where M is the combined mass of the platform and the two people, V is the recoil velocity of the platform, m is the mass of the ball, and v is the velocity of the ball.

The distance that the platform moves is given by

$$x = Vt \tag{2}$$

where t is the time that the ball is in the air. The time that the ball is in the air is given by

$$t = \frac{L}{v - V} \tag{3}$$

where L is the length of the platform, and the quantity $(v - V)$ is the velocity of the ball relative to the platform. Remember, both the ball and the platform are moving while the ball is in the air. Combining equations (2) and (3) gives

$$x = \left(\frac{V}{v - V} \right) L \tag{4}$$

From equation (1) the ratio of the velocities is $V / v = -m / M$. Equation (4) then gives

$$x = \frac{(V/v)L}{1 - (V/v)} = \frac{(-m/M)L}{1 + (m/M)} = -\frac{mL}{M + m} = -\frac{(6.0 \text{ kg})(2.0 \text{ m})}{118 \text{ kg} + 6.0 \text{ kg}} = -0.097 \text{ m}$$

The minus sign indicates that displacement of the platform is in the opposite direction to the displacement of the ball. The distance moved by the platform is the magnitude of this displacement, or $\boxed{0.097 \text{ m}}$.

---

67. **CONCEPTS** (i) The answer is the same in both cases, except that both joggers are running due north in part *a*, but one is running due north and one is running dues south in part *b*. Kinetic energy is a scalar quantity and does not depend on directional differences. Therefore, the kinetic energy of the two-jogger system is the same in both cases. (ii) Momentum is a vector. As a result, when we add Jim's momentum to Tom's momentum to get the total for the two-jogger system, we must take the directions into account. For example, we can use positive to denote north and negative to denote south. In case *a* two positive momentum vectors are added, while in case *b* one positive and one negative vector are added. Due to the partial cancelation in case *b*, the resulting total momentum will have a smaller magnitude than in case *a*.

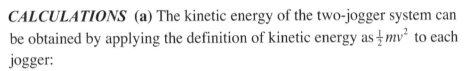

CALCULATIONS (a) The kinetic energy of the two-jogger system can be obtained by applying the definition of kinetic energy as $\frac{1}{2}mv^2$ to each jogger:

$$KE_{system} = \tfrac{1}{2}m_{Jim}v_{Jim}^2 + \tfrac{1}{2}m_{Tom}v_{Tom}^2$$

$$= \tfrac{1}{2}(90.0 \text{ kg})(4.00 \text{ m/s})^2 + \tfrac{1}{2}(55.0 \text{ kg})(4.00 \text{ m/s})^2 = \boxed{1160 \text{ J}}$$

The momentum of each jogger is his mass times his velocity, according to the definition given in Equation 7.2. Using P_{system} to denote the momentum of the two-jogger system and defining due north as the positive direction, we have

$$P_{system} = m_{Jim}v_{Jim} + m_{Tom}v_{Tom}$$

$$= (90.0 \text{ kg})(+4.00 \text{ m/s}) + (55.0 \text{ kg})(+4.00 \text{ m/s}) = \boxed{580 \text{ kg}\cdot\text{m/s}}$$

(b) In these calculations, we note that Tom's velocity is in the due south direction, so it is -4.00 m/s. Proceeding as in part (a), we find:

$$KE_{system} = \tfrac{1}{2}m_{Jim}v_{Jim}^2 + \tfrac{1}{2}m_{Tom}v_{Tom}^2$$

$$= \tfrac{1}{2}(90.0 \text{ kg})(4.00 \text{ m/s})^2 + \tfrac{1}{2}(55.0 \text{ kg})(-4.00 \text{ m/s})^2 = \boxed{1160 \text{ J}}$$

$$P_{system} = m_{Jim}v_{Jim} + m_{Tom}v_{Tom}$$

$$= (90.0 \text{ kg})(+4.00 \text{ m/s}) + (55.0 \text{ kg})(-4.00 \text{ m/s}) = \boxed{140 \text{ kg}\cdot\text{m/s}}$$

CHAPTER 8 | *ROTATIONAL KINEMATICS*

1. ***REASONING*** The average angular velocity is equal to the angular displacement divided by the elapsed time (Equation 8.2). Thus, the angular displacement of the baseball is equal to the product of the average angular velocity and the elapsed time. However, the problem gives the travel time in seconds and asks for the displacement in radians, while the angular velocity is given in revolutions per minute. Thus, we will begin by converting the angular velocity into radians per second.

SOLUTION Since 2π rad $= 1$ rev and 1 min $= 60$ s, the average angular velocity $\bar{\omega}$ (in rad/s) of the baseball is

$$\bar{\omega} = \left(\frac{330 \text{ rev}}{\text{min}}\right)\left(\frac{2\pi \text{ rad}}{1 \text{ rev}}\right)\left(\frac{1 \text{ min}}{60 \text{ s}}\right) = 35 \text{ rad/s}$$

Since the average angular velocity of the baseball is equal to the angular displacement $\Delta\theta$ divided by the elapsed time Δt, the angular displacement is

$$\Delta\theta = \bar{\omega}\Delta t = (35 \text{ rad/s})(0.60 \text{ s}) = \boxed{21 \text{ rad}} \tag{8.2}$$

---

5. ***REASONING AND SOLUTION*** Since there are 2π radians per revolution and it is stated in the problem that there are 100 grads in one-quarter of a circle, we find that the number of grads in one radian is

$$(1.00 \text{ rad})\left(\frac{1 \text{ rev}}{2\pi \text{ rad}}\right)\left(\frac{100 \text{ grad}}{0.250 \text{ rev}}\right) = \boxed{63.7 \text{ grad}}$$

---

9. ***REASONING*** Equation 8.4 $\left[\bar{\alpha} = (\omega - \omega_0)/t\right]$ indicates that the average angular acceleration is equal to the change in the angular velocity divided by the elapsed time. Since the wheel starts from rest, its initial angular velocity is $\omega_0 = 0$ rad/s. Its final angular velocity is given as $\omega = 0.24$ rad/s. Since the average angular acceleration is given as $\bar{\alpha} = 0.030$ rad/s^2, Equation 8.4 can be solved to determine the elapsed time t.

SOLUTION Solving Equation 8.4 for the elapsed time gives

$$t = \frac{\omega - \omega_0}{\bar{\alpha}} = \frac{0.24 \text{ rad/s} - 0 \text{ rad/s}}{0.030 \text{ rad/s}^2} = \boxed{8.0 \text{ s}}$$

---

17. **REASONING AND SOLUTION**
a. If the propeller is to appear stationary, each blade must move through an angle of 120° or $2\pi/3$ rad between flashes. The time required is

$$t = \frac{\theta}{\omega} = \frac{(2\pi/3)\,\text{rad}}{(16.7\,\text{rev/s})\left(\dfrac{2\pi\,\text{rad}}{1\,\text{rev}}\right)} = \boxed{2.00\times10^{-2}\,\text{s}}$$

b. The next shortest time occurs when each blade moves through an angle of 240°, or $4\pi/3$ rad, between successive flashes. This time is twice the value that we found in part a, or $\boxed{4.00\times10^{-2}\,\text{s}}$.

---

21. **REASONING** The angular displacement is given as $\theta = 0.500$ rev, while the initial angular velocity is given as $\omega_0 = 3.00$ rev/s and the final angular velocity as $\omega = 5.00$ rev/s. Since we seek the time t, we can use Equation 8.6 $\left[\theta = \frac{1}{2}(\omega_0 + \omega)t\right]$ from the equations of rotational kinematics to obtain it.

SOLUTION Solving Equation 8.6 for the time t, we find that

$$t = \frac{2\theta}{\omega_0 + \omega} = \frac{2(0.500\,\text{rev})}{3.00\,\text{rev/s} + 5.00\,\text{rev/s}} = \boxed{0.125\,\text{s}}$$

---

25. **REASONING**
a. The time t for the wheels to come to a halt depends on the initial and final velocities, ω_0 and ω, and the angular displacement θ: $\theta = \frac{1}{2}(\omega_0 + \omega)t$ (see Equation 8.6). Solving for the time yields

$$t = \frac{2\theta}{\omega_0 + \omega}$$

b. The angular acceleration α is defined as the change in the angular velocity, $\omega - \omega_0$, divided by the time t:

$$\alpha = \frac{\omega - \omega_0}{t} \qquad (8.4)$$

SOLUTION
a. Since the wheel comes to a rest, $\omega = 0$ rad/s. Converting 15.92 revolutions to radians (1 rev = 2π rad), the time for the wheel to come to rest is

$$t = \frac{2\theta}{\omega_0 + \omega} = \frac{2(+15.92\,\text{rev})\left(\dfrac{2\pi\,\text{rad}}{1\,\text{rev}}\right)}{+20.0\,\text{rad/s} + 0\,\text{rad/s}} = \boxed{10.0\,\text{s}}$$

b. The angular acceleration is

$$\alpha = \frac{\omega - \omega_0}{t} = \frac{0 \text{ rad/s} - 20.0 \text{ rad/s}}{10.0 \text{ s}} = \boxed{-2.00 \text{ rad/s}^2}$$

---

27. **REASONING** The equations of kinematics for rotational motion cannot be used directly to find the angular displacement, because the final angular velocity (not the initial angular velocity), the acceleration, and the time are known. We will combine two of the equations, Equations 8.4 and 8.6 to obtain an expression for the angular displacement that contains the three known variables.

SOLUTION The angular displacement of each wheel is equal to the average angular velocity multiplied by the time

$$\theta = \underbrace{\tfrac{1}{2}\left(\omega_0 + \omega\right)}_{\bar{\omega}} t \qquad (8.6)$$

The initial angular velocity ω_0 is not known, but it can be found in terms of the angular acceleration and time, which are known. The angular acceleration is defined as (with $t_0 = 0$ s)

$$\bar{\alpha} = \frac{\omega - \omega_0}{t} \qquad \text{or} \qquad \omega_0 = \omega - \alpha t \qquad (8.4)$$

Substituting this expression for ω_0 into Equation 8.6 gives

$$\theta = \tfrac{1}{2}\left[\underbrace{\left(\omega - \alpha t\right)}_{\omega_0} + \omega \right] t = \omega t - \tfrac{1}{2}\alpha t^2$$

$$= \left(+74.5 \text{ rad/s}\right)\left(4.50 \text{ s}\right) - \tfrac{1}{2}\left(+6.70 \text{ rad/s}^2\right)\left(4.50 \text{ s}\right)^2 = \boxed{+267 \text{ rad}}$$

---

33. **REASONING** The angular displacement of the child when he catches the horse is, from Equation 8.2, $\theta_c = \omega_c t$. In the same time, the angular displacement of the horse is, from Equation 8.7 with $\omega_0 = 0$ rad/s, $\theta_h = \tfrac{1}{2}\alpha t^2$. If the child is to catch the horse $\theta_c = \theta_h + (\pi/2)$.

SOLUTION Using the above conditions yields

$$\tfrac{1}{2}\alpha t^2 - \omega_c t + \tfrac{1}{2}\pi = 0$$

or

$$\tfrac{1}{2}(0.0100 \text{ rad/s}^2)t^2 - (0.250 \text{ rad/s})t + \tfrac{1}{2}\left(\pi \text{ rad}\right) = 0$$

The quadratic formula yields $t = 7.37$ s and 42.6 s; therefore, the shortest time needed to catch the horse is $\boxed{t = 7.37 \text{ s}}$.

---

37. **REASONING** The angular speed ω and tangential speed v_T are related by Equation 8.9 ($v_T = r\omega$), and this equation can be used to determine the radius r. However, we must remember that this relationship is only valid if we use radian measure. Therefore, it will be necessary to convert the given angular speed in rev/s into rad/s.

SOLUTION Solving Equation 8.9 for the radius gives

$$r = \frac{v_T}{\omega} = \frac{54 \text{ m/s}}{\underbrace{(47 \text{ rev/s})\left(\dfrac{2\pi \text{ rad}}{1 \text{ rev}}\right)}_{\text{Conversion from rev/s into rad/s}}} = \boxed{0.18 \text{ m}}$$

where we have used the fact that 1 rev corresponds to 2π rad to convert the given angular speed from rev/s into rad/s.

---

41. **REASONING** The tangential speed v_T of a point on the "equator" of the baseball is given by Equation 8.9 as $v_T = r\omega$, where r is the radius of the baseball and ω is its angular speed. The radius is given in the statement of the problem. The (constant) angular speed is related to that angle θ through which the ball rotates by Equation 8.2 as $\omega = \theta / t$, where we have assumed for convenience that $\theta_0 = 0$ rad when $t_0 = 0$ s. Thus, the tangential speed of the ball is

$$v_T = r\omega = r\left(\frac{\theta}{t}\right)$$

The time t that the ball is in the air is equal to the distance x it travels divided by its linear speed v, $t = x/v$, so the tangential speed can be written as

$$v_T = r\left(\frac{\theta}{t}\right) = r\left(\frac{\theta}{x/v}\right) = \frac{r\theta v}{x}$$

SOLUTION The tangential speed of a point on the equator of the baseball is

$$v_T = \frac{r\theta v}{x} = \frac{(3.67 \times 10^{-2} \text{ m})(49.0 \text{ rad})(42.5 \text{ m/s})}{16.5 \text{ m}} = \boxed{4.63 \text{ m/s}}$$

---

45. **REASONING** Since the car is traveling with a constant speed, its tangential acceleration must be zero. The radial or centripetal acceleration of the car can be found from Equation 5.2. Since the tangential acceleration is zero, the total acceleration of the car is equal to its radial acceleration.

SOLUTION
a. Using Equation 5.2, we find that the car's radial acceleration, and therefore its total acceleration, is

$$a = a_R = \frac{v_T^2}{r} = \frac{(75.0 \text{ m/s})^2}{625 \text{ m}} = \boxed{9.00 \text{ m/s}^2}$$

b. The direction of the car's total acceleration is the same as the direction of its radial acceleration. That is, the direction is $\boxed{\text{radially inward}}$.

---

47. *REASONING*
a. According to Equation 8.2, the average angular speed is equal to the magnitude of the angular displacement divided by the elapsed time. The magnitude of the angular displacement is one revolution, or 2π rad. The elapsed time is one year, expressed in seconds.

b. The tangential speed of the earth in its orbit is equal to the product of its orbital radius and its orbital angular speed (Equation 8.9).

c. Since the earth is moving on a nearly circular orbit, it has a centripetal acceleration that is directed toward the center of the orbit. The magnitude a_c of the centripetal acceleration is given by Equation 8.11 as $a_c = r\omega^2$.

SOLUTION
a. The average angular speed is

$$\omega = \bar{\omega} = \frac{\Delta\theta}{\Delta t} = \frac{2\pi \text{ rad}}{3.16\times10^7 \text{ s}} = \boxed{1.99\times10^{-7} \text{ rad/s}} \qquad (8.2)$$

b. The tangential speed of the earth in its orbit is

$$v_T = r\omega = (1.50\times10^{11} \text{ m})(1.99\times10^{-7} \text{ rad/s}) = \boxed{2.98\times10^4 \text{ m/s}} \qquad (8.9)$$

c. The centripetal acceleration of the earth due to its circular motion around the sun is

$$a_c = r\omega^2 = (1.50\times10^{11} \text{ m})(1.99\times10^{-7} \text{ rad/s})^2 = \boxed{5.94\times10^{-3} \text{ m/s}^2}$$
$$(8.11)$$

$\boxed{\text{The acceleration is directed toward the center of the orbit.}}$

---

51. **REASONING**

a. The tangential speed v_T of the sun as it orbits about the center of the Milky Way is related to the orbital radius r and angular speed ω by Equation 8.9, $v_T = r\omega$. Before we use this relation, however, we must first convert r to meters from light-years.

b. The centripetal force is the net force required to keep an object, such as the sun, moving on a circular path. According to Newton's second law of motion, the magnitude F_c of the centripetal force is equal to the product of the object's mass m and the magnitude a_c of its centripetal acceleration (see Section 5.3): $F_c = ma_c$. The magnitude of the centripetal acceleration is expressed by Equation 8.11 as $a_c = r\omega^2$, where r is the radius of the circular path and ω is the angular speed of the object.

SOLUTION

a. The radius of the sun's orbit about the center of the Milky Way is

$$r = \left(2.3 \times 10^4 \; \text{light-years}\right)\left(\frac{9.5 \times 10^{15} \; \text{m}}{1 \; \text{light-year}}\right) = 2.2 \times 10^{20} \; \text{m}$$

The tangential speed of the sun is

$$v_T = r\omega = \left(2.2 \times 10^{20} \; \text{m}\right)\left(1.1 \times 10^{-15} \; \text{rad/s}\right) = \boxed{2.4 \times 10^5 \; \text{m/s}} \qquad (8.9)$$

b. The magnitude of the centripetal force that acts on the sun is

$$\underbrace{F_c}_{\substack{\text{Centripetal} \\ \text{force}}} = ma_c = mr\omega^2$$

$$= \left(1.99 \times 10^{30} \; \text{kg}\right)\left(2.2 \times 10^{20} \; \text{m}\right)\left(1.1 \times 10^{-15} \; \text{rad/s}\right)^2 = \boxed{5.3 \times 10^{20} \; \text{N}}$$

---

53. **REASONING AND SOLUTION** From Equation 2.4, the linear acceleration of the motorcycle is

$$a = \frac{v - v_0}{t} = \frac{22.0 \; \text{m/s} - 0 \; \text{m/s}}{9.00 \; \text{s}} = 2.44 \; \text{m/s}^2$$

Since the tire rolls without slipping, the linear acceleration equals the tangential acceleration of a point on the outer edge of the tire: $a = a_T$. Solving Equation 8.13 for α gives

$$\alpha = \frac{a_T}{r} = \frac{2.44 \; \text{m/s}^2}{0.280 \; \text{m}} = \boxed{8.71 \; \text{rad/s}^2}$$

61. **REASONING** As a penny-farthing moves, both of its wheels roll without slipping. This means that the axle for each wheel moves through a linear distance (the distance through which the bicycle moves) that equals the circular arc length measured along the outer edge of the wheel. Since both axles move through the same linear distance, the circular arc length measured along the outer edge of the large front wheel must equal the circular arc length measured along the outer edge of the small rear wheel. In each case the arc length s is equal to the number n of revolutions times the circumference $2\pi r$ of the wheel (r = radius).

SOLUTION Since the circular arc length measured along the outer edge of the large front wheel must equal the circular arc length measured along the outer edge of the small rear wheel, we have

$$\underbrace{n_{Rear}\,2\pi r_{Rear}}_{\substack{\text{Arc length for rear}\\\text{wheel}}} = \underbrace{n_{Front}\,2\pi r_{Front}}_{\substack{\text{Arc length for front}\\\text{wheel}}}$$

Solving for n_{Rear} gives

$$n_{Rear} = \frac{n_{Front}\,r_{Front}}{r_{Rear}} = \frac{276\,(1.20\ \text{m})}{0.340\ \text{m}} = \boxed{974\ \text{rev}}$$

---

65. **REASONING** The tangential acceleration $\mathbf{a}_T$ of the speedboat can be found by using Newton's second law, $\mathbf{F}_T = m\mathbf{a}_T$, where $\mathbf{F}_T$ is the net tangential force. Once the tangential acceleration of the boat is known, Equation 2.4 can be used to find the tangential speed of the boat 2.0 s into the turn. With the tangential speed and the radius of the turn known, Equation 5.2 can then be used to find the centripetal acceleration of the boat.

SOLUTION
a. From Newton's second law, we obtain

$$a_T = \frac{F_T}{m} = \frac{550\ \text{N}}{220\ \text{kg}} = \boxed{2.5\ \text{m/s}^2}$$

b. The tangential speed of the boat 2.0 s into the turn is, according to Equation 2.4,

$$v_T = v_{0T} + a_T t = 5.0\ \text{m/s} + (2.5\ \text{m/s}^2)(2.0\ \text{s}) = 1.0 \times 10^1\ \text{m/s}$$

The centripetal acceleration of the boat is then

$$a_c = \frac{v_T^2}{r} = \frac{(1.0 \times 10^1\ \text{m/s})^2}{32\ \text{m}} = \boxed{3.1\ \text{m/s}^2}$$

---

67. **REASONING AND SOLUTION**   Since the angular speed of the fan decreases, the sign of the angular acceleration must be opposite to the sign for the angular velocity. Taking the angular velocity to be positive, the angular acceleration, therefore, must be a negative quantity. Using Equation 8.4 we obtain

$$\omega_0 = \omega - \alpha t = 83.8\ \text{rad/s} - (-42.0\ \text{rad/s}^2)(1.75\ \text{s}) = \boxed{157.3\ \text{rad/s}}$$

71. **REASONING AND SOLUTION**

a. From Equation 8.9, and the fact that 1 revolution = 2π radians, we obtain

$$v_T = r\omega = (0.0568 \text{ m})\left(3.50 \ \frac{\text{rev}}{\text{s}}\right)\left(\frac{2\pi \text{ rad}}{1 \text{ rev}}\right) = \boxed{1.25 \text{ m/s}}$$

b. Since the disk rotates at *constant* tangential speed,

$$v_{T1} = v_{T2} \qquad \text{or} \qquad \omega_1 r_1 = \omega_2 r_2$$

Solving for ω_2, we obtain

$$\omega_2 = \frac{\omega_1 r_1}{r_2} = \frac{(3.50 \text{ rev/s})(0.0568 \text{ m})}{0.0249 \text{ m}} = \boxed{7.98 \text{ rev/s}}$$

77. **REASONING** Since the magnitude a of the car's linear acceleration is constant, and the tires roll without slipping, the relation $a = r\alpha$ (Equation 8.13) reveals that the magnitude a of each tire's angular acceleration is also constant. Therefore, the equations of kinematics for rotational motion apply, and we will use $\theta = \omega_0 t + \frac{1}{2}\alpha t^2$ (Equation 8.7) to find the angle θ through which each tire rotates. Two kinematic variables are known (the initial angular velocity ω_0 and the time t). However, to use such an equation we also need to know the angular acceleration of each tire. To find the angular acceleration, we will use Equation 8.13.

SOLUTION Since the initial angular velocity ω_0 and the time t are known, we begin by selecting Equation 8.7, as shown above. However, at this point we do not know the angular acceleration α. Angular acceleration, being a vector, has a magnitude and a direction. Because each wheel rolls (no slipping), the magnitude of its angular acceleration is equal to a/r (Equation 8.13), where a is the magnitude of the car's linear acceleration and r is the radius of each wheel. As the car accelerates to the right, the tires rotate faster and faster in the clockwise, or negative direction, so the angular acceleration is also negative. Inserting a minus sign into the expression a/r to denote the direction of the angular acceleration gives

$$\alpha = -\frac{a}{r}$$

All the variables on the right side of this equation are known, so we can substitute this expression for α into Equation 8.7:

$$\theta = \omega_0 t + \tfrac{1}{2}\alpha t^2 = \omega_0 t + \tfrac{1}{2}\left(-\frac{a}{r}\right)t^2$$

$$= (0 \text{ rad/s})(20.0 \text{ s}) + \tfrac{1}{2}\left(-\frac{0.800 \text{ m/s}^2}{0.330 \text{ m}}\right)(20.0\text{s})^2 = \boxed{-485 \text{ rad}}$$

The minus sign indicates that wheel has rotated in the clockwise direction.

---

79. ***CONCEPTS*** (i) Yes. Recall than an object has an acceleration if its velocity is changing in time. The velocity has two attributes, a magnitude (or speed) and a direction. In this instance, the speed is not changing since it is steady at 32 m/s. However, the *direction* of the velocity is changing continually as the car moves on the circular road. A change in direction gives rise to the centripetal acceleration $\vec{a}_c$, which is directed toward the center of the circular path. (ii) Yes. When the car's angular speed ω decreases, for example, its tangential speed v_T also decreases. This is because they are related by $v_T = r\omega$, where r is the radius of the circular path (the road). A decreasing tangential speed v_T, in turn, means that the car has a tangential acceleration $\vec{a}_T$. The direction of the tangential acceleration $\vec{a}_T$ must be opposite to that of the tangential velocity $\vec{v}_T$, because the tangential speed is decreasing. (Note: If the two vectors were in the same direction, the tangential speed would be increasing.)

CALCULATIONS (a) The angular speed ω of the car is equal to its tangential speed $v_T = 32$ m/s (the speed indicated by the speedometer) divided by the radius $r = 390$ m of the circular road:

$$\omega = \frac{v_T}{r} = \frac{32 \text{ m/s}}{390 \text{ m}} = \boxed{8.2\times10^{-2} \text{ rad/s}}$$

(b) The acceleration is the centripetal acceleration and arises because the tangential velocity is changing direction as the car travels around the circular path. The magnitude of the centripetal acceleration is

$$a_C = r\omega^2 = (390 \text{ m})(8.2\times10^{-2} \text{ rad/s})^2 = \boxed{2.6 \text{ m/s}^2}$$

As always, the centripetal acceleration is directed toward the center of the circle.

(c) The magnitude of the tangential acceleration is related to the magnitude of the angular acceleration through $a_T = r\alpha$. The angular acceleration arises because the brakes reduce the angular speed of the car to $\omega = 4.9\times10^{-2}$ rad/s in a time $t = 4.0$ s. The angular acceleration is given by Equation 8.4 as

$$\alpha = \frac{\omega - \omega_0}{t} = \frac{(4.9\times10^{-2} \text{ rad/s}) - (8.2\times10^{-2} \text{ rad/s})}{4.0 \text{ s}} = -8.3\times10^{-3} \text{ rad/s}^2$$

The magnitude of the angular acceleration is 8.3×10^{-3} rad/s^2. Therefore, the magnitude of the tangential acceleration is

$$\alpha_T = r\alpha = (390 \text{ m})(8.3 \times 10^{-3} \text{ rad/s}^2) = \boxed{3.2 \text{ m/s}^2}$$

Since the car is slowing down, the tangential acceleration is directed opposite to the direction of the tangential velocity.

CHAPTER 9 | *ROTATIONAL DYNAMICS*

1. **REASONING** The drawing shows the wheel as it rolls to the right, so the torque applied by the engine is assumed to be clockwise about the axis of rotation. The force of static friction that the ground applies to the wheel is labeled as f_s. This force produces a counterclockwise torque τ about the axis of rotation, which is given by Equation 9.1 as $\tau = f_s l$, where l is the lever arm. Using this relation we can find the magnitude f_s of the static frictional force.

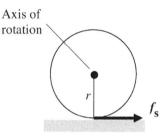

SOLUTION The countertorque is given as $\tau = f_s l$, where f_s is the magnitude of the static frictional force and l is the lever arm. The lever arm is the distance between the line of action of the force and the axis of rotation; in this case the lever arm is just the radius r of the tire. Solving for f_s gives

$$f_s = \frac{\tau}{l} = \frac{295 \text{ N·m}}{0.350 \text{ m}} = \boxed{843 \text{ N}}$$

3. **REASONING** According to Equation 9.1, we have

$$\text{Magnitude of torque} = Fl$$

where F is the magnitude of the applied force and l is the lever arm. From the figure in the text, the lever arm is given by $l = (0.28 \text{ m}) \sin 50.0°$. Since both the magnitude of the torque and l are known, Equation 9.1 can be solved for F.

SOLUTION Solving Equation 9.1 for F, we have

$$F = \frac{\text{Magnitude of torque}}{l} = \frac{45 \text{ N·m}}{(0.28 \text{ m}) \sin 50.0°} = \boxed{2.1 \times 10^2 \text{ N}}$$

5. **REASONING** To calculate the torques, we need to determine the lever arms for each of the forces. These lever arms are shown in the following drawings:

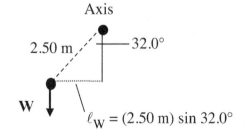 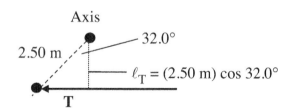

SOLUTION

a. Using Equation 9.1, we find that the magnitude of the torque due to the weight **W** is

$$\text{Magnitude of torque} = Wl_{\text{W}} = (10\,200\text{ N})(2.5\text{ m})\sin 32° = \boxed{13\,500\text{ N}\cdot\text{m}}$$

b. Using Equation 9.1, we find that the magnitude of the torque due to the thrust **T** is

$$\text{Magnitude of torque} = Tl_{\text{T}} = (62\,300\text{ N})(2.5\text{ m})\cos 32° = \boxed{132\,000\text{ N}\cdot\text{m}}$$

---

7. **REASONING AND SOLUTION** The torque produced by each force of magnitude F is given by Equation 9.1, $\tau = Fl$, where l is the lever arm and the torque is positive since each force causes a counterclockwise rotation. In each case, the torque produced by the couple is equal to the sum of the individual torques produced by each member of the couple.

a. When the axis passes through point A, the torque due to the force at A is zero. The lever arm for the force at C is L. Therefore, taking counterclockwise as the positive direction, we have

$$\tau = \tau_A + \tau_C = 0 + FL = \boxed{FL}$$

b. Each force produces a counterclockwise rotation. The magnitude of each force is F and each force has a lever arm of $L/2$. Taking counterclockwise as the positive direction, we have

$$\tau = \tau_A + \tau_C = F\left(\frac{L}{2}\right) + F\left(\frac{L}{2}\right) = \boxed{FL}$$

c. When the axis passes through point C, the torque due to the force at C is zero. The lever arm for the force at A is L. Therefore, taking counterclockwise as the positive direction, we have

$$\tau = \tau_A + \tau_C = FL + 0 = \boxed{FL}$$

Note that the value of the torque produced by the couple is the same in all three cases; in other words, when the couple acts on the tire wrench, the couple produces a torque that does *not* depend on the location of the axis.

---

13. **REASONING** The drawing shows the bridge and the four forces that act on it: the upward force $\mathbf{F}_1$ exerted on the left end by the support, the force due to the weight $\mathbf{W}_h$ of the hiker, the weight $\mathbf{W}_b$ of the bridge, and the upward force $\mathbf{F}_2$ exerted on the right side by the support. Since the bridge is in equilibrium, the sum of the torques about any axis of rotation must be zero $\left(\Sigma\tau = 0\right)$, and the sum of the forces in the vertical direction must be zero $\left(\Sigma F_y = 0\right)$. These two conditions will allow us to determine the magnitudes of $\mathbf{F}_1$ and $\mathbf{F}_2$.

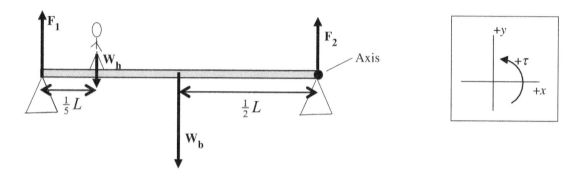

SOLUTION
a. We will begin by taking the axis of rotation about the right end of the bridge. The torque produced by $\mathbf{F}_2$ is zero, since its lever arm is zero. When we set the sum of the torques equal to zero, the resulting equation will have only one unknown, $\mathbf{F}_1$, in it. Setting the sum of the torques produced by the three forces equal to zero gives

$$\Sigma\tau = -F_1 L + W_h\left(\tfrac{4}{5}L\right) + W_b\left(\tfrac{1}{2}L\right) = 0$$

Algebraically eliminating the length L of the bridge from this equation and solving for F_1 gives

$$F_1 = \tfrac{4}{5}W_h + \tfrac{1}{2}W_b = \tfrac{4}{5}(985\text{ N}) + \tfrac{1}{2}(3610\text{ N}) = \boxed{2590\text{ N}}$$

b. Since the bridge is in equilibrium, the sum of the forces in the vertical direction must be zero:

$$\Sigma F_y = F_1 - W_h - W_b + F_2 = 0$$

Solving for F_2 gives

$$F_2 = -F_1 + W_h + W_b = -2590\text{ N} + 985\text{ N} + 3610\text{ N} = \boxed{2010\text{ N}}$$

17. **REASONING** Multiple-Concept Example 8 discusses the static stability factor (SSF) and rollover. In that example, it is determined that the maximum speed v at which a vehicle can negotiate a curve of radius r is related to the SSF according to $v = \sqrt{rg\left(\text{SSF}\right)}$. No value is

given for the radius of the turn. However, by applying this result separately to the sport utility vehicle (SUV) and to the sports car (SC), we will be able to eliminate r algebraically and determine the maximum speed at which the sports car can negotiate the curve without rolling over.

SOLUTION Applying $v = \sqrt{rg\,(\text{SSF})}$ to each vehicle, we obtain

$$v_{\text{SUV}} = \sqrt{rg\,(\text{SSF})_{\text{SUV}}} \quad \text{and} \quad v_{\text{SC}} = \sqrt{rg\,(\text{SSF})_{\text{SC}}}$$

Dividing these two expressions gives

$$\frac{v_{\text{SC}}}{v_{\text{SUV}}} = \frac{\sqrt{rg\,(\text{SSF})_{\text{SC}}}}{\sqrt{rg\,(\text{SSF})_{\text{SUV}}}} \quad \text{or} \quad v_{\text{SC}} = v_{\text{SUV}}\sqrt{\frac{(\text{SSF})_{\text{SC}}}{(\text{SSF})_{\text{SUV}}}} = (18 \text{ m/s})\sqrt{\frac{1.4}{0.80}} = \boxed{24 \text{ m/s}}$$

---

23. **REASONING** The drawing shows the forces acting on the board, which has a length L. The ground exerts the vertical normal force **V** on the lower end of the board. The maximum force of static friction has a magnitude of $\mu_s V$ and acts horizontally on the lower end of the board. The weight **W** acts downward at the board's center. The vertical wall applies a force **P** to the upper end of the board, this force being perpendicular to the wall since the wall is smooth (i.e., there is no friction along the wall). We take upward and to the right as our positive directions. Then, since the horizontal forces balance to zero, we have

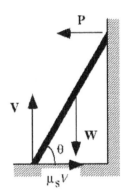

$$\mu_s V - P = 0 \qquad (1)$$

The vertical forces also balance to zero giving

$$V - W = 0 \qquad (2)$$

Using an axis through the lower end of the board, we express the fact that the torques balance to zero as

$$PL \sin\theta - W\left(\frac{L}{2}\right)\cos\theta = 0 \qquad (3)$$

Equations (1), (2), and (3) may then be combined to yield an expression for θ.

SOLUTION Rearranging Equation (3) gives

$$\tan\theta = \frac{W}{2P} \qquad (4)$$

But, $P = \mu_s V$ according to Equation (1), and $W = V$ according to Equation (2). Substituting these results into Equation (4) gives

$$\tan \theta = \frac{V}{2\mu_s V} = \frac{1}{2\mu_s}$$

Therefore,

$$\theta = \tan^{-1}\left(\frac{1}{2\mu_s}\right) = \tan^{-1}\left[\frac{1}{2(0.650)}\right] = \boxed{37.6°}$$

---

25. ***REASONING*** The following drawing shows the beam and the five forces that act on it: the horizontal and vertical components S_x and S_y that the wall exerts on the left end of the beam, the weight W_b of the beam, the force due to the weight W_c of the crate, and the tension **T** in the cable. The beam is uniform, so its center of gravity is at the center of the beam, which is where its weight can be assumed to act. Since the beam is in equilibrium, the sum of the torques about any axis of rotation must be zero $(\Sigma\tau = 0)$, and the sum of the forces in the horizontal and vertical directions must be zero $\left(\Sigma F_x = 0, \Sigma F_y = 0\right)$. These three conditions will allow us to determine the magnitudes of S_x, S_y, and **T**.

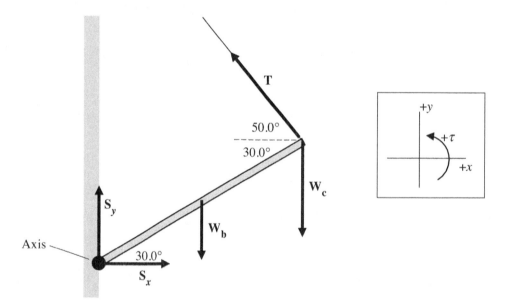

SOLUTION
a. We will begin by taking the axis of rotation to be at the left end of the beam. Then the torques produced by S_x and S_y are zero, since their lever arms are zero. When we set the sum of the torques equal to zero, the resulting equation will have only one unknown, T, in it. Setting the sum of the torques produced by the three forces equal to zero gives (with L equal to the length of the beam)

$$\Sigma\tau = -W_b\left(\tfrac{1}{2}L\cos 30.0°\right) - W_c\left(L\cos 30.0°\right) + T\left(L\sin 80.0°\right) = 0$$

Algebraically eliminating L from this equation and solving for T gives

$$T = \frac{W_b\left(\frac{1}{2}\cos 30.0°\right) + W_c\left(\cos 30.0°\right)}{\sin 80.0°}$$

$$= \frac{(1220 \text{ N})\left(\frac{1}{2}\cos 30.0°\right) + (1960 \text{ N})\left(\cos 30.0°\right)}{\sin 80.0°} = \boxed{2260 \text{ N}}$$

b. Since the beam is in equilibrium, the sum of the forces in the vertical direction is zero:

$$\Sigma F_y = +S_y - W_b - W_c + T\sin 50.0° = 0$$

Solving for S_y gives

$$S_y = W_b + W_c - T\sin 50.0° = 1220 \text{ N} + 1960 \text{ N} - (2260 \text{ N})\sin 50.0° = \boxed{1450 \text{ N}}$$

The sum of the forces in the horizontal direction must also be zero:

$$\Sigma F_x = +S_x - T\cos 50.0° = 0$$

so that

$$S_x = T\cos 50.0° = (2260 \text{ N})\cos 50.0° = \boxed{1450 \text{ N}}$$

---

29. **REASONING** The drawing shows the forces acting on the board, which has a length L. Wall 2 exerts a normal force P_2 on the lower end of the board. The maximum force of static friction that wall 2 can apply to the lower end of the board is $\mu_s P_2$ and is directed upward in the drawing. The weight W acts downward at the board's center. Wall 1 applies a normal force P_1 to the upper end of the board. We take upward and to the right as our positive directions.

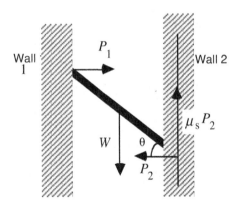

SOLUTION Then, since the horizontal forces balance to zero, we have

$$P_1 - P_2 = 0 \qquad (1)$$

The vertical forces also balance to zero:

$$\mu_s P_2 - W = 0 \qquad (2)$$

Using an axis through the lower end of the board, we now balance the torques to zero:

$$W\left(\frac{L}{2}\right)(\cos\theta) - P_1 L(\sin\theta) = 0 \qquad (3)$$

Rearranging Equation (3) gives

$$\tan \theta = \frac{W}{2P_1} \qquad (4)$$

But $W = \mu_s P_2$ according to Equation (2), and $P_2 = P_1$ according to Equation (1). Therefore, $W = \mu_s P_1$, which can be substituted in Equation (4) to show that

$$\tan \theta = \frac{\mu_s P_1}{2P_1} = \frac{\mu_s}{2} = \frac{0.98}{2}$$

or

$$\theta = \tan^{-1}(0.49) = 26°$$

From the drawing at the right,

$$\cos \theta = \frac{d}{L}$$

Therefore, the longest board that can be propped between the two walls is

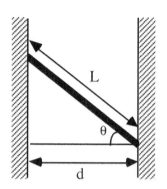

$$L = \frac{d}{\cos \theta} = \frac{1.5 \text{ m}}{\cos 26°} = \boxed{1.7 \text{ m}}$$

---

37. **REASONING**

a. The angular acceleration α is defined as the change, $\omega - \omega_0$, in the angular velocity divided by the elapsed time t (see Equation 8.4). Since all these variables are known, we can determine the angular acceleration directly from this definition.

b. The magnitude τ of the torque is defined by Equation 9.1 as the product of the magnitude F of the force and the lever arm l. The lever arm is the radius of the cylinder, which is known. Since there is only one torque acting on the cylinder, the magnitude of the force can be obtained by using Newton's second law for rotational motion, $\Sigma \tau = Fl = I\alpha$.

SOLUTION

a. From Equation 8.4 we have that $\alpha = (\omega - \omega_0)/t$. We are given that $\omega_0 = 76.0$ rad/s, $\omega = \frac{1}{2}\omega_0 = 38.0$ rad/s, and $t = 6.40$ s, so

$$\alpha = \frac{\omega - \omega_0}{t} = \frac{38.0 \text{ rad/s} - 76.0 \text{ rad/s}}{6.40 \text{ s}} = -5.94 \text{ rad/s}^2$$

The magnitude of the angular acceleration is $\boxed{5.94 \text{ rad/s}^2}$.

b. Using Newton's second law for rotational motion, we have that $\Sigma \tau = Fl = I\alpha$. Thus, the magnitude of the force is

$$F = \frac{I\alpha}{1} = \frac{\left(0.615 \text{ kg} \cdot \text{m}^2\right)\left(5.94 \text{ rad/s}^2\right)}{0.0830 \text{ m}} = \boxed{44.0 \text{ N}}$$

39. **REASONING** The figure below shows eight particles, each one located at a different corner of an imaginary cube. As shown, if we consider an axis that lies along one edge of the cube, two of the particles lie on the axis, and for these particles $r = 0$. The next four particles closest to the axis have $r = 1$, where l is the length of one edge of the cube. The remaining two particles have $r = d$, where d is the length of the diagonal along any one of the faces. From the Pythagorean theorem, $d = \sqrt{\ell^2 + \ell^2} = \ell\sqrt{2}$.

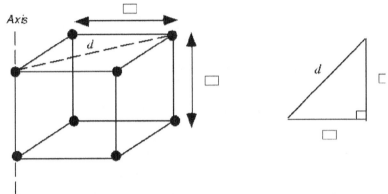

According to Equation 9.6, the moment of inertia of a system of particles is given by $I = \sum mr^2$.

SOLUTION Direct application of Equation 9.6 gives

$$I = \sum mr^2 = 4(ml^2) + 2(md^2) = 4(ml^2) + 2(2ml^2) = 8ml^2$$

or

$$I = 8(0.12 \text{ kg})(0.25 \text{ m})^2 = \boxed{0.060 \text{ kg} \cdot \text{m}^2}$$

41. **REASONING** The drawing shows the two identical sheets and the axis of rotation for each.

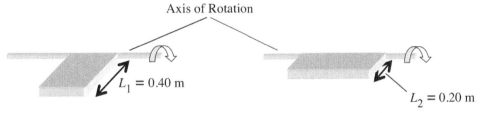

Axis of Rotation

$L_1 = 0.40 \text{ m}$

$L_2 = 0.20 \text{ m}$

The time t it takes for each sheet to reach its final angular velocity depends on the angular acceleration α of the sheet. This relation is given by Equation 8.4 as $t = \left(\omega - \omega_0\right)/\alpha$, where ω and ω_0 are the final and initial angular velocities, respectively. We know that $\omega_0 = 0$ rad/s in each case and that the final angular velocities are the same. The angular acceleration can be determined by using Newton's second law for rotational motion, Equation 9.7, as $\alpha = \tau/I$, where τ is the torque applied to a sheet and I is its moment of inertia.

SOLUTION Substituting the relation $\alpha = \tau / I$ into $t = (\omega - \omega_0) / \alpha$ gives

$$t = \frac{(\omega - \omega_0)}{\alpha} = \frac{(\omega - \omega_0)}{\left(\dfrac{\tau}{I}\right)} = \frac{I(\omega - \omega_0)}{\tau}$$

The time it takes for each sheet to reach its final angular velocity is:

$$t_{\text{Left}} = \frac{I_{\text{Left}}(\omega - \omega_0)}{\tau} \qquad \text{and} \qquad t_{\text{Right}} = \frac{I_{\text{Right}}(\omega - \omega_0)}{\tau}$$

The moments of inertia I of the left and right sheets about the axes of rotation are given by the following relations, where M is the mass of each sheet (see Table 9.1 and the drawings above): $I_{\text{Left}} = \frac{1}{3} M L_1^2$ and $I_{\text{Right}} = \frac{1}{3} M L_2^2$. Note that the variables M, ω, ω_0, and τ are the same for both sheets. Dividing the time-expression for the right sheet by that for the left sheet gives

$$\frac{t_{\text{Right}}}{t_{\text{Left}}} = \frac{\dfrac{I_{\text{Right}}(\omega - \omega_0)}{\tau}}{\dfrac{I_{\text{Left}}(\omega - \omega_0)}{\tau}} = \frac{I_{\text{Right}}}{I_{\text{Left}}} = \frac{\frac{1}{3} M L_2^2}{\frac{1}{3} M L_1^2} = \frac{L_2^2}{L_1^2}$$

Solving this expression for t_{Right} yields

$$t_{\text{Right}} = t_{\text{Left}} \left(\frac{L_2^2}{L_1^2} \right) = (8.0 \text{ s}) \frac{(0.20 \text{ m})^2}{(0.40 \text{ m})^2} = \boxed{2.0 \text{ s}}$$

---

45. **REASONING**   The angular acceleration of the bicycle wheel can be calculated from Equation 8.4. Once the angular acceleration is known, Equation 9.7 can be used to find the net torque caused by the brake pads. The normal force can be calculated from the torque using Equation 9.1.

SOLUTION   The angular acceleration of the wheel is, according to Equation 8.4,

$$\alpha = \frac{\omega - \omega_0}{t} = \frac{3.7 \text{ rad/s} - 13.1 \text{ rad/s}}{3.0 \text{ s}} = -3.1 \text{ rad/s}^2$$

If we assume that all the mass of the wheel is concentrated in the rim, we may treat the wheel as a hollow cylinder. From Table 9.1, we know that the moment of inertia of a hollow cylinder of mass m and radius r about an axis through its center is $I = mr^2$. The net torque that acts on the wheel due to the brake pads is, therefore,

$$\sum \tau = I\alpha = (mr^2)\alpha \tag{1}$$

From Equation 9.1, the net torque that acts on the wheel due to the action of the two brake pads is

$$\sum \tau = -2f_k l \tag{2}$$

where f_k is the kinetic frictional force applied to the wheel by each brake pad, and $l = 0.33$ m is the lever arm between the axle of the wheel and the brake pad (see the drawing in the text). The factor of 2 accounts for the fact that there are two brake pads. The minus sign arises because the net torque must have the same sign as the angular acceleration. The kinetic frictional force can be written as (see Equation 4.8)

$$f_k = \mu_k F_N \tag{3}$$

where μ_k is the coefficient of kinetic friction and F_N is the magnitude of the normal force applied to the wheel by each brake pad. Combining Equations (1), (2), and (3) gives

$$-2(\mu_k F_N)l = (mr^2)\alpha$$

$$F_N = \frac{-mr^2\alpha}{2\mu_k l} = \frac{-(1.3 \text{ kg})(0.33 \text{ m})^2(-3.1 \text{ rad/s}^2)}{2(0.85)(0.33 \text{ m})} = \boxed{0.78 \text{ N}}$$

---

49. **REASONING AND SOLUTION**
   a. The tangential speed of each object is given by Equation 8.9, $v_T = r\omega$. Therefore,

For object 1: $\qquad\qquad v_{T1} = (2.00 \text{ m})(6.00 \text{ rad/s}) = \boxed{12.0 \text{ m/s}}$

For object 2: $\qquad\qquad v_{T2} = (1.50 \text{ m})(6.00 \text{ rad/s}) = \boxed{9.00 \text{ m/s}}$

For object 3: $\qquad\qquad v_{T3} = (3.00 \text{ m})(6.00 \text{ rad/s}) = \boxed{18.0 \text{ m/s}}$

   b. The total kinetic energy of this system can be calculated by computing the sum of the kinetic energies of each object in the system. Therefore,

$$\text{KE} = \tfrac{1}{2}m_1 v_1^2 + \tfrac{1}{2}m_2 v_2^2 + \tfrac{1}{2}m_3 v_3^2$$

$$\text{KE} = \tfrac{1}{2}\left[(6.00 \text{ kg})(12.0 \text{ m/s})^2 + (4.00 \text{ kg})(9.00 \text{ m/s})^2 + (3.00 \text{ kg})(18.0 \text{ m/s})^2 \right] = \boxed{1.08 \times 10^3 \text{ J}}$$

   c. The total moment of inertia of this system can be calculated by computing the sum of the moments of inertia of each object in the system. Therefore,

$$I = \sum mr^2 = m_1 r_1^2 + m_2 r_2^2 + m_3 r_3^2$$

$$I = (6.00 \text{ kg})(2.00 \text{ m})^2 + (4.00 \text{ kg})(1.50 \text{ m})^2 + (3.00 \text{ kg})(3.00 \text{ m})^2 = \boxed{60.0 \text{ kg} \cdot \text{m}^2}$$

d. The rotational kinetic energy of the system is, according to Equation 9.9,

$$KE_R = \tfrac{1}{2} I \omega^2 = \tfrac{1}{2}(60.0 \text{ kg} \cdot \text{m}^2)(6.00 \text{ rad}/\text{s})^2 = \boxed{1.08 \times 10^3 \text{ J}}$$

This agrees, as it should, with the result for part (b).

---

51. ***REASONING*** The kinetic energy of the flywheel is given by Equation 9.9. The moment of inertia of the flywheel is the same as that of a solid disk, and, according to Table 9.1 in the text, is given by $I = \tfrac{1}{2} MR^2$. Once the moment of inertia of the flywheel is known, Equation 9.9 can be solved for the angular speed ω in rad/s. This quantity can then be converted to rev/min.

SOLUTION Solving Equation 9.9 for ω, we obtain,

$$\omega = \sqrt{\frac{2(KE_R)}{I}} = \sqrt{\frac{2(KE_R)}{\tfrac{1}{2} MR^2}} = \sqrt{\frac{4(1.2 \times 10^9 \text{ J})}{(13 \text{ kg})(0.30 \text{ m})^2}} = 6.4 \times 10^4 \text{ rad}/\text{s}$$

Converting this answer into rev/min, we find that

$$\omega = \left(6.4 \times 10^4 \text{ rad/s}\right)\left(\frac{1 \text{ rev}}{2\pi \text{ rad}}\right)\left(\frac{60 \text{ s}}{1 \text{ min}}\right) = \boxed{6.1 \times 10^5 \text{ rev/min}}$$

---

57. ***REASONING AND SOLUTION*** The conservation of energy gives

$$mgh + (1/2) mv^2 + (1/2) I\omega^2 = (1/2) mv_0^2 + (1/2) I\omega_0^2$$

If the ball rolls without slipping, $\omega = v/R$ and $\omega_0 = v_0/R$. We also know that $I = (2/5) mR^2$. Substitution of the last two equations into the first and rearrangement gives

$$v = \sqrt{v_0^2 - \tfrac{10}{7} gh} = \sqrt{(3.50 \text{ m/s})^2 - \tfrac{10}{7}(9.80 \text{ m/s}^2)(0.760 \text{ m})} = \boxed{1.3 \text{ m/s}}$$

59. **REASONING** Let the two disks constitute the system. Since there are no external torques acting on the system, the principle of conservation of angular momentum applies. Therefore we have $L_{\text{initial}} = L_{\text{final}}$, or

$$I_A \omega_A + I_B \omega_B = (I_A + I_B)\omega_{\text{final}}$$

This expression can be solved for the moment of inertia of disk B.

SOLUTION Solving the above expression for I_B, we obtain

$$I_B = I_A\left(\frac{\omega_{\text{final}} - \omega_A}{\omega_B - \omega_{\text{final}}}\right) = (3.4\ \text{kg}\cdot\text{m}^2)\left[\frac{-2.4\ \text{rad/s} - 7.2\ \text{rad/s}}{-9.8\ \text{rad/s} - (-2.4\ \text{rad/s})}\right] = \boxed{4.4\ \text{kg}\cdot\text{m}^2}$$

---

69. **REASONING** In both parts of the problem, the magnitude of the torque is given by Equation 9.1 as the magnitude F of the force times the lever arm l. In part (a), the lever arm is just the distance of 0.55 m given in the drawing. However, in part (b), the lever arm is less than the given distance and must be expressed using trigonometry as $l = (0.55\ \text{m})\sin\theta$. See the drawing at the right.

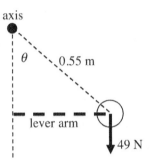

SOLUTION
a. Using Equation 9.1, we find that

$$\text{Magnitude of torque} = Fl = (49\ \text{N})(0.55\ \text{m}) = \boxed{27\ \text{N}\cdot\text{m}}$$

b. Again using Equation 9.1, this time with a lever arm of $l = (0.55\ \text{m})\sin\theta$, we obtain

$$\text{Magnitude of torque} = 15\ \text{N}\cdot\text{m} = Fl = (49\ \text{N})(0.55\ \text{m})\sin\theta$$

$$\sin\theta = \frac{15\ \text{N}\cdot\text{m}}{(49\ \text{N})(0.55\ \text{m})} \quad \text{or} \quad \theta = \sin^{-1}\left[\frac{15\ \text{N}\cdot\text{m}}{(49\ \text{N})(0.55\ \text{m})}\right] = \boxed{34°}$$

---

73. **REASONING** The rotational analog of Newton's second law is given by Equation 9.7, $\sum\tau = I\alpha$. Since the person pushes on the outer edge of one section of the door with a force **F** that is directed perpendicular to the section, the torque exerted on the door has a magnitude of FL, where the lever arm L is equal to the width of one section. Once the moment of inertia is known, Equation 9.7 can be solved for the angular acceleration α.

The moment of inertia of the door relative to the rotation axis is $I = 4I_{\text{P}}$, where I_{P} is the moment of inertia for one section. According to Table 9.1, we find $I_{\text{P}} = \frac{1}{3}ML^2$, so that the rotational inertia of the door is $I = \frac{4}{3}ML^2$.

SOLUTION Solving Equation 9.7 for α, and using the expression for I determined above, we have

$$\alpha = \frac{FL}{\frac{4}{3}ML^2} = \frac{F}{\frac{4}{3}ML} = \frac{68\ \text{N}}{\frac{4}{3}(85\ \text{kg})(1.2\ \text{m})} = \boxed{0.50\ \text{rad/s}^2}$$

---

77. **REASONING** When the modules pull together, they do so by means of forces that are internal. These pulling forces, therefore, do not create a net external torque, and the angular momentum of the system is conserved. In other words, it remains constant. We will use the conservation of angular momentum to obtain a relationship between the initial and final angular speeds. Then, we will use Equation 8.9 ($v = r\omega$) to relate the angular speeds ω_0 and ω_{f} to the tangential speeds v_0 and v_{f}.

SOLUTION Let L be the initial length of the cable between the modules and ω_0 be the initial angular speed. Relative to the center-of-mass axis, the initial momentum of inertia of the two-module system is $I_0 = 2M(L/2)^2$, according to Equation 9.6. After the modules pull together, the length of the cable is $L/2$, the final angular speed is ω_{f}, and the momentum of inertia is $I_{\text{f}} = 2M(L/4)^2$. The conservation of angular momentum indicates that

$$\underbrace{I_{\text{f}}\omega_{\text{f}}}_{\substack{\text{Final angular}\\\text{momentum}}} = \underbrace{I_0\omega_0}_{\substack{\text{Initial angular}\\\text{momentum}}}$$

$$\left[2M\left(\frac{L}{4}\right)^2\right]\omega_{\text{f}} = \left[2M\left(\frac{L}{2}\right)^2\right]\omega_0$$

$$\omega_{\text{f}} = 4\omega_0$$

According to Equation 8.9, $\omega_{\text{f}} = v_{\text{f}}/(L/4)$ and $\omega_0 = v_0/(L/2)$. With these substitutions, the result that $\omega_{\text{f}} = 4\omega_0$ becomes

$$\frac{v_{\text{f}}}{L/4} = 4\left(\frac{v_0}{L/2}\right) \quad \text{or} \quad v_{\text{f}} = 2v_0 = 2(17\ \text{m/s}) = \boxed{34\ \text{m/s}}$$

81. **REASONING**

To determine the angular acceleration of the dual pulley and the tension in the cable attached to the crate, we will apply Newton's second law to the pulley and the crate separately. Four external forces act on the dual pulley, as its free-body diagram in the figure shows. These are (1) the tension $\vec{T}_1$ in the cable connected to the motor, (2) the tension $\vec{T}_2$ in the cable attached to the crate, (3) the pulley's weight $\vec{W}_P$, and (4) the reaction force $\vec{P}$ exerted on the dual pulley by the axle. The

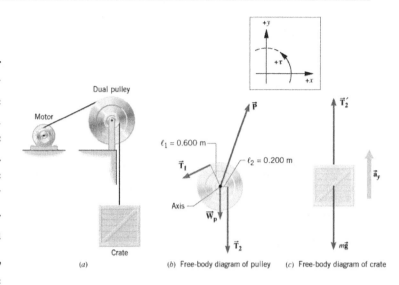

force $\vec{P}$ arises because the two cables and the pulley's weight pull the pulley down and to the left into the axle, and the axle pushes back, thus keeping the pulley in place. The net torque that results from these forces obeys Newton's second law for rotational motion (Equation 9.7). Two external forces act on the crate, as its free-body diagram in figure indicates. These are (1) the cable tension $\vec{T}_2{}'$ and (2) the weight $m\vec{g}$ of the crate. The net force that results from these forces obeys Newton's second law for translational motion (Equation 4.2b).

Using the lever arms ℓ_1 and ℓ_2 shown in the figure, we can apply the second law to the rotational motion of the dual pulley:

$$\sum \tau = T_1\ell_1 - T_2\ell_2 = I\alpha \qquad (1)$$

Note that the forces $\vec{P}$ and $\vec{W}_P$ have zero lever arms, since their lines of action pass through the axis of rotation. Thus, these forces contribute nothing to the net torque.

Applying Newton's second law to the upward translational motion of the crate gives:

$$\sum F_y = T_2{}' - mg = ma_y$$

Note that the magnitude of the tension in the cable between the crate and the pulley is $T_2' = T_2$, so that we can solve the above equation for the tension and obtain:

$$T_2' = T_2 = mg + ma_y$$

Because the cable attached to the crate rolls on the pulley without slipping, the linear acceleration a_y of the crate is related to the angular acceleration α of the pulley via $a_y = r\alpha$ (Equation 8.13), where $r = \ell_2$. Thus, we have $a_y = \ell_2 \alpha$. We can now substitute our expressions for T_2 and a_y into Equation (1):

$$I\alpha = T_1\ell_1 - T_2\ell_2 = T_1\ell_1 - (mg + ma_y)\ell_2 = T_1\ell_1 - (mg + \ell_2\alpha)\ell_2$$

Solving for α gives:

$$\alpha = \frac{T_1\ell_1 - mg\ell_2}{I + m\ell_2^2} = \frac{(2150 \text{ N})(0.600 \text{ m}) - (451 \text{ kg})(9.80 \text{ m/s}^2)(0.200 \text{ m})}{46.0 \text{ kg} \cdot \text{m}^2 + (451 \text{ kg})(0.200 \text{ m})^2} = \boxed{6.3 \text{ rad/s}^2}$$

Using this value for α and Equation (1), we can now solve for T_2:

$$T_2 = \frac{T_1\ell_1 - I\alpha}{\ell_2} = \frac{(2150 \text{ N})(0.600 \text{ m}) - (46.0 \text{ kg} \cdot \text{m}^2)(6.3 \text{ rad/s}^2)}{0.200 \text{ m}} = \boxed{5.00 \times 10^3 \text{ N}}$$

83. SSM   *CONCEPTS* (i) The solid sphere has a moment of inertia of $\frac{2}{5}MR^2$, while the shell has a moment of inertia of $\frac{2}{3}MR^2$. Since the masses and radii of the spheres are the same, it follows that the shell has the greater moment of inertia. The reason is that more of the mass of the shell is located farther from the rotational axis than is the case for the solid sphere. In the solid sphere, some of the mass is located close to the axis and, therefore, does not contribute as much to the moment of inertia. (ii) Newton's second law for rotation specifies that the angular acceleration is $\alpha = (\Sigma\tau)/I$, where $\Sigma\tau$ is the net external torque and I is the moment of inertia. Since the moment of inertia is in the denominator, the angular acceleration is smaller when I is greater. Because it has the greater moment of inertia, the shell has the angular acceleration with the smaller magnitude. (iii) Since the angular acceleration of the shell has the smaller magnitude, the shell requires a longer time for the deceleration to reduce its angular velocity to zero.

CALCULATIONS According to the equations of rotational kinematics,

$$\alpha = \frac{\omega - \omega_0}{t - t_0}$$

where ω and ω_0 are the final and initial angular velocities, respectively. Taking the initial time to be $t_0 = 0$, we write

$$t = \frac{\omega - \omega_0}{\alpha}$$

From Newton's second law for rotation, the angular acceleration is given by $\alpha = (\Sigma\tau)/I$, and substituting this into the previous equation results in

$$t = \frac{\omega - \omega_0}{(\Sigma\tau)/I} = \frac{I(\omega - \omega_0)}{\Sigma\tau}$$

In applying this result, we arbitrarily choose the direction of the initial rotation to be positive. With this choice, the torque must be negative, since it causes a deceleration. Using the proper moments of inertia, we find the following times for the spheres to come to a halt:

Solid $$t = \frac{I(\omega - \omega_0)}{\Sigma\tau} = \frac{\frac{2}{5}MR^2(\omega - \omega_0)}{\Sigma\tau}$$ **sphere:**

$$= \frac{\frac{2}{5}(1.5 \text{ kg})(0.20 \text{ m})^2 \left[(0 \text{ rad/s}) - (24 \text{ rad/s})\right]}{-0.12 \text{ N} \cdot \text{m}} = \boxed{4.8 \text{ s}}$$

Spherical $$t = \frac{I(\omega - \omega_0)}{\Sigma\tau} = \frac{\frac{2}{3}MR^2(\omega - \omega_0)}{\Sigma\tau}$$

$$= \frac{\frac{2}{3}(1.5 \text{ kg})(0.20 \text{ m})^2 \left[(0 \text{ rad/s}) - (24 \text{ rad/s})\right]}{-0.12 \text{ N} \cdot \text{m}} = \boxed{8.0 \text{ s}}$$

CHAPTER 10 | SIMPLE HARMONIC MOTION AND ELASTICITY

1. **REASONING AND SOLUTION** Using Equation 10.1, we first determine the spring constant:

$$k = \frac{F_x^{\text{Applied}}}{x} = \frac{89.0 \text{ N}}{0.0191 \text{ m}} = 4660 \text{ N/m}$$

Again using Equation 10.1, we find that the force needed to compress the spring by 0.0508 m is

$$F_x^{\text{Applied}} = kx = (4660 \text{ N/m})(0.0508 \text{ m}) = \boxed{237 \text{ N}}$$

---

5. **REASONING** The weight of the person causes the spring in the scale to compress. The amount x of compression, according to Equation 10.1, depends on the magnitude F_x^{Applied} of the applied force and the spring constant k.

SOLUTION

a. Since the applied force is equal to the person's weight, the spring constant is

$$k = \frac{F_x^{\text{Applied}}}{x} = \frac{670 \text{ N}}{0.79 \times 10^{-2} \text{ m}} = \boxed{8.5 \times 10^4 \text{ N/m}} \qquad (10.1)$$

b. When another person steps on the scale, it compresses by 0.34 cm. The weight (or applied force) that this person exerts on the scale is

$$F_x^{\text{Applied}} = kx = \left(8.5 \times 10^4 \text{ N/m}\right)\left(0.34 \times 10^{-2} \text{ m}\right) = \boxed{290 \text{ N}} \qquad (10.1)$$

---

9. **REASONING AND SOLUTION** The force that acts on the block is given by Newton's Second law, $F_x = ma_x$ (Equation 4.2a). Since the block has a constant acceleration, the acceleration is given by Equation 2.8 with $v_0 = 0$ m/s; that is, $a_x = 2d/t^2$, where d is the distance through which the block is pulled. Therefore, the force that acts on the block is given by

$$F_x = ma_x = \frac{2md}{t^2}$$

The force acting on the block is the restoring force of the spring. Thus, according to Equation 10.2, $F_x = -kx$, where k is the spring constant and x is the displacement. Solving Equation 10.2 for x and using the expression above for F_x, we obtain

$$x = -\frac{F_x}{k} = -\frac{2md}{kt^2} = -\frac{2(7.00\,\text{kg})(4.00\,\text{m})}{(415\,\text{N/m})(0.750\,\text{s})^2} = -0.240\,\text{m}$$

The amount that the spring stretches is $\boxed{0.240\,\text{m}}$.

---

11. **REASONING**   When the ball is whirled in a horizontal circle of radius r at speed v, the centripetal force is provided by the restoring force of the spring. From Equation 5.3, the magnitude of the centripetal force is mv^2/r, while the magnitude of the restoring force is kx (see Equation 10.2). Thus,

$$\frac{mv^2}{r} = kx \tag{1}$$

The radius of the circle is equal to $(L_0 + \Delta L)$, where L_0 is the unstretched length of the spring and ΔL is the amount that the spring stretches. Equation (1) becomes

$$\frac{mv^2}{L_0 + \Delta L} = k\,\Delta L \tag{1'}$$

If the spring were attached to the ceiling and the ball were allowed to hang straight down, motionless, the net force must be zero: $mg - kx = 0$, where $-kx$ is the restoring force of the spring. If we let Δy be the displacement of the spring in the vertical direction, then

$$mg = k\Delta y$$

Solving for Δy, we obtain

$$\Delta y = \frac{mg}{k} \tag{2}$$

SOLUTION   According to equation (1') above, the spring constant k is given by

$$k = \frac{mv^2}{\Delta L(L_0 + \Delta L)}$$

Substituting this expression for k into equation (2) gives

$$\Delta y = \frac{mg\Delta L(L_0 + \Delta L)}{mv^2} = \frac{g\Delta L(L_0 + \Delta L)}{v^2}$$

or

$$\Delta y = \frac{(9.80\,\text{m/s}^2)(0.010\,\text{m})(0.200\,\text{m} + 0.010\,\text{m})}{(3.00\,\text{m/s})^2} = \boxed{2.29 \times 10^{-3}\,\text{m}}$$

15. **REASONING** The frequency f of the eardrum's vibration is related to its angular frequency ω via $\omega = 2\pi f$ (Equation 10.6). The maximum speed during vibration is given by $v_{max} = A\omega$ (Equation 10.8). We will find the frequency f in part a from Equations 10.6 and 10.8. In part b, we will find the maximum acceleration that the eardrum undergoes with the aid of $a_{max} = A\omega^2$ (Equation 10.10).

SOLUTION
a. Substituting $\omega = 2\pi f$ (Equation 10.6) into $v_{max} = A\omega$ (Equation 10.8) and solving for the frequency f, we obtain

$$v_{max} = A\omega = A(2\pi f) \qquad \text{or} \qquad f = \frac{v_{max}}{2\pi A} = \frac{2.9 \times 10^{-3} \text{ m/s}}{2\pi (6.3 \times 10^{-7} \text{ m})} = \boxed{730 \text{ Hz}}$$

b. Substituting $\omega = 2\pi f$ (Equation 10.6) into $a_{max} = A\omega^2$ (Equation 10.10) yields the maximum acceleration of the vibrating eardrum:

$$a_{max} = A\omega^2 = A(2\pi f)^2 = (6.3 \times 10^{-7} \text{ m})[2\pi(730 \text{ Hz})]^2 = \boxed{13 \text{ m/s}^2}$$

---

21. **REASONING** The frequency of vibration of the spring is related to the added mass m by Equations 10.6 and 10.11:

$$f = \frac{1}{2\pi}\sqrt{\frac{k}{m}} \qquad (1)$$

The spring constant can be determined from Equation 10.1.

SOLUTION Since the spring stretches by 0.018 m when a 2.8-kg object is suspended from its end, the spring constant is, according to Equation 10.1,

$$k = \frac{F_x^{\text{Applied}}}{x} = \frac{mg}{x} = \frac{(2.8 \text{ kg})(9.80 \text{ m/s}^2)}{0.018 \text{ m}} = 1.52 \times 10^3 \text{ N/m}$$

Solving Equation (1) for m, we find that the mass required to make the spring vibrate at 3.0 Hz is

$$m = \frac{k}{4\pi^2 f^2} = \frac{1.52 \times 10^3 \text{ N/m}}{4\pi^2 (3.0 \text{ Hz})^2} = \boxed{4.3 \text{ kg}}$$

29. **_REASONING_** Since air resistance is negligible, we can apply the principle of conservation of mechanical energy, which indicates that the total mechanical energy of the block and the spring is the same at the instant it comes to a momentary halt on the spring and at the instant the block is dropped. Gravitational potential energy is one part of the total mechanical energy, and Equation 6.5 indicates that it is mgh for a block of mass m located at a height h relative to an arbitrary zero level. This dependence on h will allow us to determine the height at which the block was dropped.

SOLUTION The conservation of mechanical energy states that the final total mechanical energy E_f is equal to the initial total mechanical energy E_0. The expression for the total mechanical energy for an object on a spring is given by Equation 10.14, so that we have

$$\underbrace{\tfrac{1}{2}mv_f^2 + \tfrac{1}{2}I\omega_f^2 + mgh_f + \tfrac{1}{2}ky_f^2}_{E_f} = \underbrace{\tfrac{1}{2}mv_0^2 + \tfrac{1}{2}I\omega_0^2 + mgh_0 + \tfrac{1}{2}ky_0^2}_{E_0}$$

The block does not rotate, so the angular speeds ω_f and ω_0 are zero. Since the block comes to a momentary halt on the spring and is dropped from rest, the translational speeds v_f and v_0 are also zero. Because the spring is initially unstrained, the initial displacement y_0 of the spring is likewise zero. Thus, the above expression can be simplified as follows:

$$mgh_f + \tfrac{1}{2}ky_f^2 = mgh_0$$

The block was dropped at a height of $h_0 - h_f$ above the compressed spring. Solving the simplified energy-conservation expression for this quantity gives

$$h_0 - h_f = \frac{ky_f^2}{2mg} = \frac{(450 \text{ N/m})(0.025 \text{ m})^2}{2(0.30 \text{ kg})(9.80 \text{ m/s}^2)} = \boxed{0.048 \text{ m or 4.8 cm}}$$

---

33. **_REASONING_** The only force that acts on the block along the line of motion is the force due to the spring. Since the force due to the spring is a conservative force, the principle of conservation of mechanical energy applies. Initially, when the spring is unstrained, all of the mechanical energy is kinetic energy, $(1/2)mv_0^2$. When the spring is fully compressed, all of the mechanical energy is in the form of elastic potential energy, $(1/2)kx_{max}^2$, where x_{max}, the maximum compression of the spring, is the amplitude A. Therefore, the statement of energy conservation can be written as

$$\tfrac{1}{2}mv_0^2 = \tfrac{1}{2}kA^2$$

This expression may be solved for the amplitude A.

SOLUTION Solving for the amplitude A, we obtain

$$A = \sqrt{\frac{mv_0^2}{k}} = \sqrt{\frac{(1.00 \times 10^{-2} \text{ kg})(8.00 \text{ m/s})^2}{124 \text{ N/m}}} = \boxed{7.18 \times 10^{-2} \text{ m}}$$

37. **REASONING** The angular frequency ω (in rad/s) is given by $\omega = \sqrt{k/m}$ (Equation 10.11), where k is the spring constant and m is the mass of the object. However, we are given neither k nor m. Instead, we are given information about how much the spring is compressed and the launch speed of the object. Once launched, the object has kinetic energy, which is related to its speed. Before launching, the spring/object system has elastic potential energy, which is related to the amount by which the spring is compressed. This suggests that we apply the principle of conservation of mechanical energy in order to use the given information. This principle indicates that the total mechanical energy of the system is the same after the object is launched as it is before the launch. The resulting equation will provide us with the value of k/m that we need in order to determine the angular frequency from $\omega = \sqrt{k/m}$.

SOLUTION The conservation of mechanical energy states that the final total mechanical energy E_f is equal to the initial total mechanical energy E_0. The expression for the total mechanical energy for a spring/mass system is given by Equation 10.14, so that we have

$$\underbrace{\tfrac{1}{2}mv_f^2 + \tfrac{1}{2}I\omega_f^2 + mgh_f + \tfrac{1}{2}kx_f^2}_{E_f} = \underbrace{\tfrac{1}{2}mv_0^2 + \tfrac{1}{2}I\omega_0^2 + mgh_0 + \tfrac{1}{2}kx_0^2}_{E_0}$$

Since the object does not rotate, the angular speeds ω_f and ω_0 are zero. Since the object is initially at rest, the initial translational speed v_0 is also zero. Moreover, the motion takes place horizontally, so that the final height h_f is the same as the initial height h_0. Lastly, the spring is unstrained after the launch, so that x_f is zero. Thus, the above expression can be simplified as follows:

$$\tfrac{1}{2}mv_f^2 = \tfrac{1}{2}kx_0^2 \quad \text{or} \quad \frac{k}{m} = \frac{v_f^2}{x_0^2}$$

Substituting this result into Equation 10.11 shows that

$$\omega = \sqrt{\frac{k}{m}} = \sqrt{\frac{v_f^2}{x_0^2}} = \frac{v_f}{x_0} = \frac{1.50 \text{ m/s}}{0.0620 \text{ m}} = \boxed{24.2 \text{ rad/s}}$$

41. **REASONING** Using the principle of conservation of mechanical energy, the initial elastic potential energy stored in the elastic bands must be equal to the sum of the kinetic energy and the gravitational potential energy of the performer at the point of ejection:

$$\frac{1}{2}kx^2 = \frac{1}{2}mv_0^2 + mgh$$

where v_0 is the speed of the performer at the point of ejection and, from the figure at the right, $h = x \sin \theta$. Thus,

$$\frac{1}{2}kx^2 = \frac{1}{2}mv_0^2 + mgx \sin \theta \qquad (1)$$

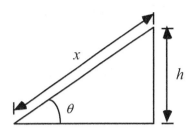

From the horizontal motion of the performer

$$v_{0x} = v_0 \cos\theta \qquad (2)$$

where

$$v_{0x} = \frac{s_x}{t} \qquad (3)$$

and $s_x = 26.8$ m. Combining equations (2) and (3) gives

$$v_0 = \frac{s_x}{t(\cos\theta)}$$

Equation (1) becomes:

$$\frac{1}{2}kx^2 = \frac{1}{2}m\frac{s_x^2}{t^2\cos^2\theta} + mgx \sin\theta$$

This expression can be solved for k, the spring constant of the firing mechanism.

SOLUTION Solving for k yields:

$$k = m\left(\frac{s_x}{xt\cos\theta}\right)^2 + \frac{2mg(\sin\theta)}{x}$$

$$k = (70.0\ \text{kg})\left[\frac{26.8\ \text{m}}{(3.00\ \text{m})(2.14\ \text{s})(\cos 40.0°)}\right]^2$$

$$+\frac{2(70.0\ \text{kg})(9.80\ \text{m/s}^2)(\sin 40.0°)}{3.00\ \text{m}} = \boxed{2.37\times10^3\ \text{N/m}}$$

---

51. **REASONING** When the tow truck pulls the car out of the ditch, the cable stretches and a tension exists in it. This tension is the force that acts on the car. The amount ΔL that the cable stretches depends on the tension F, the length L_0 and cross-sectional area A of the cable, as well as Young's modulus Y for steel. All of these quantities are given in the statement of the problem, except for Young's modulus, which can be found by consulting Table 10.1.

SOLUTION Solving Equation 10.17, $F = Y\left(\dfrac{\Delta L}{L_0}\right)A$, for the change in length, we have

$$\Delta L = \frac{FL_0}{AY} = \frac{(890\ \text{N})(9.1\ \text{m})}{\pi\left(0.50\times10^{-2}\ \text{m}\right)^2\left(2.0\times10^{11}\ \text{N/m}^2\right)} = \boxed{5.2\times10^{-4}\ \text{m}}$$

---

53. *REASONING AND SOLUTION* According to Equation 10.20, it follows that

$$\Delta P = -B\frac{\Delta V}{V_0} = -\left(2.6\times10^{10}\ \text{N/m}^2\right)\frac{-1.0\times10^{-10}\ \text{m}^3}{1.0\times10^{-6}\ \text{m}^3} = 2.6\times10^6\ \text{N/m}^2 \qquad (10.20)$$

where we have expressed the volume V_0 of the cube at the ocean's surface as $V_0 = \left(1.0\times10^{-2}\ \text{m}\right)^3 = 1.0\times10^{-6}\ \text{m}^3$.

Since the pressure increases by $1.0\times10^4\ \text{N/m}^2$ per meter of depth, the depth is

$$\frac{2.6\times10^6\ \text{N/m}^2}{1.0\times10^4\ \dfrac{\text{N/m}^2}{\text{m}}} = \boxed{260\ \text{m}}$$

---

59. *REASONING AND SOLUTION* The shearing stress is equal to the force per unit area applied to the rivet. Thus, when a shearing stress of 5.0×10^8 Pa is applied to each rivet, the force experienced by each rivet is

$$F = (\text{Stress})A = (\text{Stress})(\pi r^2) = (5.0\times10^8\ \text{Pa})\left[\pi(5.0\times10^{-3}\ \text{m})^2\right] = 3.9\times10^4\ \text{N}$$

Therefore, the maximum tension T that can be applied to each beam, assuming that each rivet carries one-fourth of the total load, is $4F = \boxed{1.6\times10^5\ \text{N}}$.

---

65. *REASONING* Our approach is straightforward. We will begin by writing Equation 10.17 $\left[F = Y\left(\dfrac{\Delta L}{L_0}\right)A\right]$ as it applies to the composite rod. In so doing, we will use subscripts for only those variables that have different values for the composite rod and the aluminum and tungsten sections. Thus, we note that the force applied to the end of the composite rod (see Figure 10.27) is also applied to each section of the rod, with the result that the magnitude F of the force has no subscript. Similarly, the cross-sectional area A is the same for the composite rod and for the aluminum and tungsten sections. Next, we will express the change

$\Delta L_{\text{Composite}}$ in the length of the composite rod as the sum of the changes in lengths of the aluminum and tungsten sections. Lastly, we will take into account that the initial length of the composite rod is twice the initial length of either of its two sections and thereby simply our equation algebraically to the point that we can solve it for the effective value of Young's modulus that applies to the composite rod.

SOLUTION Applying Equation 10.17 to the composite rod, we obtain

$$F = Y_{\text{Composite}} \left(\frac{\Delta L_{\text{Composite}}}{L_{0,\,\text{Composite}}} \right) A \tag{1}$$

Since the change $\Delta L_{\text{Composite}}$ in the length of the composite rod is the sum of the changes in lengths of the aluminum and tungsten sections, we have $\Delta L_{\text{Composite}} = \Delta L_{\text{Aluminum}} + \Delta L_{\text{Tungsten}}$. Furthermore, the changes in length of each section can be expressed using Equation 10.17 $\left(\Delta L = \dfrac{FL_0}{YA} \right)$, so that

$$\Delta L_{\text{Composite}} = \Delta L_{\text{Aluminum}} + \Delta L_{\text{Tungsten}} = \frac{FL_{0,\,\text{Aluminum}}}{Y_{\text{Aluminum}} A} + \frac{FL_{0,\,\text{Tungsten}}}{Y_{\text{Tungsten}} A}$$

Substituting this result into Equation (1) gives

$$F = \left(\frac{Y_{\text{Composite}} A}{L_{0,\,\text{Composite}}} \right) \Delta L_{\text{Composite}} = \left(\frac{Y_{\text{Composite}} A}{L_{0,\,\text{Composite}}} \right) \left(\frac{FL_{0,\,\text{Aluminum}}}{Y_{\text{Aluminum}} A} + \frac{FL_{0,\,\text{Tungsten}}}{Y_{\text{Tungsten}} A} \right)$$

$$1 = Y_{\text{Composite}} \left(\frac{L_{0,\,\text{Aluminum}}}{L_{0,\,\text{Composite}} Y_{\text{Aluminum}}} + \frac{L_{0,\,\text{Tungsten}}}{L_{0,\,\text{Composite}} Y_{\text{Tungsten}}} \right)$$

In this result we now use the fact that $L_{0,\,\text{Aluminum}}/L_{0,\,\text{Composite}} = L_{0,\,\text{Tungsten}}/L_{0,\,\text{Composite}} = 1/2$ and obtain

$$1 = Y_{\text{Composite}} \left(\frac{1}{2Y_{\text{Aluminum}}} + \frac{1}{2Y_{\text{Tungsten}}} \right)$$

Solving for $Y_{\text{Composite}}$ shows that

$$Y_{\text{Composite}} = \frac{2Y_{\text{Tungsten}} Y_{\text{Aluminum}}}{Y_{\text{Tungsten}} + Y_{\text{Aluminum}}} = \frac{2\left(3.6 \times 10^{11} \text{ N/m}^2\right)\left(6.9 \times 10^{10} \text{ N/m}^2\right)}{\left(3.6 \times 10^{11} \text{ N/m}^2\right) + \left(6.9 \times 10^{10} \text{ N/m}^2\right)} = \boxed{1.2 \times 10^{11} \text{ N/m}^2}$$

The values for Y_{Tungsten} and Y_{Aluminum} have been taken from Table 10.1.

69. ***REASONING AND SOLUTION*** Strain $= \Delta L/L_0 = F/(YA)$ where $F = mg$ and $A = \pi r^2$. Setting the strain for the spider thread equal to the strain for the wire

$$\underbrace{\frac{F}{Y A}}_{\substack{\text{Spider} \\ \text{thread}}} = \underbrace{\frac{F'}{Y' A'}}_{\substack{\text{Aluminum} \\ \text{wire}}} \quad \text{so that} \quad \frac{F}{Y r^2} = \frac{F'}{Y' r'^2}$$

Thus,

$$r'^2 = \frac{F' Y r^2}{F Y'}$$

Taking the value for Young's modulus Y' for aluminum from Table 10.1, we find that

$$r' = \sqrt{\frac{(95 \text{ kg})(9.80 \text{ m/s}^2)(4.5 \times 10^9 \text{ Pa})(13 \times 10^{-6} \text{ m})^2}{(1.0 \times 10^{-3} \text{ kg})(9.80 \text{ m/s}^2)(6.9 \times 10^{10} \text{ Pa})}} = \boxed{1.0 \times 10^{-3} \text{ m}}$$

73. ***REASONING*** Equation 10.20 can be used to find the fractional change in volume of the brass sphere when it is exposed to the Venusian atmosphere. Once the fractional change in volume is known, it can be used to calculate the fractional change in radius.

SOLUTION According to Equation 10.20, the fractional change in volume is

$$\frac{\Delta V}{V_0} = -\frac{\Delta P}{B} = -\frac{8.9 \times 10^6 \text{ Pa}}{6.7 \times 10^{10} \text{ Pa}} = -1.33 \times 10^{-4}$$

Here, we have used the fact that $\Delta P = 9.0 \times 10^6 \text{ Pa} - 1.0 \times 10^5 \text{ Pa} = 8.9 \times 10^6 \text{ Pa}$, and we have taken the value for the bulk modulus B of brass from Table 10.3. The initial volume of the sphere is $V_0 = \frac{4}{3}\pi r_0^3$. If we assume that the change in the radius of the sphere is very small relative to the initial radius, we can think of the sphere's change in volume as the addition or subtraction of a spherical shell of volume ΔV, whose radius is r_0 and whose thickness is Δr. Then, the change in volume of the sphere is equal to the volume of the shell and is given by $\Delta V = \left(4\pi r_0^2\right)\Delta r$. Combining the expressions for V_0 and ΔV, and solving for $\Delta r / r_0$, we have

$$\frac{\Delta r}{r_0} = \left(\frac{1}{3}\right)\frac{\Delta V}{V_0}$$

Therefore,

$$\frac{\Delta r}{r_0} = \frac{1}{3}\left(-1.33\times10^{-4}\right) = \boxed{-4.4\times10^{-5}}$$

---

77. **REASONING** Each spring supports one-quarter of the total mass m_{total} of the system (the empty car plus the four passengers), or $\frac{1}{4}m_{total}$. The mass $m_{one\ passenger}$ of one passenger is equal to $\frac{1}{4}m_{total}$ minus one-quarter of the mass $m_{empty\ car}$ of the empty car:

$$m_{one\ passenger} = \frac{1}{4}m_{total} - \frac{1}{4}m_{empty\ car} \tag{1}$$

The mass of the empty car is known. Since the car and its passengers oscillate up and down in simple harmonic motion, the angular frequency ω of oscillation is related to the spring constant k and the mass $\frac{1}{4}m_{total}$ supported by each spring by:

$$\omega = \sqrt{\frac{k}{\frac{1}{4}m_{total}}} \tag{10.11}$$

Solving this expression for $\frac{1}{4}m_{total}$ $\left(\frac{1}{4}m_{total} = k/\omega^2\right)$ and substituting the result into Equation (1) gives

$$m_{one\ passenger} = \frac{k}{\omega^2} - \frac{1}{4}m_{empty\ car} \tag{2}$$

The angular frequency ω is inversely related to the period T of oscillation by $\omega = 2\pi/T$ (see Equation 10.4). Substituting this expression for ω into Equation (2) yields

$$m_{one\ passenger} = \frac{k}{\left(\dfrac{2\pi}{T}\right)^2} - \frac{1}{4}m_{empty\ car}$$

SOLUTION The mass of one of the passengers is

$$m_{one\ passenger} = \frac{k}{\left(\dfrac{2\pi}{T}\right)^2} - \frac{1}{4}m_{empty\ car} = \frac{1.30\times10^5\ \text{N/m}}{\left(\dfrac{2\pi}{0.370\ \text{s}}\right)^2} - \frac{1}{4}\left(1560\ \text{kg}\right) = \boxed{61\ \text{kg}}$$

---

93. **CONCEPTS** (i) The amplitude is the distance from the midpoint of the motion to either the highest or lowest point. Thus, the amplitude is one-half the vertical distance between the highest and lowest points of the motion. (ii) The time for the diver to complete one motional cycle is defined as the period. In one cycle, the diver moves downward from the highest point to the lowest point and then moves upward and returns to the highest point. In a time equal to one-quarter of a period, the diver completes one-quarter of this cycle and, therefore, is halfway between the highest and lowest points. His speed is momentarily zero at the highest and lowest points and is a maximum at the halfway point. (iii) No. The period is the time to complete one cycle, and it is equal to the distance traveled during one cycle divided by the average speed. If the amplitude doubles, the distance also doubles. However, the average speed also doubles. We can verify this by examining Equation 10.7, which gives the diver's velocity as $v_y = -A\omega \sin \omega t$. The speed is the magnitude of this value, or $A\omega \sin \omega t$.

Since the speed is proportional to the amplitude A, the speed at every point in the cycle also doubles when the amplitude doubles. Thus the average speed doubles. However, the period, being the distance divided by the average speed, does not change.

CALCULATIONS (a) Since the amplitude A is one-half the vertical distance between the highest and lowest points of the motion,

$$A = \tfrac{1}{2}(0.30 \text{ m}) = \boxed{0.15 \text{ m}}$$

(b) When the diver is halfway between the highest and lowest points, his speed is a maximum. The maximum speed of an object vibrating in simple harmonic motion is $v_{max} = A\omega$, where A is the amplitude of the motion and ω is the angular frequency. The angular frequency can be determined from $\omega = \sqrt{k/m}$, where k is the effective spring constant of the diving board and m is the mass of the diver. The maximum speed is

$$v_{max} = A\omega = A\sqrt{\frac{k}{m}} = (0.15 \text{ m})\sqrt{\frac{4100 \text{ N/m}}{75 \text{ kg}}} = \boxed{1.1 \text{ m/s}}$$

(c) The period is the same, regardless of the amplitude of the motion. We know that the period T and the angular speed ω are related by $T = 2\pi/\omega$, where $\omega = \sqrt{k/m}$. Thus the period can be written as

$$T = \frac{2\pi}{\omega} = \frac{2\pi}{\sqrt{k/m}} = 2\pi\sqrt{\frac{m}{k}} = 2\pi\sqrt{\frac{75 \text{ kg}}{4100 \text{ N/m}}} = \boxed{0.85 \text{ s}}$$

CHAPTER 11 | *FLUIDS*

1. ***REASONING*** According to Equation 4.5, the pillar's weight is $W = mg$. Equation 11.1 can be solved for the mass m to show that the pillar's mass is $m = \rho V$. The volume V of the cylindrical pillar is its height times its circular cross-sectional area.

   SOLUTION Expressing the weight as $W = mg$ (Equation 4.5) and substituting $m = \rho V$ (Equation 11.1) for the mass give

   $$W = mg = (\rho V) g \qquad (1)$$

   The volume of the pillar is $V = h \pi r^2$, where h is the height and r is the radius of the pillar. Substituting this expression for the volume into Equation (1), we find that the weight is

   $$W = (\rho V) g = \left[\rho (h \pi r^2) \right] g = (2.2 \times 10^3 \text{ kg/m}^3)(2.2 \text{ m}) \pi (0.50 \text{ m})^2 (9.80 \text{ m/s}^2) = 3.7 \times 10^4 \text{ N}$$

   Converting newtons (N) to pounds (lb) gives

   $$W = (3.7 \times 10^4 \text{ N}) \left(\frac{0.2248 \text{ lb}}{1 \text{ N}} \right) = \boxed{8.3 \times 10^3 \text{ lb}}$$

3. ***REASONING*** Equation 11.1 can be used to find the volume occupied by 1.00 kg of silver. Once the volume is known, the area of a sheet of silver of thickness d can be found from the fact that the volume is equal to the area of the sheet times its thickness.

   SOLUTION Solving Equation 11.1 for V and using a value of $\rho = 10\,500 \text{ kg/m}^3$ for the density of silver (see Table 11.1), we find that the volume of 1.00 kg of silver is

   $$V = \frac{m}{\rho} = \frac{1.00 \text{ kg}}{10\,500 \text{ kg/m}^3} = 9.52 \times 10^{-5} \text{ m}^3$$

   The area of the silver, is, therefore,

   $$A = \frac{V}{d} = \frac{9.52 \times 10^{-5} \text{ m}^3}{3.00 \times 10^{-7} \text{ m}} = \boxed{317 \text{ m}^2}$$

7. **REASONING** According to the definition of density ρ given in Equation 11.1, the mass m of a substance is $m = \rho V$, where V is the volume. We will use this equation and the fact that the mass of the water and the gold are equal to find our answer. To convert from a volume in cubic meters to a volume in gallons, we refer to the inside of the front cover of the text to find that $1\ \text{gal} = 3.785 \times 10^{-3}\ \text{m}^3$.

REASONING Using Equation 11.1, we find that

$$\rho_{\text{Water}} V_{\text{Water}} = \rho_{\text{Gold}} V_{\text{Gold}} \quad \text{or} \quad V_{\text{Water}} = \frac{\rho_{\text{Gold}} V_{\text{Gold}}}{\rho_{\text{Water}}}$$

Using the fact that $1\ \text{gal} = 3.785 \times 10^{-3}\ \text{m}^3$ and densities for gold and water from Table 11.1, we find

$$V_{\text{Water}} = \frac{\rho_{\text{Gold}} V_{\text{Gold}}}{\rho_{\text{Water}}}$$

$$= \frac{(19\,300\ \text{kg/m}^3)(0.15\ \text{m})(0.050\ \text{m})(0.050\ \text{m})}{(1000\ \text{kg/m}^3)}\left(\frac{1\ \text{gal}}{3.785 \times 10^{-3}\ \text{m}^3}\right) = \boxed{1.9\ \text{gal}}$$

---

11. **REASONING** Since the inside of the box is completely evacuated; there is no air to exert an upward force on the lid from the inside. Furthermore, since the weight of the lid is negligible, there is only one force that acts on the lid; the downward force caused by the air pressure on the outside of the lid. In order to pull the lid off the box, one must supply a force that is at least equal in magnitude and opposite in direction to the force exerted on the lid by the outside air.

SOLUTION According to Equation 11.3, pressure is defined as $P = F / A$; therefore, the magnitude of the force on the lid due to the air pressure is

$$F = (0.85 \times 10^5\ \text{N/m}^2)(1.3 \times 10^{-2}\ \text{m}^2) = \boxed{1.1 \times 10^3\ \text{N}}$$

---

15. **REASONING** The cap is in equilibrium, so the sum of all the forces acting on it must be zero. There are three forces in the vertical direction: the force $\mathbf{F}_{\text{inside}}$ due to the gas pressure inside the bottle, the force $\mathbf{F}_{\text{outside}}$ due to atmospheric pressure outside the bottle, and the force $\mathbf{F}_{\text{thread}}$ that the screw thread exerts on the cap. By setting the sum of these forces to zero, and using the relation $F = PA$, where P is the pressure and A is the area of the cap, we can determine the magnitude of the force that the screw threads exert on the cap.

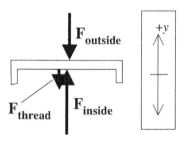

SOLUTION The drawing shows the free-body diagram of the cap and the three vertical forces that act on it. Since the cap is in equilibrium, the net force in the vertical direction must be zero.

$$\Sigma F_y = -F_{\text{thread}} + F_{\text{inside}} - F_{\text{outside}} = 0 \qquad \text{(4.9b)}$$

Solving this equation for F_{thread}, and using the fact that force equals pressure times area, $F = PA$ (Equation 11.3), we have

$$F_{\text{thread}} = F_{\text{inside}} - F_{\text{outside}} = P_{\text{inside}} A - P_{\text{outside}} A$$

$$= \left(P_{\text{inside}} - P_{\text{outside}}\right) A = \left(1.80 \times 10^5 \text{ Pa} - 1.01 \times 10^5 \text{ Pa}\right)\left(4.10 \times 10^{-4} \text{ m}^2\right) = \boxed{32 \text{ N}}$$

---

23. **REASONING** Since the faucet is closed, the water in the pipe may be treated as a static fluid. The gauge pressure P_2 at the faucet on the first floor is related to the gauge pressure P_1 at the faucet on the second floor by Equation 11.4, $P_2 = P_1 + \rho g h$.

SOLUTION
a. Solving Equation 11.4 for P_1, we find the gauge pressure at the second-floor faucet is

$$P_1 = P_2 - \rho g h = 1.90 \times 10^5 \text{ Pa} - (1.00 \times 10^3 \text{ kg/m}^3)(9.80 \text{ m/s}^2)(6.50 \text{ m}) = \boxed{1.26 \times 10^5 \text{ Pa}}$$

b. If the second faucet were placed at a height h above the first-floor faucet so that the gauge pressure P_1 at the second faucet were zero, then no water would flow from the second faucet, even if it were open. Solving Equation 11.4 for h when P_1 equals zero, we obtain

$$h = \frac{P_2 - P_1}{\rho g} = \frac{1.90 \times 10^5 \text{ Pa} - 0}{(1.00 \times 10^3 \text{ kg/m}^3)(9.80 \text{ m/s}^2)} = \boxed{19.4 \text{ m}}$$

---

27. **REASONING** The pressure P at a distance h beneath the water surface at the vented top of the water tower is $P = P_{\text{atm}} + \rho g h$ (Equation 11.4). We note that the value for h in this expression is different for the two houses and must take into account the diameter of the spherical reservoir in each case.

SOLUTION
a. The pressure at the level of house A is given by Equation 11.4 as $P_A = P_{\text{atm}} + \rho g h_A$. Now the height h_A consists of the 15.0 m plus the diameter d of the tank. We first calculate the radius of the tank, from which we can infer d. Since the tank is spherical, its full mass is given by $M = \rho V = \rho[(4/3)\pi r^3]$. Therefore,

$$r^3 = \frac{3M}{4\pi\rho} \quad \text{or} \quad r = \left(\frac{3M}{4\pi\rho}\right)^{1/3} = \left[\frac{3(5.25\times10^5\,\text{kg})}{4\pi(1.000\times10^3\,\text{kg/m}^3)}\right]^{1/3} = 5.00\,\text{m}$$

Therefore, the diameter of the tank is 10.0 m, and the height h_A is given by

$$h_A = 10.0\,\text{m} + 15.0\,\text{m} = 25.0\,\text{m}$$

According to Equation 11.4, the gauge pressure in house A is, therefore,

$$P_A - P_{\text{atm}} = \rho g h_A = (1.000\times10^3\,\text{kg/m}^3)(9.80\,\text{m/s}^2)(25.0\,\text{m}) = \boxed{2.45\times10^5\,\text{Pa}}$$

b. The pressure at house B is $P_B = P_{\text{atm}} + \rho g h_B$, where

$$h_B = 15.0\,\text{m} + 10.0\,\text{m} - 7.30\,\text{m} = 17.7\,\text{m}$$

According to Equation 11.4, the gauge pressure in house B is

$$P_B - P_{\text{atm}} = \rho g h_B = (1.000\times10^3\,\text{kg/m}^3)(9.80\,\text{m/s}^2)(17.7\,\text{m}) = \boxed{1.73\times10^5\,\text{Pa}}$$

---

29. ***REASONING AND SOLUTION*** The pressure at the bottom of the container is

$$P = P_{\text{atm}} + \rho_w g h_w + \rho_m g h_m$$

We want $P = 2P_{\text{atm}}$, and we know $h = h_w + h_m = 1.00$ m. Using the above and rearranging gives

$$h_m = \frac{P_{\text{atm}} - \rho_w g h}{(\rho_m - \rho_w)g} = \frac{1.01\times10^5\,\text{Pa} - (1.00\times10^3\,\text{kg/m}^3)(9.80\,\text{m/s}^2)(1.00\,\text{m})}{(13.6\times10^3\,\text{kg/m}^3 - 1.00\times10^3\,\text{kg/m}^3)(9.80\,\text{m/s}^2)} = \boxed{0.74\,\text{m}}$$

---

33. ***REASONING*** According to Equation 11.4, the initial pressure at the bottom of the pool is $P_0 = (P_{\text{atm}})_0 + \rho g h$, while the final pressure is $P_f = (P_{\text{atm}})_f + \rho g h$. Therefore, the change in pressure at the bottom of the pool is

$$\Delta P = P_f - P_0 = \left[(P_{\text{atm}})_f + \rho g h\right] - \left[(P_{\text{atm}})_0 + \rho g h\right] = (P_{\text{atm}})_f - (P_{\text{atm}})_0$$

According to Equation 11.3, $F = PA$, the change in the force at the bottom of the pool is

$$\Delta F = (\Delta P)A = \left[(P_{\text{atm}})_f - (P_{\text{atm}})_0\right]A$$

SOLUTION   Direct substitution of the data given in the problem into the expression above yields

$$\Delta F = \left(765 \text{ mm Hg} - 755 \text{ mm Hg}\right)(12 \text{ m})(24 \text{ m}) \left(\frac{133 \text{ Pa}}{1.0 \text{ mm Hg}}\right) = \boxed{3.8 \times 10^5 \text{ N}}$$

Note that the conversion factor 133 Pa = 1.0 mm Hg is used to convert mm Hg to Pa.

---

41.   **REASONING**   According to Archimedes principle, the buoyant force that acts on the block is equal to the weight of the water that is displaced by the block. The block displaces an amount of water V, where V is the volume of the block. Therefore, the weight of the water displaced by the block is $W = mg = \left(\rho_{\text{water}} V\right) g$.

SOLUTION   The buoyant force that acts on the block is, therefore,

$$F = \rho_{\text{water}} V g = (1.00 \times 10^3 \text{ kg/m}^3)(0.10 \text{ m} \times 0.20 \text{ m} \times 0.30 \text{ m})(9.80 \text{ m/s}^2) = \boxed{59 \text{ N}}$$

---

45.   **REASONING**   According to Equation 11.1, the density of the life jacket is its mass divided by its volume. The volume is given. To obtain the mass, we note that the person wearing the life jacket is floating, so that the upward-acting buoyant force balances the downward-acting weight of the person and life jacket. The magnitude of the buoyant force is the weight of the displaced water, according to Archimedes' principle. We can express each of the weights as mg (Equation 4.5) and then relate the mass of the displaced water to the density of water and the displaced volume by using Equation 11.1.

SOLUTION   According to Equation 11.1, the density of the life jacket is

$$\rho_{\text{J}} = \frac{m_{\text{J}}}{V_{\text{J}}} \tag{1}$$

Since the person wearing the life jacket is floating, the upward-acting buoyant force F_{B} balances the downward-acting weight W_{P} of the person and the weight W_{J} of the life jacket. The buoyant force has a magnitude that equals the weight $W_{\text{H}_2\text{O}}$ of the displaced water, as stated by Archimedes' principle. Thus, we have

$$F_{\text{B}} = W_{\text{H}_2\text{O}} = W_{\text{P}} + W_{\text{J}} \tag{2}$$

In Equation (2), we can use Equation 4.5 to express each weight as mass m times the magnitude g of the acceleration due to gravity. Then, the mass of the water can be expressed as $m_{\text{H}_2\text{O}} = \rho_{\text{H}_2\text{O}} V_{\text{H}_2\text{O}}$ (Equation 11.1). With these substitutions, Equation (2) becomes

$$m_{H_2O}\,g = m_P\,g + m_J\,g \quad \text{or} \quad (\rho_{H_2O} V_{H_2O})g = m_P\,g + m_J\,g$$

Solving this result for m_J shows that

$$m_J = \rho_{H_2O} V_{H_2O} - m_P$$

Substituting this result into Equation (1) and noting that the volume of the displaced water is $V_{H_2O} = 3.1 \times 10^{-2} \text{ m}^3 + 6.2 \times 10^{-2} \text{ m}^3$ gives

$$\rho_J = \frac{\rho_{H_2O} V_{H_2O} - m_P}{V_J} = \frac{\left(1.00 \times 10^3 \text{ kg/m}^3\right)\left(3.1 \times 10^{-2} \text{ m}^3 + 6.2 \times 10^{-2} \text{ m}^3\right) - 81 \text{ kg}}{3.1 \times 10^{-2} \text{ m}^3} = \boxed{390 \text{ kg/m}^3}$$

---

47. **REASONING AND SOLUTION**   The buoyant force exerted by the water must at least equal the weight of the logs plus the weight of the people,

$$F_B = W_L + W_P$$

$$\rho_w g V = \rho_L g V + W_P$$

Now the volume of logs needed is

$$V = \frac{M_P}{\rho_W - \rho_L} = \frac{4(80.0 \text{ kg})}{1.00 \times 10^3 \text{ kg/m}^3 - 725 \text{ kg/m}^3} = 1.16 \text{ m}^3$$

The volume of one log is

$$V_L = \pi(8.00 \times 10^{-2} \text{ m})^2(3.00 \text{ m}) = 6.03 \times 10^{-2} \text{ m}^3$$

The number of logs needed is

$$N = V/V_L = (1.16)/(6.03 \times 10^{-2}) = 19.2$$

Therefore, at least 20 logs are needed.

---

51. **REASONING** The height of the cylinder that is in the oil is given by $h_{oil} = V_{oil}/(\pi r^2)$, where V_{oil} is the volume of oil displaced by the cylinder and r is the radius of the cylinder. We must, therefore, find the volume of oil displaced by the cylinder. After the oil is poured in, the buoyant force that acts on the cylinder is equal to the sum of the weight of the water displaced by the cylinder and the weight of the oil displaced by the cylinder. Therefore, the

magnitude of the buoyant force is given by $F = \rho_{water} g V_{water} + \rho_{oil} g V_{oil}$. Since the cylinder floats in the fluid, the net force that acts on the cylinder must be zero. Therefore, the buoyant force that supports the cylinder must be equal to the weight of the cylinder, or

$$\rho_{water} g V_{water} + \rho_{oil} g V_{oil} = mg$$

where m is the mass of the cylinder. Substituting values into the expression above leads to

$$V_{water} + (0.725)V_{oil} = 7.00 \times 10^{-3} \text{ m}^3 \tag{1}$$

From the figure in the text, $V_{cylinder} = V_{water} + V_{oil}$. Substituting values into the expression for $V_{cylinder}$ gives

$$V_{water} + V_{oil} = 8.48 \times 10^{-3} \text{ m}^3 \tag{2}$$

Subtracting Equation (1) from Equation (2) yields $V_{oil} = 5.38 \times 10^{-3} \text{ m}^3$.

SOLUTION The height of the cylinder that is in the oil is, therefore,

$$h_{oil} = \frac{V_{oil}}{\pi r^2} = \frac{5.38 \times 10^{-3} \text{ m}^3}{\pi (0.150 \text{ m})^2} = \boxed{7.6 \times 10^{-2} \text{ m}}$$

---

55. ***REASONING AND SOLUTION*** The mass flow rate Q_{mass} is the amount of fluid mass that flows per unit time. Therefore,

$$Q_{mass} = \frac{m}{t} = \frac{\rho V}{t} = \frac{(1030 \text{ kg/m}^3)(9.5 \times 10^{-4} \text{ m}^3)}{6.0 \text{ h}} \left(\frac{1.0 \text{ h}}{3600 \text{ s}} \right) = \boxed{4.5 \times 10^{-5} \text{ kg/s}}$$

---

61. ***REASONING AND SOLUTION*** Using Bernoulli's equation, we have

$$\Delta P = P_1 - P_2 = (1/2)\rho v_2{}^2 - (1/2)\rho v_1{}^2 = (1/2)(1.29 \text{ kg/m}^3)[(8.5 \text{ m/s})^2 - (1.1 \text{ m/s})^2]$$

$$\Delta P = \boxed{46 \text{ Pa}}$$

The air flows from high pressure to low pressure (from lower to higher velocity), so it $\boxed{\text{enters at } B \text{ and exits at } A}$.

63. ***REASONING AND SOLUTION*** Let the speed of the air below and above the wing be given by v_1 and v_2, respectively. According to Equation 11.12, the form of Bernoulli's equation with $y_1 = y_2$, we have

$$P_1 - P_2 = \tfrac{1}{2}\rho \left(v_2^2 - v_1^2\right) = \frac{1.29 \text{ kg/m}^3}{2}\left[(251 \text{ m/s})^2 - (225 \text{ m/s})^2\right] = 7.98 \times 10^3 \text{ Pa}$$

From Equation 11.3, the lifting force is, therefore,

$$F = \left(P_1 - P_2\right)A = (7.98 \times 10^3 \text{ Pa})(24.0 \text{ m}^2) = \boxed{1.92 \times 10^5 \text{ N}}$$

---

69. ***REASONING*** Since the pressure difference is known, Bernoulli's equation can be used to find the speed v_2 of the gas in the pipe. Bernoulli's equation also contains the unknown speed v_1 of the gas in the Venturi meter; therefore, we must first express v_1 in terms of v_2. This can be done by using Equation 11.9, the equation of continuity.

SOLUTION
a. From the equation of continuity (Equation 11.9) it follows that $v_1 = \left(A_2 / A_1\right)v_2$. Therefore,

$$v_1 = \frac{0.0700 \text{ m}^2}{0.0500 \text{ m}^2}v_2 = (1.40)v_2$$

Substituting this expression into Bernoulli's equation (Equation 11.12), we have

$$P_1 + \tfrac{1}{2}\rho(1.40 \, v_2)^2 = P_2 + \tfrac{1}{2}\rho v_2^2$$

Solving for v_2, we obtain

$$v_2 = \sqrt{\frac{2(P_2 - P_1)}{\rho\left[(1.40)^2 - 1\right]}} = \sqrt{\frac{2(120 \text{ Pa})}{(1.30 \text{ kg/m}^3)\left[(1.40)^2 - 1\right]}} = \boxed{14 \text{ m/s}}$$

b. According to Equation 11.10, the volume flow rate is

$$Q = A_2 v_2 = (0.0700 \text{ m}^2)(14 \text{ m/s}) = \boxed{0.98 \text{ m}^3/\text{s}}$$

---

75. ***REASONING AND SOLUTION*** Bernoulli's equation (Equation 11.12) is

$$P_1 + \tfrac{1}{2}\rho v_1^2 = P_2 + \tfrac{1}{2}\rho v_2^2$$

If we let the right hand side refer to the air above the plate, and the left hand side refer to the air below the plate, then $v_1 = 0$ m/s, since the air below the plate is stationary. We wish to find v_2 for the situation illustrated in part b of the figure shown in the text. Solving the equation above for v_2 (with $v_2 = v_{2b}$ and $v_1 = 0$) gives

$$v_{2b} = \sqrt{\frac{2(P_1 - P_2)}{\rho}} \qquad (1)$$

In Equation (1), P_1 is atmospheric pressure and P_2 must be determined. We must first consider the situation in part a of the text figure.

The figure at the right shows the forces that act on the rectangular plate in part a of the text drawing. F_1 is the force exerted on the plate from the air below the plate, and F_2 is the force exerted on the plate from the air above the plate. Applying Newton's second law, we have (taking "up" to be the positive direction),

$$F_1 - F_2 - mg = 0$$

$$F_1 - F_2 = mg$$

Thus, the difference in pressures exerted by the air on the plate in part a of the drawing is

$$P_1 - P_2 = \frac{F_1 - F_2}{A} = \frac{mg}{A} \qquad (2)$$

where A is the area of the plate. From Bernoulli's equation (Equation 11.12) we have, with $v_2 = v_{2a}$ and $v_{1a} = 0$ m/s,

$$P_1 - P_2 = \tfrac{1}{2}\rho v_{2a}^2 \qquad (3)$$

where v_{2a} is the speed of the air along the top of the plate in part a of the text drawing. Combining Equations (2) and (3) we have

$$\frac{mg}{A} = \tfrac{1}{2}\rho v_{2a}^2 \qquad (4)$$

The following figure, on the left, shows the forces that act on the plate in part b of the text drawing. The notation is the same as that used when the plate was horizontal (part a of the text figure). The following figure, on the right, shows the same forces resolved into components along the plate and perpendicular to the plate.

Applying Newton's second law we have $F_1 - F_2 - mg \sin \theta = 0$, or

$$F_1 - F_2 = mg \sin \theta$$

Thus, the difference in pressures exerted by the air on the plate in part *b* of the text figure is

$$P_1 - P_2 = \frac{F_1 - F_2}{A} = \frac{mg \sin \theta}{A}$$

Using Equation (4) above,

$$P_1 - P_2 = \tfrac{1}{2} \rho v_{2a}^2 \sin \theta$$

Thus, Equation (1) becomes

$$v_{2b} = \sqrt{\frac{2\left(\tfrac{1}{2} \rho v_{2a}^2 \sin \theta\right)}{\rho}}$$

Therefore,

$$v_{2b} = \sqrt{v_{2a}^2 \sin \theta} = \sqrt{(11.0 \text{ m/s})^2 \sin 30.0^\circ} = \boxed{7.78 \text{ m/s}}$$

---

77. ***REASONING*** AND ***SOLUTION*** The Reynold's number, Re, can be written as Re $= 2\bar{v}\rho R/\eta$. To find the average speed $\bar{v}$,

$$\bar{v} = \frac{(\text{Re})\eta}{2\rho R} = \frac{(2000)\left(4.0 \times 10^{-3} \text{ Pa} \cdot \text{s}\right)}{2\left(1060 \text{ kg/m}^3\right)\left(8.0 \times 10^{-3} \text{ m}\right)} = \boxed{0.5 \text{ m/s}}$$

---

81. ***REASONING*** The volume flow rate Q of a viscous fluid flowing through a pipe of radius R is given by Equation 11.14 as $Q = \dfrac{\pi R^4 \left(P_2 - P_1\right)}{8\eta L}$, where $P_2 - P_1$ is the pressure difference

between the ends of the pipe, L is the length of the pipe, and η is the viscosity of the fluid. Since all the variables are known except L, we can use this relation to find it.

SOLUTION Solving Equation 11.14 for the pipe length, we have

$$L = \frac{\pi R^4 \left(P_2 - P_1\right)}{8\eta Q} = \frac{\pi \left(5.1\times10^{-3} \text{ m}\right)^4 \left(1.8\times10^3 \text{ Pa}\right)}{8\left(1.0\times10^{-3} \text{ Pa}\cdot\text{s}\right)\left(2.8\times10^{-4} \text{ m}^3/\text{s}\right)} = \boxed{1.7 \text{ m}}$$

---

85. **REASONING** As the depth h increases, the pressure increases according to Equation 11.4 $(P_2 = P_1 + \rho g h)$. In this equation, P_1 is the pressure at the shallow end, P_2 is the pressure at the deep end, and ρ is the density of water $(1.00 \times 10^3 \text{ kg/m}^3$, see Table 11.1). We seek a value for the pressure at the deep end minus the pressure at the shallow end.

SOLUTION Using Equation 11.4, we find

$$P_{\text{Deep}} = P_{\text{Shallow}} + \rho g h \quad \text{or} \quad P_{\text{Deep}} - P_{\text{Shallow}} = \rho g h$$

The drawing at the right shows that a value for h can be obtained from the 15-m length of the pool by using the tangent of the 11° angle:

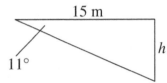

$$\tan 11° = \frac{h}{15 \text{ m}} \quad \text{or} \quad h = (15 \text{ m}) \tan 11°$$

$$P_{\text{Deep}} - P_{\text{Shallow}} = \rho g (15 \text{ m}) \tan 11°$$

$$= \left(1.00\times10^3 \text{ kg/m}^3\right)\left(9.80 \text{ m/s}^2\right)(15 \text{ m})\tan 11° = \boxed{2.9\times10^4 \text{ Pa}}$$

---

87. **REASONING** The buoyant force exerted on the balloon by the air must be equal in magnitude to the weight of the balloon and its contents (load and hydrogen). The magnitude of the buoyant force is given by $\rho_{\text{air}} V g$. Therefore,

$$\rho_{\text{air}} V g = W_{\text{load}} + \rho_{\text{hydrogen}} V g$$

where, since the balloon is spherical, $V = (4/3)\pi r^3$. Making this substitution for V and solving for r, we obtain

$$r = \left[\frac{3W_{\text{load}}}{4\pi g(\rho_{\text{air}} - \rho_{\text{hydrogen}})}\right]^{1/3}$$

SOLUTION Direct substitution of the data given in the problem yields

$$r = \left[\frac{3(5750 \text{ N})}{4\pi(9.80 \text{ m/s}^2)(1.29 \text{ kg/m}^3 - 0.0899 \text{ kg/m}^3)} \right]^{1/3} = \boxed{4.89 \text{ m}}$$

---

95. **REASONING AND SOLUTION**
a. The volume flow rate is given by Equation 11.10. Assuming that the line has a circular cross section, $A = \pi r^2$, we have

$$Q = Av = (\pi r^2)v = \pi(0.0065 \text{ m})^2(1.2 \text{ m/s}) = \boxed{1.6 \times 10^{-4} \text{ m}^3/\text{s}}$$

b. The volume flow rate calculated in part (a) above is the flow rate for all twelve holes. Therefore, the volume flow rate through one of the twelve holes is

$$Q_{hole} = \frac{Q}{12} = (\pi r_{hole}^2)v_{hole}$$

Solving for v_{hole} we have

$$v_{hole} = \frac{Q}{12\pi r_{hole}^2} = \frac{1.6 \times 10^{-4} \text{ m}^3/\text{s}}{12\pi(4.6 \times 10^{-4} \text{ m})^2} = \boxed{2.0 \times 10^1 \text{ m/s}}$$

---

99. **REASONING** Let the length of the tube be denoted by L, and let the length of the liquid be denoted by ℓ. When the tube is whirled in a circle at a constant angular speed about an axis through one end, the liquid collects at the other end and experiences a centripetal force given by (see Equation 8.11, and use $F = ma$) $F = mr\omega^2 = mL\omega^2$.

Since there is no air in the tube, this is the only radial force experienced by the liquid, and it results in a pressure of

$$P = \frac{F}{A} = \frac{mL\omega^2}{A}$$

where A is the cross-sectional area of the tube. The mass of the liquid can be expressed in terms of its density ρ and volume V: $m = \rho V = \rho A\ell$. The pressure may then be written as

$$P = \frac{\rho A\ell L\omega^2}{A} = \rho\ell L\omega^2 \tag{1}$$

If the tube were completely filled with liquid and allowed to hang vertically, the pressure at the bottom of the tube (that is, where $h = L$) would be given by

$$P = \rho g L \tag{2}$$

SOLUTION According to the statement of the problem, the quantities calculated by Equations (1) and (2) are equal, so that $\rho \ell L \omega^2 = \rho g L$. Solving for ω gives

$$\omega = \sqrt{\frac{g}{\ell}} = \sqrt{\frac{9.80 \text{ m/s}^2}{0.0100 \text{ m}}} = \boxed{31.3 \text{ rad/s}}$$

---

105. **SSM** **REASONING** Bernoulli's equation (Equation 11.12) may be used to find the pressure difference between two points in a fluid moving horizontally. However, in order to use this relation we need to know the speed of the blood in the enlarged region of the artery, as well as the speed in the normal section. We can obtain the speed in the enlarged region by using the equation of continuity (Equation 11.9), which relates it to the speed in the normal region and the cross-sectional areas of the two parts.

In the form pertinent to horizontal flow, Bernoulli's equation is given by Equation 11.12:

$$P_1 + \tfrac{1}{2}\rho v_1^2 = P_2 + \tfrac{1}{2}\rho v_2^2$$

Rearranging this expression gives the following: $P_2 - P_1 = \tfrac{1}{2}\rho(v_1^2 - v_2^2)$ (1).

For an incompressible fluid like blood, the equation of continuity is given by Equation 11.9 as

$$A_1 v_1 = A_2 v_2 \quad \text{or} \quad v_2 = \left(\frac{A_1}{A_2}\right) v_1. \quad \text{This expression for } v_2 \text{ can now be substituted into}$$

Equation (1):

$$P_2 - P_1 = \tfrac{1}{2}\rho(v_1^2 - v_2^2) = \tfrac{1}{2}\rho\left\{ v_1^2 - \left[\left(\frac{A_1}{A_2}\right)v_1\right]^2 \right\}$$

Since it is given that $A_2 = 1.7 A_1$, this result for $P_2 - P_1$ reveals that

$$P_2 - P_1 = \tfrac{1}{2}\rho\left\{ v_1^2 - \left[\left(\frac{A_1}{A_2}\right)v_1\right]^2 \right\} = \tfrac{1}{2}\rho v_1^2\left(1 - \frac{1}{1.7^2}\right)$$

$$= \tfrac{1}{2}(1060 \text{ kg/m}^3)(0.40 \text{ m/s})^2\left(1 - \frac{1}{1.7^2}\right) = \boxed{55 \text{ Pa}}$$

The positive answer indicates that P_2 is greater than P_1. The excess pressure puts added stress on the already weakened tissue of the arterial wall at the aneurysm.

---

107. SSM *CONCEPTS* (i) The downward-acting weight is balanced by the upward-acting buoyant force F_B that the water applies to the ball. (ii) The buoyant force acting on the father's beach ball is greater, since it must balance his greater weight. (iii) According to Archimedes' principle, the magnitude of the buoyant force equals the weight of the fluid that the ball displaces. Since the ball is completely submerged, it displaces a volume of water that equals the ball's volume. The weight of this volume of water is the magnitude of the buoyant force. (iv) The father's beach ball has the larger volume and the larger radius. This follows because a larger buoyant force acts on that ball. For the buoyant force to be larger, that ball must displace a greater volume of water, according to Archimedes' principle. Therefore, the volume of that ball is larger, since both balls are completely submerged.

CALCULATIONS Since the balls are in equilibrium, the net force acting on each of them must be zero. Therefore, taking upward as the positive direction, we have

$$\underbrace{\sum F}_{\substack{\text{Net} \\ \text{Force}}} = F_B - W = 0$$

Archimedes' principle specifies that the magnitude of the buoyant force is the weight of the water displaced by the ball. Using the definition of density, the mass of the displaced water is $m = \rho V$, where $\rho = 1.00 \times 10^3$ kg/m^3 is the density of water and V is the volume displaced. Since all of the ball is submerged (and assumed to remain spherical)

$$V = \tfrac{4}{3}\pi r^3 \quad \text{and} \quad W = mg = \rho\left(\tfrac{4}{3}\pi r^3\right)g$$

With this value for the buoyant force, the force equation becomes

$$F_B - W = \rho\left(\tfrac{4}{3}\pi r^3\right)g - W = 0$$

Solving for the radius r we find:

Father: $r = \sqrt[3]{\dfrac{3W}{4\pi\rho g}} = \sqrt[3]{\dfrac{3(830\ \text{N})}{4\pi\left(1.00\times10^3\ \text{kg/m}^3\right)\left(9.80\ \text{m/s}^2\right)}} = \boxed{0.27\ \text{m}}$

Daughter: $r = \sqrt[3]{\dfrac{3W}{4\pi\rho g}} = \sqrt[3]{\dfrac{3(340\ \text{N})}{4\pi\left(1.00\times10^3\ \text{kg/m}^3\right)\left(9.80\ \text{m/s}^2\right)}} = \boxed{0.20\ \text{m}}$

The radius of the father's beach ball is larger, as expected.

CHAPTER 12 | *TEMPERATURE AND HEAT*

1. ***REASONING***

   a. According to the discussion in Section 12.1, the size of a Fahrenheit degree is smaller than that of a Celsius degree by a factor of $\frac{5}{9}$; thus, $1\ \text{F}° = \frac{5}{9}\text{C}°$. This factor will be used to find the temperature difference in Fahrenheit degrees.

   b. The size of one kelvin is identical to that of one Celsius degree (see Section 12.2), $1\ \text{K} = 1\ \text{C}°$. Thus, the temperature difference, expressed in kelvins, is the same at that expressed in Celsius degrees.

   SOLUTION

   a. The difference in the two temperatures is $34\ °\text{C} - 3\ °\text{C} = 31\ \text{C}°$. This difference, expressed in Fahrenheit degrees, is

   $$\text{Temperature difference} = 31\ \text{C}° = \left(31\ \cancel{\text{C}°}\right)\left(\frac{1\ \text{F}°}{\frac{5}{9}\cancel{\text{C}°}}\right) = \boxed{56\ \text{F}°}$$

   b. Since $1\ \text{K} = 1\ \text{C}°$, the temperature difference, expressed in kelvins, is

   $$\text{Temperature difference} = 31\ \text{C}° = \left(31\ \cancel{\text{C}°}\right)\left(\frac{1\ \text{K}}{1\ \cancel{\text{C}°}}\right) = \boxed{31\ \text{K}}$$

5. ***REASONING AND SOLUTION***

   a. The Kelvin temperature and the temperature on the Celsius scale are related by Equation 12.1: $T = T_\text{c} + 273.15$, where T is the Kelvin temperature and T_c is the Celsius temperature. Therefore, a temperature of 77 K on the Celsius scale is

   $$T_\text{c} = T - 273.15 = 77\ \text{K} - 273.15\ \text{K} = \boxed{-196\ °\text{C}}$$

   b. The temperature of $-196\ °\text{C}$ is 196 Celsius degrees *below* the ice point of $0\ °\text{C}$. Since $1\ \text{C}° = \frac{9}{5}\ \text{F}°$, this number of Celsius degrees corresponds to

   $$196\ \cancel{\text{C}°}\left(\frac{\frac{9}{5}\ \text{F}°}{1\ \cancel{\text{C}°}}\right) = 353\ \text{F}°$$

   Subtracting 353 Fahrenheit degrees from the ice point of $32.0\ °\text{F}$ on the Fahrenheit scale gives a Fahrenheit temperature of $\boxed{-321\ °\text{F}}$.

9. ***REASONING AND SOLUTION*** We begin by noting that on the Rankine scale the difference between the steam point and the ice point temperatures is

$$671.67 \ ^\circ R - 491.67 \ ^\circ R = 180.00 \ ^\circ R$$

Thus, the Rankine and Fahrenheit degrees are the same size, since the difference between the steam point and ice point temperatures is also 180 degrees on the Fahrenheit scale. The difference in the ice points of the two scales is $491.67 - 32.00 = 459.67$. To get a Rankine from a Fahrenheit temperature, this amount must be added, so $\boxed{T_R = T_F + 459.67}$.

---

11. ***REASONING AND SOLUTION*** Using Equation 12.2 and the value for the coefficient of thermal expansion of steel given in Table 12.1, we find that the linear expansion of the aircraft carrier is

$$\Delta L = \alpha L_0 \Delta T = (12 \times 10^{-6} \ C^{\circ -1})(370 \ m)(21 \ ^\circ C - 2.0 \ ^\circ C) = \boxed{0.084 \ m} \qquad (12.2)$$

---

15. ***REASONING AND SOLUTION*** The change in the coin's diameter is $\Delta d = \alpha d_0 \Delta T$, according to Equation 12.2. Solving for α gives

$$\alpha = \frac{\Delta d}{d_0 \Delta T} = \frac{2.3 \times 10^{-5} \ m}{(1.8 \times 10^{-2} \ m)(75 \ C^\circ)} = \boxed{1.7 \times 10^{-5} \ (C^\circ)^{-1}} \qquad (12.2)$$

---

21. ***REASONING AND SOLUTION*** Recall that $\omega = 2\pi / T$ (Equation 10.6), where ω is the angular frequency of the pendulum and T is the period. Using this fact and Equation 10.16, we know that the period of the pendulum before the temperature rise is given by $T_1 = 2\pi \sqrt{L_0 / g}$, where L_0 is the length of the pendulum. After the temperature has risen, the period becomes (using Equation 12.2), $T_2 = 2\pi \sqrt{\left[L_0 + \alpha L_0 \Delta T \right] / g}$. Dividing these expressions, solving for T_2, and taking the coefficient of thermal expansion of brass from Table 12.1, we find that

$$T_2 = T_1 \sqrt{1 + \alpha \Delta T} = (2.0000 \ s) \sqrt{1 + (19 \times 10^{-6} / C^\circ)(140 \ C^\circ)} = \boxed{2.0027 \ s}$$

---

27. ***REASONING AND SOLUTION*** Let $L_0 = 0.50$ m and L be the true length of the line at 40.0 °C. The ruler has expanded an amount

$$\Delta L_r = L - L_0 = \alpha_r L_0 \Delta T_r \qquad (1)$$

The copper plate must shrink by an amount

$$\Delta L_{\text{p}} = L_0 - L = \alpha_{\text{p}} L \Delta T_{\text{p}} \tag{2}$$

Eliminating L from Equations (1) and (2), solving for ΔT_{p}, and using values of the coefficients of thermal expansion for copper and steel from Table 12.1, we find that

$$\Delta T_{\text{p}} = \frac{-\alpha_{\text{r}} \Delta T_{\text{r}}}{\alpha_{\text{p}} \left(1 + \alpha_{\text{r}} \Delta T_{\text{r}}\right)}$$

$$= \frac{-\left[12 \times 10^{-6} \ (\text{C}^\circ)^{-1}\right](40.0 \ ^\circ\text{C} - 20.0 \ ^\circ\text{C})}{\left[17 \times 10^{-6} \ (\text{C}^\circ)^{-1}\right]\left\{1 + \left[12 \times 10^{-6} \ (\text{C}^\circ)^{-1}\right](40.0 \ ^\circ\text{C} - 20.0 \ ^\circ\text{C})\right\}} = -14 \ ^\circ\text{C}$$

Therefore, $T_{\text{p}} = 40.0 \ ^\circ\text{C} - 14 \ \text{C}^\circ = \boxed{26 \ ^\circ\text{C}}$.

---

31. **REASONING AND SOLUTION** The volume V_0 of an object changes by an amount ΔV when its temperature changes by an amount ΔT; the mathematical relationship is given by Equation 12.3: $\Delta V = \beta V_0 \Delta T$. Thus, the volume of the kettle at 24 °C can be found by solving Equation 12.3 for V_0. According to Table 12.1, the coefficient of volumetric expansion for copper is $51 \times 10^{-6} \ (\text{C}^\circ)^{-1}$. Solving Equation 12.3 for V_0, we have

$$V_0 = \frac{\Delta V}{\beta \Delta T} = \frac{1.2 \times 10^{-5} \ \text{m}^3}{[51 \times 10^{-6} \ (\text{C}^\circ)^{-1}](100 \ ^\circ\text{C} - 24 \ ^\circ\text{C})} = \boxed{3.1 \times 10^{-3} \ \text{m}^3}$$

---

35. **REASONING** When the temperature increases, both the gasoline and the tank cavity expand. If they were to expand by the same amount, there would be no overflow. However, the gasoline expands more than the tank cavity, and the overflow volume is the amount of gasoline expansion minus the amount of the tank cavity expansion. The tank cavity expands as if it were solid steel.

SOLUTION The coefficients of volumetric expansion β_{g} and β_{s} for gasoline and steel are available in Table 12.1. According to Equation 12.3, the volume expansion of the gasoline is

$$\Delta V_{\text{g}} = \beta_{\text{g}} V_0 \Delta T = \left[950 \times 10^{-6} \ (\text{C}^\circ)^{-1}\right](20.0 \ \text{gal})\left[(35 \ \text{C}^\circ) - (17 \ \text{C}^\circ)\right] = 0.34 \ \text{gal}$$

while the volume of the steel tank expands by an amount

$$\Delta V_{\text{s}} = \beta_{\text{s}} V_0 \Delta T = \left[36 \times 10^{-6} \ (\text{C}^\circ)^{-1}\right](20.0 \ \text{gal})\left[(35 \ \text{C}^\circ) - (17 \ \text{C}^\circ)\right] = 0.013 \ \text{gal}$$

The amount of gasoline that spills out is

$$\Delta V_g - \Delta V_s = \boxed{0.33 \text{ gal}}$$

---

39. **REASONING** In order to keep the water from expanding as its temperature increases from 15 to 25 °C, the atmospheric pressure must be increased to compress the water as it tries to expand. The magnitude of the pressure change ΔP needed to compress a substance by an amount ΔV is, according to Equation 10.20, $\Delta P = B(\Delta V / V_0)$. The ratio $\Delta V / V_0$ is, according to Equation 12.3, $\Delta V / V_0 = \beta \Delta T$. Combining these two equations yields

$$\Delta P = B\beta\Delta T$$

SOLUTION Taking the value for the coefficient of volumetric expansion β for water from Table 12.1, we find that the change in atmospheric pressure that is required to keep the water from expanding is

$$\Delta P = (2.2 \times 10^9 \text{ N/m}^2)\left[207 \times 10^{-6} \ (\text{C}°)^{-1} \right](25 \text{ °C} - 15 \text{ °C})$$

$$= \left(4.6 \times 10^6 \text{ Pa} \right)\left(\frac{1 \text{ atm}}{1.01 \times 10^5 \text{ Pa}} \right) = \boxed{45 \text{ atm}}$$

---

41. **REASONING** The cavity that contains the liquid in either Pyrex thermometer expands according to Equation 12.3, $\Delta V_g = \beta_g V_0 \Delta T$. On the other hand, the volume of mercury expands by an amount $\Delta V_m = \beta_m V_0 \Delta T$, while the volume of alcohol expands by an amount $\Delta V_a = \beta_a V_0 \Delta T$. Therefore, the net change in volume for the mercury thermometer is

$$\Delta V_m - \Delta V_g = (\beta_m - \beta_g)V_0\Delta T$$

while the net change in volume for the alcohol thermometer is

$$\Delta V_a - \Delta V_g = (\beta_a - \beta_g)V_0\Delta T$$

In each case, this volume change is related to a movement of the liquid into a cylindrical region of the thermometer with volume $\pi r^2 h$, where r is the radius of the region and h is the height of the region. For the mercury thermometer, therefore,

$$h_m = \frac{(\beta_m - \beta_g)V_0\Delta T}{\pi r^2}$$

Similarly, for the alcohol thermometer

$$h_a = \frac{(\beta_a - \beta_g)V_0 \Delta T}{\pi r^2}$$

These two expressions can be combined to give the ratio of the heights, h_a/h_m.

SOLUTION Taking the values for the coefficients of volumetric expansion for methyl alcohol, Pyrex glass, and mercury from Table 12.1, we divide the two expressions for the heights of the liquids in the thermometers and find that

$$\frac{h_a}{h_m} = \frac{\beta_a - \beta_g}{\beta_m - \beta_g} = \frac{1200 \times 10^{-6} \ (C°)^{-1} - 9.9 \times 10^{-6} \ (C°)^{-1}}{182 \times 10^{-6} \ (C°)^{-1} - 9.9 \times 10^{-6} \ (C°)^{-1}} = 6.9$$

Therefore, the degree marks are $\boxed{6.9 \text{ times further apart}}$ on the alcohol thermometer than on the mercury thermometer.

---

43. **REASONING** Since there is no heat lost or gained by the system, the heat lost by the water in cooling down must be equal to the heat gained by the thermometer in warming up. The heat Q lost or gained by a substance is given by Equation 12.4 as $Q = cm\Delta T$, where c is the specific heat capacity, m is the mass, and ΔT is the change in temperature. Thus, we have that

$$\underbrace{c_{H_2O}\, m_{H_2O}\, \Delta T_{H_2O}}_{\text{Heat lost by water}} = \underbrace{c_{\text{therm}}\, m_{\text{therm}}\, \Delta T_{\text{therm}}}_{\text{Heat gained by thermometer}}$$

We can use this equation to find the temperature of the water before the insertion of the thermometer.

SOLUTION Solving the equation above for ΔT_{H_2O}, and using the value of c_{H_2O} from Table 12.2, we have

$$\Delta T_{H_2O} = \frac{c_{\text{therm}}\, m_{\text{therm}}\, \Delta T_{\text{therm}}}{c_{H_2O}\, m_{H_2O}}$$

$$= \frac{\left[815 \ \text{J}/(\text{kg} \cdot \text{C}°)\right](31.0 \ \text{g})(41.5 \ °\text{C} - 12.0 \ °\text{C})}{\left[4186 \ \text{J}/(\text{kg} \cdot \text{C}°)\right](119 \ \text{g})} = 1.50 \ \text{C}°$$

The temperature of the water before the insertion of the thermometer was

$$T = 41.5 \ °\text{C} + 1.50 \ \text{C}° = \boxed{43.0 \ °\text{C}}$$

49. **REASONING**   Let the system be comprised only of the metal forging and the oil. Then, according to the principle of energy conservation, the heat lost by the forging equals the heat gained by the oil, or $Q_{metal} = Q_{oil}$. According to Equation 12.4, the heat lost by the forging is $Q_{metal} = c_{metal}m_{metal}(T_{0metal} - T_{eq})$, where T_{eq} is the final temperature of the system at thermal equilibrium. Similarly, the heat gained by the oil is given by $Q_{oil} = c_{oil}m_{oil}(T_{eq} - T_{0oil})$.

SOLUTION

$$Q_{metal} = Q_{oil}$$

$$c_{metal}m_{metal}(T_{0metal} - T_{eq}) = c_{oil}m_{oil}(T_{eq} - T_{0oil})$$

Solving for T_{0metal}, we have

$$T_{0metal} = \frac{c_{oil}m_{oil}(T_{eq} - T_{0oil})}{c_{metal}m_{metal}} + T_{eq}$$

or

$$T_{0metal} = \frac{[2700 \text{ J/(kg} \cdot \text{C}°)](710 \text{ kg})(47 \text{ °C} - 32 \text{ °C})}{[430 \text{ J/(kg} \cdot \text{C}°)](75 \text{ kg})} + 47 \text{ °C} = \boxed{940 \text{ °C}}$$

---

55. **REASONING AND SOLUTION**   As the rock falls through a distance h, its initial potential energy $m_{rock}gh$ is converted into kinetic energy. This kinetic energy is then converted into heat when the rock is brought to rest in the pail. If we ignore the heat absorbed by the pail, the principle of conservation of energy indicates that

$$m_{rock}gh = c_{rock}m_{rock}\Delta T + c_{water}m_{water}\Delta T$$

where we have used Equation 12.4 to express the heat absorbed by the rock and the water. Table 12.2 gives the specific heat capacity of the water. Solving for ΔT yields

$$\Delta T = \frac{m_{rock}gh}{c_{rock}m_{rock} + c_{water}m_{water}}$$

Substituting values yields

$$\Delta T = \frac{(0.20 \text{ kg})(9.80 \text{ m/s}^2)(15 \text{ m})}{[1840 \text{ J/(kg} \cdot \text{C}°)](0.20 \text{ kg}) + [4186 \text{ J/(kg} \cdot \text{C}°)](0.35 \text{ kg})} = \boxed{0.016 \text{ C}°}$$

---

57. **REASONING**   Heat Q_1 must be added to raise the temperature of the aluminum in its solid phase from 130 °C to its melting point at 660 °C. According to Equation 12.4, $Q_1 = cm\Delta T$.

The specific heat c of aluminum is given in Table 12.2. Once the solid aluminum is at its melting point, additional heat Q_2 must be supplied to change its phase from solid to liquid. The additional heat required to melt or liquefy the aluminum is $Q_2 = mL_f$, where L_f is the latent heat of fusion of aluminum. Therefore, the total amount of heat which must be added to the aluminum in its solid phase to liquefy it is

$$Q_{total} = Q_1 + Q_2 = m(c\Delta T + L_f)$$

SOLUTION Substituting values, we obtain

$$Q_{total} = (0.45 \text{ kg}) \left\{ [9.00\times10^2 \text{ J/(kg·C°)}](660 \text{ °C} - 130 \text{ °C}) + 4.0\times10^5 \text{ J/kg} \right\} = \boxed{3.9\times10^5 \text{ J}}$$

---

61. ***REASONING*** From the conservation of energy, the heat lost by the mercury is equal to the heat gained by the water. As the mercury loses heat, its temperature decreases; as the water gains heat, its temperature rises to its boiling point. Any remaining heat gained by the water will then be used to vaporize the water.

According to Equation 12.4, the heat lost by the mercury is $Q_{mercury} = (cm\Delta T)_{mercury}$. The heat required to vaporize the water is, from Equation 12.5, $Q_{vap} = (m_{vap}L_v)_{water}$. Thus, the total amount of heat gained by the water is $Q_{water} = (cm\Delta T)_{water} + (m_{vap}L_v)_{water}$.

SOLUTION

$$Q_{\substack{\text{lost by} \\ \text{mercury}}} = Q_{\substack{\text{gained by} \\ \text{water}}}$$

$$(cm\Delta T)_{mercury} = (cm\Delta T)_{water} + (m_{vap}L_v)_{water}$$

where $\Delta T_{mercury} = (205 \text{ °C} - 100.0 \text{ °C})$ and $\Delta T_{water} = (100.0 \text{ °C} - 80.0 \text{ °C})$. The specific heats of mercury and water are given in Table 12.2, and the latent heat of vaporization of water is given in Table 12.3. Solving for the mass of the water that vaporizes gives

$$m_{vap} = \frac{c_{mercury}m_{mercury}\Delta T_{mercury} - c_{water}m_{water}\Delta T_{water}}{(L_v)_{water}}$$

$$= \frac{[139 \text{ J/(kg·C°)}](2.10 \text{ kg})(105 \text{ C°}) - [4186 \text{ J/(kg·C°)}](0.110 \text{ kg})(20.0 \text{ C°})}{22.6\times10^5 \text{ J/kg}}$$

$$= \boxed{9.49\times10^{-3} \text{ kg}}$$

67. **REASONING** All of the heat generated by friction goes into the ice, so this heat provides the heat needed for melting to occur. Since the surface on which the block slides is horizontal, the gravitational potential energy does not change, and energy conservation dictates that the heat generated by friction equals the amount by which the kinetic energy decreases or $Q_{\text{Friction}} = \frac{1}{2} M v_0^2 - \frac{1}{2} M v^2$, where v_0 and v are, respectively, the initial and final speeds and M is the mass of the block. In reality, the mass of the block decreases as the melting proceeds. However, only a very small amount of ice melts, so we may consider M to be essentially constant at its initial value. The heat Q needed to melt a mass m of water is given by Equation 12.5 as $Q = mL_f$, where L_f is the latent heat of fusion. Thus, by equating Q_{Friction} to mL_f and solving for m, we can determine the mass of ice that melts.

SOLUTION Equating Q_{Friction} to mL_f and solving for m gives

$$Q_{\text{Friction}} = \tfrac{1}{2} M v_0^2 - \tfrac{1}{2} M v^2 = mL_f$$

$$m = \frac{M\left(v_0^2 - v^2\right)}{2L_f} = \frac{(42 \text{ kg})\left[(7.3 \text{ m/s})^2 - (3.5 \text{ m/s})^2\right]}{2\left(33.5 \times 10^4 \text{ J/kg}\right)} = \boxed{2.6 \times 10^{-3} \text{ kg}}$$

We have taken the value for the latent heat of fusion for water from Table 12.3.

---

69. **REASONING** The system is comprised of the unknown material, the glycerin, and the aluminum calorimeter. From the principle of energy conservation, the heat gained by the unknown material is equal to the heat lost by the glycerin and the calorimeter. The heat gained by the unknown material is used to melt the material and then raise its temperature from the initial value of $-25.0\,°C$ to the final equilibrium temperature of $T_{\text{eq}} = 20.0\,°C$.

SOLUTION

$$\underset{\substack{\text{unknown}}}{Q_{\text{gained by}}} = \underset{\substack{\text{glycerine}}}{Q_{\text{lost by}}} + \underset{\substack{\text{calorimeter}}}{Q_{\text{lost by}}}$$

$$m_u L_f + c_u m_u \Delta T_u = c_{gl} m_{gl} \Delta T_{gl} + c_{al} m_{al} \Delta T_{al}$$

Taking values for the specific heat capacities of glycerin and aluminum from Table 12.2, we have

$$(0.10 \text{ kg}) L_f + [160 \text{ J/(kg} \cdot \text{C}°)](0.10 \text{ kg})(45.0 \text{ C}°) = [2410 \text{ J/(kg} \cdot \text{C}°)](0.100 \text{ kg})(7.0 \text{ C}°)$$
$$+ [9.0 \times 10^2 \text{ J/(kg} \cdot \text{C}°)](0.150 \text{ kg})(7.0 \text{ C}°)$$

Solving for L_f yields,

$$L_f = \boxed{1.9 \times 10^4 \text{ J/kg}}$$

71. **REASONING** In order to melt, the bullet must first heat up to 327.3 °C (its melting point) and then undergo a phase change. According to Equation 12.4, the amount of heat necessary to raise the temperature of the bullet to 327.3 °C is $Q = cm(327.3 \text{ °C} - 30.0 \text{ °C})$, where m is the mass of the bullet. The amount of heat required to melt the bullet is given by $Q_{melt} = mL_f$, where L_f is the latent heat of fusion of lead.

The lead bullet melts completely when it comes to a sudden halt; all of the kinetic energy of the bullet is converted into heat; therefore,

$$KE = Q + Q_{melt}$$

$$\tfrac{1}{2}mv^2 = cm(327.3 \text{ °C} - 30.0 \text{ °C}) + mL_f$$

The value for the specific heat c of lead is given in Table 12.2, and the value for the latent heat of fusion L_f of lead is given in Table 12.3. This expression can be solved for v, the minimum speed of the bullet for such an event to occur. Note that the mass m of the bullet appears in each term on both sides of the expression and, therefore, is eliminated algebraically from the solution.

SOLUTION Solving for v, we find that the minimum speed of the lead bullet is

$$v = \sqrt{2L_f + 2c\,(327.3 \text{ °C} - 30.0 \text{ °C})}$$

$$v = \sqrt{2(2.32 \times 10^4 \text{ J/kg}) + 2\left[128 \text{ J/(kg} \cdot \text{C°)}\right](327.3 \text{ °C} - 30.0 \text{ °C})} = \boxed{3.50 \times 10^2 \text{ m/s}}$$

75. **REASONING** The definition of percent relative humidity is given by Equation 12.6 as follows:

$$\text{Percent relative humidity} = \frac{\text{Partial pressure of water vapor}}{\substack{\text{Equilibrium vapor pressure of} \\ \text{water at the existing temperature}}} \times 100$$

Using R to denote the percent relative humidity, P to denote the partial pressure of water vapor, and P_V to denote the equilibrium vapor pressure of water at the existing temperature, we can write Equation 12.6 as

$$R = \frac{P}{P_V} \times 100$$

The partial pressure of water vapor P is the same at the two given temperatures. The relative humidity is not the same at the two temperatures, however, because the equilibrium vapor pressure P_V is different at each temperature, with values that are available from the vapor pressure curve given with the problem statement. To determine the ratio R_{10}/R_{40}, we will apply Equation 12.6 at each temperature.

SOLUTION Using Equation 12.6 and reading the values of $P_{V,\,10}$ and $P_{V,\,40}$ from the vapor pressure curve given with the problem statement, we find

$$\frac{R_{10}}{R_{40}}=\frac{P/P_{V,\,10}}{P/P_{V,\,40}}=\frac{P_{V,\,40}}{P_{V,\,10}}=\frac{7200\text{ Pa}}{1300\text{ Pa}}=\boxed{5.5}$$

81. ***REASONING*** Since her glasses (at 10 °C) steam up when she enters the room, the partial pressure of water vapor in the air in the room must be greater or equal to the vapor pressure of water at 10 °C. At 10 °C the equilibrium vapor pressure is 1250 Pa (see the vapor pressure curve for water that accompanies Problem 75). The problem asks for the smallest possible value of the relative humidity of the room. Therefore, we take the partial pressure of water vapor in the air in the room to be 1250 Pa.

SOLUTION According to Equation 12.6, the relative humidity is

$$\text{Percent relative humidity}=\frac{\text{Partial pressure of water vapor}}{\text{Equilibrium vapor pressure of water at the existing temperature}}\times100$$

At 25 °C, the equilibrium vapor pressure is 3200 Pa (see the vapor pressure curve for water that accompanies Problem 75). Using Equation 12.6 we find for the smallest possible relative humidity that

$$\text{Percent relative humidity}=\frac{\text{Partial pressure of water vapor}}{\text{Equilibrium vapor pressure of water at the existing temperature}}\times100=\frac{1250\text{ Pa}}{3200\text{ Pa}}\times100=\boxed{39\%}$$

87. ***REASONING*** From the principle of conservation of energy, the heat lost by the coin must be equal to the heat gained by the liquid nitrogen. The heat lost by the silver coin is, from Equation 12.4, $Q=c_{coin}m_{coin}\Delta T_{coin}$ (see Table 12.2 for the specific heat capacity of silver). If the liquid nitrogen is at its boiling point, –195.8 °C, then the heat gained by the nitrogen will cause it to change phase from a liquid to a vapor. The heat gained by the liquid nitrogen is $Q=m_{nitrogen}L_v$, where $m_{nitrogen}$ is the mass of liquid nitrogen that vaporizes, and L_v is the latent heat of vaporization for nitrogen (see Table 12.3).

SOLUTION

$$Q_{\substack{\text{lost by}\\\text{coin}}}=Q_{\substack{\text{gained by}\\\text{nitrogen}}}$$

$$c_{coin}m_{coin}\Delta T_{coin}=m_{nitrogen}L_v$$

Solving for the mass of the nitrogen that vaporizes, we obtain

$$m_{\text{nitrogen}} = \frac{c_{\text{coin}} m_{\text{coin}} \Delta T_{\text{coin}}}{L_v}$$

$$= \frac{[235 \text{ J/(kg} \cdot \text{C}°)](1.5 \times 10^{-2} \text{ kg})[25 \text{ }°\text{C} - (-195.8 \text{ }°\text{C})]}{2.00 \times 10^5 \text{ J/kg}} = \boxed{3.9 \times 10^{-3} \text{ kg}}$$

---

91. ***REASONING*** The increase ΔV in volume is given by Equation 12.3 as $\Delta V = \beta V_0 \Delta T$, where β is the coefficient of volume expansion, V_0 is the initial volume, and ΔT is the increase in temperature. The lead and quartz objects experience the same change in volume. Therefore, we can use Equation 12.3 to express the two volume changes and set them equal. We will solve the resulting equation for ΔT_{Quartz}.

SOLUTION Recognizing that the lead and quartz objects experience the same change in volume and expressing that change with Equation 12.3, we have

$$\underbrace{\beta_{\text{Lead}} V_0 \Delta T_{\text{Lead}}}_{\Delta V_{\text{Lead}}} = \underbrace{\beta_{\text{Quartz}} V_0 \Delta T_{\text{Quartz}}}_{\Delta V_{\text{Quartz}}}$$

In this result V_0 is the initial volume of each object. Solving for ΔT_{Quartz} and taking values for the coefficients of volume expansion for lead and quartz from Table 12.1 gives

$$\Delta T_{\text{Quartz}} = \frac{\beta_{\text{Lead}} \Delta T_{\text{Lead}}}{\beta_{\text{Quartz}}} = \frac{\left[87 \times 10^{-6} \left(\text{C}°\right)^{-1}\right](4.0 \text{ C}°)}{1.5 \times 10^{-6} \left(\text{C}°\right)^{-1}} = \boxed{230 \text{ C}°}$$

---

95. ***REASONING*** According to the statement of the problem, the initial state of the system is comprised of the ice and the steam. From the principle of energy conservation, the heat lost by the steam equals the heat gained by the ice, or $Q_{\text{steam}} = Q_{\text{ice}}$. When the ice and the steam are brought together, the steam immediately begins losing heat to the ice. An amount $Q_{1(\text{lost})}$ is released as the temperature of the steam drops from 130 °C to 100 °C, the boiling point of water. Then an amount of heat $Q_{2(\text{lost})}$ is released as the steam condenses into liquid water at 100 °C. The remainder of the heat lost by the "steam" $Q_{3(\text{lost})}$ is the heat that is released as the water at 100 °C cools to the equilibrium temperature of $T_{\text{eq}} = 50.0 \text{ }°\text{C}$. According to Equation 12.4, $Q_{1(\text{lost})}$ and $Q_{3(\text{lost})}$ are given by

$$Q_{1(\text{lost})} = c_{\text{steam}} m_{\text{steam}} (T_{\text{steam}} - 100.0 \text{ }°\text{C}) \quad \text{and} \quad Q_{3(\text{lost})} = c_{\text{water}} m_{\text{steam}} (100.0 \text{ }°\text{C} - T_{\text{eq}})$$

$Q_{2(\text{lost})}$ is given by $Q_{2(\text{lost})} = m_{\text{steam}}L_{\text{v}}$, where L_{v} is the latent heat of vaporization of water. The total heat lost by the steam has three effects on the ice. First, a portion of this heat $Q_{1(\text{gained})}$ is used to raise the temperature of the ice to its melting point at 0.00 °C. Then, an amount of heat $Q_{2(\text{gained})}$ is used to melt the ice completely (we know this because the problem states that after thermal equilibrium is reached the liquid phase is present at 50.0 °C). The remainder of the heat $Q_{3(\text{gained})}$ gained by the "ice" is used to raise the temperature of the resulting liquid at 0.0 °C to the final equilibrium temperature. According to Equation 12.4, $Q_{1(\text{gained})}$ and $Q_{3(\text{gained})}$ are given by

$$Q_{1(\text{gained})} = c_{\text{ice}}m_{\text{ice}}(0.00\ °\text{C} - T_{\text{ice}}) \quad \text{and} \quad Q_{3(\text{gained})} = c_{\text{water}}m_{\text{ice}}(T_{\text{eq}} - 0.00\ °\text{C})$$

$Q_{2(\text{gained})}$ is given by $Q_{2(\text{gained})} = m_{\text{ice}}L_{\text{f}}$, where L_{f} is the latent heat of fusion of ice.

SOLUTION According to the principle of energy conservation, we have

$$Q_{\text{steam}} = Q_{\text{ice}}$$

$$Q_{1(\text{lost})} + Q_{2(\text{lost})} + Q_{3(\text{lost})} = Q_{1(\text{gained})} + Q_{2(\text{gained})} + Q_{3(\text{gained})}$$

or

$$c_{\text{steam}}m_{\text{steam}}(T_{\text{steam}} - 100.0\ °\text{C}) + m_{\text{steam}}L_{\text{v}} + c_{\text{water}}m_{\text{steam}}(100.0\ °\text{C} - T_{\text{eq}})$$

$$= c_{\text{ice}}m_{\text{ice}}(0.00\ °\text{C} - T_{\text{ice}}) + m_{\text{ice}}L_{\text{f}} + c_{\text{water}}m_{\text{ice}}(T_{\text{eq}} - 0.00\ °\text{C})$$

Values for specific heats are given in Table 12.2, and values for the latent heats are given in Table 12.3. Solving for the ratio of the masses gives

$$\frac{m_{\text{steam}}}{m_{\text{ice}}} = \frac{c_{\text{ice}}(0.00\ °\text{C} - T_{\text{ice}}) + L_{\text{f}} + c_{\text{water}}(T_{\text{eq}} - 0.00\ °\text{C})}{c_{\text{steam}}(T_{\text{steam}} - 100.0\ °\text{C}) + L_{\text{v}} + c_{\text{water}}(100.0\ °\text{C} - T_{\text{eq}})}$$

$$= \frac{\left[2.00\times10^3\ \text{J/}(\text{kg}\cdot\text{C}°)\right]\left[0.0\ °\text{C}-(-10.0°\text{C})\right] + 33.5\times10^4\ \text{J/kg} + \left[4186\ \text{J/}(\text{kg}\cdot\text{C}°))\right](50.0\ °\text{C}-0.0\ °\text{C})}{\left[2020\ \text{J/}(\text{kg}\cdot\text{C}°)\right](130\ °\text{C}-100.0\ °\text{C}) + 22.6\times10^5\ \text{J/kg} + \left[4186\ \text{J/}(\text{kg}\cdot\text{C}°)\right](100.0\ °\text{C}-50.0\ °\text{C})}$$

or

$$\frac{m_{\text{steam}}}{m_{\text{ice}}} = \boxed{0.223}$$

99. **REASONING** When heat Q is supplied to the block, its temperature changes by an amount ΔT. The relation between Q and ΔT is given by

$$Q = cm\Delta T \qquad (12.4)$$

where c is the specific heat capacity and m is the mass. When the temperature of the block changes by an amount ΔT, the change ΔV in its volume is given by Equation 12.3 as $\Delta V = \beta V_0 \Delta T$, where β is the coefficient of volume expansion and V_0 is the initial volume of the block. Solving for ΔT gives

$$\Delta T = \frac{\Delta V}{\beta V_0}$$

Substituting this expression for ΔT into Equation 12.4 gives

$$Q = cm\Delta T = cm\left(\frac{\Delta V}{\beta V_0}\right)$$

SOLUTION The heat supplied to the block is

$$Q = cm\left(\frac{\Delta V}{\beta V_0}\right) = \frac{\left[750\ \text{J/(kg·C°)}\right](130\ \text{kg})(1.2\times10^{-5}\ \text{m}^3)}{\left[6.4\times10^{-5}\ (\text{C°})^{-1}\right](4.6\times10^{-2}\ \text{m}^3)} = \boxed{4.0\times10^5\ \text{J}}$$

---

103. **REASONING** As the water heats up, its volume increases. According to the relation $\Delta V = \beta V_0 \Delta T$ (Equation 12.3), the change ΔV in volume depends on the change ΔT in temperature. The change in temperature, in turn, depends on the amount of heat Q absorbed by the water and on the mass m of water being heated, since $Q = mc\Delta T$ (Equation 12.4). To evaluate the mass, we recognize that it depends on the density ρ and the initial volume V_0, since $\rho = m/V_0$ (Equation 11.1). These three relations will be used to determine the change in volume of the water.

SOLUTION When the temperature of the water changes by an amount ΔT, the volume of the water changes bay an amount ΔV, as given by Equation 12.3: $\Delta V = \beta V_0 \Delta T$, where V_0 is the initial volume and β is the coefficient of volume expansion for water. Both V_0 and ΔT are unknown, which we will deal with below.

When heat Q is supplied to the water, the water temperature changes by an amount ΔT. The relations between Q and ΔT is given by Equation 12.4 as $Q = mc\Delta T$, where c is the specific heat capacity and m is the mass of the water. Solving this equation for ΔT gives:

$$\Delta T = \frac{Q}{cm}$$

which can be substituted into Equation 12.3. The value for Q is known, and we now need to consider the mass m.

The mass m and initial volume V_0 of the water are related to the mass density ρ by Equation 11.1 as $\rho = m/V_0$. Solving this equation for the mass yields: $m = \rho V_0$.

This expression for m can be substituted into our result for ΔT above:

$$\Delta V = \beta V_0 \Delta T = \beta V_0 \left(\frac{Q}{cm} \right) = \beta V_0 \left[\frac{Q}{c(\rho V_0)} \right] = \frac{\beta Q}{c\rho}$$

Notice that the initial volume V_0 of the water is eliminated algebraically, so it does not appear in the final expression for ΔV. Taking values of $\beta = 207 \times 10^{-6}$ $(C°)^{-1}$, $c = 4186$ J/(kg·C°), and $\rho = 1.000 \times 10^3$ kg/m^3, we find

$$\Delta V = \frac{\beta Q}{c\rho} = \frac{[207 \times 10^{-6} \ (C°)^{-1}](2.00 \times 10^9 \ J)}{[4186 \ J/(kg·C°)](1.000 \times 10^3 \ kg/m^3)} = \boxed{9.89 \times 10^{-2} \ m^3}$$

---

105. **CONCEPTS** (i) Consider an extreme example. Suppose a cup and a swimming pool are filled with water. For the same heat input, would you intuitively expect the temperature of the cup to rise more than the temperature of the pool? Yes, because the cup has less mass. Another way to arrive at this conclusion is to solve the equation $Q = cm\Delta T$ for the change in temperature: $\Delta T = Q/cm$. Since the heat Q and the specific heat capacity c are the same for objects A and B, ΔT is inversely proportional to the mass m. So the object with the smaller mass undergoes the larger temperature change. Therefore, object A has a greater temperature change than object B. (ii) Objects B and C are made from different materials. To see how the rise in temperature depends of the type of material, let's examine the equation $\Delta T = Q/cm$. Since the heat Q and the mass m are the same for objects B and C, ΔT is inversely proportional to the specific heat capacity c. So the object with the smaller specific heat undergoes the larger temperature change. The heat capacities of copper and glass are 387 and 840 J/(kg·C°), respectively. Since object B is made from copper, which has the smaller specific heat capacity, it has the greater temperature change.

CALCULATIONS Using the equation $\Delta T = Q/cm$, the temperature change for each object is:

$$\Delta T_A = \frac{Q}{c_A m_A} = \frac{14 \ J}{\left[387 \ J/(kg·C°) \right]\left(2.0 \times 10^{-3} \ kg \right)} = \boxed{18 \ C°}$$

$$\Delta T_{\mathrm{B}} = \frac{Q}{c_{\mathrm{B}} m_{\mathrm{B}}} = \frac{14 \text{ J}}{\left[387 \text{ J}/(\text{kg} \cdot \text{C}°) \right] \left(6.0 \times 10^{-3} \text{ kg} \right)} = \boxed{6.0 \text{ C}°}$$

$$\Delta T_{\mathrm{C}} = \frac{Q}{c_{\mathrm{C}} m_{\mathrm{C}}} = \frac{14 \text{ J}}{\left[840 \text{ J}/(\text{kg} \cdot \text{C}°) \right] \left(6.0 \times 10^{-3} \text{ kg} \right)} = \boxed{2.8 \text{ C}°}$$

As anticipated, $\Delta T_{\mathrm{A}} > \Delta T_{\mathrm{B}} > \Delta T_{\mathrm{C}}$.

CHAPTER 13 │ *THE TRANSFER OF HEAT*

3. ***REASONING AND SOLUTION*** According to Equation 13.1, the heat per second lost is

$$\frac{Q}{t} = \frac{kA\,\Delta T}{L} = \frac{[0.040\ \text{J/(s·m·C}^{\circ})]\,(1.6\ \text{m}^2)(25\ \text{C}^{\circ})}{2.0\times10^{-3}\ \text{m}} = \boxed{8.0\times10^2\ \text{J/s}}$$

where the value for the thermal conductivity k of wool has been taken from Table 13.1.

---

5. ***REASONING*** The heat conducted through the iron poker is given by Equation 13.1, $Q = (kA\,\Delta T)t\,/\,L$. If we assume that the poker has a circular cross-section, then its cross-sectional area is $A = \pi r^2$. Table 13.1 gives the thermal conductivity of iron as $79\ \text{J}/(\text{s·m·C}^{\circ})$.

SOLUTION The amount of heat conducted from one end of the poker to the other in 5.0 s is, therefore,

$$Q = \frac{(kA\,\Delta T)t}{L} = \frac{\left[79\ \text{J/(s·m·C}^{\circ})\right]\pi\left(5.0\times10^{-3}\ \text{m}\right)^2\left(502\ ^{\circ}\text{C} - 26\ ^{\circ}\text{C}\right)(5.0\ \text{s})}{1.2\ \text{m}} = \boxed{12\ \text{J}}$$

---

7. ***REASONING AND SOLUTION*** Values for the thermal conductivities of Styrofoam and air are given in Table 11.1. The conductance of an 0.080 mm thick sample of Styrofoam of cross-sectional area A is

$$\frac{k_{\text{s}}A}{L_{\text{s}}} = \frac{\left[0.010\ \text{J/(s·m·C}^{\circ})\right]A}{0.080\times10^{-3}\ \text{m}} = [125\ \text{J/(s·m}^2\text{·C}^{\circ})]\,A$$

The conductance of a 3.5 mm thick sample of air of cross-sectional area A is

$$\frac{k_{\text{a}}A}{L_{\text{a}}} = \frac{\left[0.0256\ \text{J/(s·m·C}^{\circ})\right]A}{3.5\times10^{-3}\ \text{m}} = [7.3\ \text{J/(s·m}^2\text{·C}^{\circ})]\,A$$

Dividing the conductance of Styrofoam by the conductance of air for samples of the same cross-sectional area A, gives

$$\frac{[125\ \text{J/(s·m}^2\text{·C}^{\circ})]\,A}{[7.3\ \text{J/(s·m}^2\text{·C}^{\circ})]\,A} = 17$$

Therefore, the body can adjust the conductance of the tissues beneath the skin by a factor of 17 .

17. **REASONING** The heat Q required to change liquid water at 100.0 °C into steam at 100.0 °C is given by the relation $Q = mL_v$ (Equation 12.5), where m is the mass of the water and L_v is the latent heat of vaporization. The heat required to vaporize the water is conducted through the bottom of the pot and the stainless steel plate. The amount of heat conducted in a time t is given by $Q = \dfrac{(kA\Delta T)t}{L}$ (Equation 13.1), where k is the thermal conductivity, A and L are the cross-sectional area and length, and ΔT is the temperature difference. We will use these two relations to find the temperatures at the aluminum-steel interface and at the steel surface in contact with the heating element.

SOLUTION
a. Substituting Equation 12.5 into Equation 13.1 and solving for ΔT, we have

$$\Delta T = \frac{QL}{kAt} = \frac{(mL_v)L}{kAt}$$

The thermal conductivity k_{Al} of aluminum can be found in Table 13.1, and the latent heat of vaporization for water can be found in Table 12.3. The temperature difference ΔT_{Al} between the aluminum surfaces is

$$\Delta T_{Al} = \frac{(mL_v)L}{k_{Al}At} = \frac{(0.15\ \text{kg})(22.6\times10^5\ \text{J/kg})(3.1\times10^{-3}\,\text{m})}{[240\ \text{J/(s}\cdot\text{m}\cdot\text{C°})](0.015\ \text{m}^2)(240\ \text{s})} = 1.2\ \text{C°}$$

The temperature at the aluminum-steel interface is $T_{\text{Al-Steel}} = 100.0\ °C + \Delta T_{Al} = \boxed{101.2\ °C}$.

b. Using the thermal conductivity k_{ss} of stainless steel from Table 13.1, we find that the temperature difference ΔT_{ss} between the stainless steel surfaces is

$$\Delta T_{ss} = \frac{(mL_v)L}{k_{ss}At} = \frac{(0.15\ \text{kg})(22.6\times10^5\ \text{J/kg})(1.4\times10^{-3}\,\text{m})}{[14\ \text{J/(s}\cdot\text{m}\cdot\text{C°})](0.015\ \text{m}^2)(240\ \text{s})} = 9.4\ \text{C°}$$

The temperature at the steel-burner interface is $T = 101.2\ °C + \Delta T_{ss} = \boxed{110.6\ °C}$.

___

19. **REASONING** Heat flows along the rods via conduction, so that Equation 13.1 applies: $Q = \dfrac{(kA\Delta T)t}{L}$, where Q is the amount of heat that flows in a time t, k is the thermal conductivity of the material from which a rod is made, A is the cross-sectional area of the rod, and ΔT is the difference in temperature between the ends of a rod. In arrangement a, this expression applies to each rod and ΔT has the same value of $\Delta T = T_W - T_C$. The total heat Q' is the sum of the heats through each rod. In arrangement b, the situation is more complicated. We will use the fact that the same heat flows through each rod to determine the

temperature at the interface between the rods and then use this temperature to determine ΔT and the heat flow through either rod.

SOLUTION For arrangement a, we apply Equation 13.1 to each rod and obtain for the total heat that

$$Q' = Q_1 + Q_2 = \frac{k_1 A (T_W - T_C) t}{L} + \frac{k_2 A (T_W - T_C) t}{L} = \frac{(k_1 + k_2) A (T_W - T_C) t}{L} \tag{1}$$

For arrangement b, we use T to denote the temperature at the interface between the rods and note that the same heat flows through each rod. Thus, using Equation 13.1 to express the heat flowing in each rod, we have

$$\underbrace{\frac{k_1 A (T_W - T) \cancel{t}}{\cancel{L}}}_{\substack{\text{Heat flowing} \\ \text{through rod 1}}} = \underbrace{\frac{k_2 A (T - T_C) \cancel{t}}{\cancel{L}}}_{\substack{\text{Heat flowing} \\ \text{through rod 2}}} \quad \text{or} \quad k_1 (T_W - T) = k_2 (T - T_C)$$

Solving this expression for the temperature T gives

$$T = \frac{k_1 T_W + k_2 T_C}{k_1 + k_2} \tag{2}$$

Applying Equation 13.1 to either rod in arrangement b and using Equation (2) for the interface temperature, we can determine the heat Q that is flowing. Choosing rod 2, we find that

$$Q = \frac{k_2 A (T - T_C) t}{L} = \frac{k_2 A \left(\dfrac{k_1 T_W + k_2 T_C}{k_1 + k_2} - T_C \right) t}{L}$$

$$= \frac{k_2 A \left(\dfrac{k_1 T_W - k_1 T_C}{k_1 + k_2} \right) t}{L} = \frac{k_2 A k_1 (T_W - T_C) t}{L (k_1 + k_2)} \tag{3}$$

Using Equations (1) and (3), we obtain for the desired ratio that

$$\frac{Q'}{Q} = \frac{\dfrac{(k_1 + k_2) A (T_W - T_C) t}{L}}{\dfrac{k_2 A k_1 (T_W - T_C) t}{L (k_1 + k_2)}} = \frac{(k_1 + k_2) A \cancel{(T_W - T_C)} \cancel{t} \cancel{L} (k_1 + k_2)}{\cancel{L} k_2 A k_1 \cancel{(T_W - T_C)} \cancel{t}} = \frac{(k_1 + k_2)^2}{k_2 k_1}$$

Using the fact that $k_2 = 2k_1$, we obtain

$$\frac{Q'}{Q} = \frac{(k_1 + k_2)^2}{k_2 k_1} = \frac{(k_1 + 2k_1)^2}{2k_1 k_1} = \boxed{4.5}$$

21. **REASONING** The radiant energy Q radiated by the sun is given by $Q = e\sigma T^4 At$ (Equation 13.2), where e is the emissivity, σ is the Stefan-Boltzmann constant, T is its temperature (in Kelvins), A is the surface area of the sun, and t is the time. The radiant energy emitted per second is $Q/t = e\sigma T^4 A$. Solving this equation for T gives the surface temperature of the sun.

SOLUTION The radiant power produced by the sun is $Q/t = 3.9 \times 10^{26}$ W. The surface area of a sphere of radius r is $A = 4\pi r^2$. Since the sun is a perfect blackbody, $e = 1$. Solving Equation 13.2 for the surface temperature of the sun gives

$$T = \sqrt[4]{\frac{Q/t}{e\sigma 4\pi r^2}} = \sqrt[4]{\frac{3.9 \times 10^{26} \text{ W}}{(1)\left[5.67 \times 10^{-8} \text{ J}/\left(\text{s} \cdot \text{m}^2 \cdot \text{K}^4\right)\right] 4\pi \left(6.96 \times 10^8 \text{ m}\right)^2}} = \boxed{5800 \text{ K}}$$

23. **REASONING** The radiant energy Q absorbed by the person's head is given by $Q = e\sigma T^4 At$ (Equation 13.2), where e is the emissivity, σ is the Stefan-Boltzmann constant, T is the Kelvin temperature of the environment surrounding the person ($T = 28 \text{ °C} + 273 = 301$ K), A is the area of the head that is absorbing the energy, and t is the time. The radiant energy absorbed per second is $Q/t = e\sigma T^4 A$.

SOLUTION
a. The radiant energy absorbed per second by the person's head when it is covered with hair ($e = 0.85$) is

$$\frac{Q}{t} = e\sigma T^4 A = (0.85)\left[5.67 \times 10^{-8} \text{ J}/\left(\text{s} \cdot \text{m}^2 \cdot \text{K}^4\right)\right](301 \text{ K})^4 \left(160 \times 10^{-4} \text{ m}^2\right) = \boxed{6.3 \text{ J/s}}$$

b. The radiant energy absorbed per second by a bald person's head ($e = 0.65$) is

$$\frac{Q}{t} = e\sigma T^4 A = (0.65)\left[5.67 \times 10^{-8} \text{ J}/\left(\text{s} \cdot \text{m}^2 \cdot \text{K}^4\right)\right](301 \text{ K})^4 \left(160 \times 10^{-4} \text{ m}^2\right) = \boxed{4.8 \text{ J/s}}$$

25. **REASONING** According to the discussion in Section 13.3, the net power P_{net} radiated by the person is $P_{net} = e\sigma A\left(T^4 - T_0^4\right)$, where e is the emissivity, σ is the Stefan-Boltzmann constant, A is the surface area, and T and T_0 are the temperatures of the person and the environment, respectively. Since power is the change in energy per unit time (see Equation 6.10b), the time t required for the person to emit the energy Q contained in the dessert is $t = Q/P_{net}$.

SOLUTION The time required to emit the energy from the dessert is

$$t = \frac{Q}{P_{net}} = \frac{Q}{e\sigma A\left(T^4 - T_0^4\right)}$$

The energy is $Q = (260 \text{ Calories}) \left(\dfrac{4186 \text{ J}}{1 \text{ Calorie}} \right)$, and the Kelvin temperatures are

$T = 36\,°\text{C} + 273 = 309 \text{ K}$ and $T_0 = 21\,°\text{C} + 273 = 294 \text{ K}$. The time is

$$t = \frac{(260 \text{ Calories}) \left(\dfrac{4186 \text{ J}}{1 \text{ Calorie}} \right)}{(0.75) \left[5.67 \times 10^{-8} \text{ J}/\left(\text{s} \cdot \text{m}^2 \cdot \text{K}^4 \right) \right] \left(1.3 \text{ m}^2 \right) \left[(309 \text{ K})^4 - (294 \text{ K})^4 \right]} = \boxed{1.2 \times 10^4 \text{ s}}$$

---

33. **REASONING** The total radiant power emitted by an object that has a Kelvin temperature T, surface area A, and emissivity e can be found by rearranging Equation 13.2, the Stefan-Boltzmann law: $Q = e\sigma T^4 A t$. The emitted power is $P = Q/t = e\sigma T^4 A$. Therefore, when the original cylinder is cut perpendicular to its axis into N smaller cylinders, the ratio of the power radiated by the pieces to that radiated by the original cylinder is

$$\frac{P_{\text{pieces}}}{P_{\text{original}}} = \frac{e\sigma T^4 A_2}{e\sigma T^4 A_1} \tag{1}$$

where A_1 is the surface area of the original cylinder, and A_2 is the sum of the surface areas of all N smaller cylinders. The surface area of the original cylinder is the sum of the surface area of the ends and the surface area of the cylinder body; therefore, if L and r represent the length and cross-sectional radius of the original cylinder, with $L = 10r$,

$$A_1 = (\text{area of ends}) + (\text{area of cylinder body})$$

$$= 2(\pi r^2) + (2\pi r)L = 2(\pi r^2) + (2\pi r)(10r) = 22\pi r^2$$

When the original cylinder is cut perpendicular to its axis into N smaller cylinders, the total surface area A_2 is

$$A_2 = N2(\pi r^2) + (2\pi r)L = N2(\pi r^2) + (2\pi r)(10r) = (2N + 20)\pi r^2$$

Substituting the expressions for A_1 and A_2 into Equation (1), we obtain the following expression for the ratio of the power radiated by the N pieces to that radiated by the original cylinder

$$\frac{P_{\text{pieces}}}{P_{\text{original}}} = \frac{e\sigma T^4 A_2}{e\sigma T^4 A_1} = \frac{(2N + 20)\pi r^2}{22\pi r^2} = \frac{N + 10}{11}$$

SOLUTION Since the total radiant power emitted by the N pieces is twice that emitted by the original cylinder, $P_{\text{pieces}} / P_{\text{original}} = 2$, we have $(N + 10)/11 = 2$. Solving this expression for N gives $N = 12$. Therefore, $\boxed{\text{there are 12 smaller cylinders}}$.

35. **REASONING** The heat transferred in a time t is given by Equation 13.1, $Q = (k\,A\,\Delta T)t\,/\,L$. If the same amount of heat per second is conducted through the two plates, then $(Q/t)_{\text{al}} = (Q/t)_{\text{st}}$. Using Equation 13.1, this becomes

$$\frac{k_{\text{al}}A\,\Delta T}{L_{\text{al}}} = \frac{k_{\text{st}}A\,\Delta T}{L_{\text{st}}}$$

This expression can be solved for L_{st}.

SOLUTION Solving for L_{st} gives

$$L_{\text{st}} = \frac{k_{\text{st}}}{k_{\text{al}}}L_{\text{al}} = \frac{14\ \text{J/(s}\cdot\text{m}\cdot\text{C}°)}{240\ \text{J/(s}\cdot\text{m}\cdot\text{C}°)}(0.035\ \text{m}) = \boxed{2.0\times10^{-3}\ \text{m}}$$

---

43. **REASONING** The rate at which heat is conducted along either rod is given by Equation 13.1, $Q/t = (k\,A\,\Delta T)/L$. Since both rods conduct the same amount of heat per second, we have

$$\frac{k_s A_s\ \Delta T}{L_s} = \frac{k_i A_i\ \Delta T}{L_i} \tag{1}$$

Since the same temperature difference is maintained across both rods, we can algebraically cancel the ΔT terms. Because both rods have the same mass, $m_s = m_i$; in terms of the densities of silver and iron, the statement about the equality of the masses becomes $\rho_s(L_s A_s) = \rho_i(L_i A_i)$, or

$$\frac{A_s}{A_i} = \frac{\rho_i L_i}{\rho_s L_s} \tag{2}$$

Equations (1) and (2) may be combined to find the ratio of the lengths of the rods. Once the ratio of the lengths is known, Equation (2) can be used to find the ratio of the cross-sectional areas of the rods. If we assume that the rods have circular cross sections, then each has an area of $A = \pi r^2$. Hence, the ratio of the cross-sectional areas can be used to find the ratio of the radii of the rods.

SOLUTION
a. Solving Equation (1) for the ratio of the lengths and substituting the right hand side of Equation (2) for the ratio of the areas, we have

$$\frac{L_s}{L_i} = \frac{k_s A_s}{k_i A_i} = \frac{k_s(\rho_i L_i)}{k_i(\rho_s L_s)} \quad \text{or} \quad \left(\frac{L_s}{L_i}\right)^2 = \frac{k_s \rho_i}{k_i \rho_s}$$

Solving for the ratio of the lengths, we have

$$\frac{L_s}{L_i} = \sqrt{\frac{k_s \rho_i}{k_i \rho_s}} = \sqrt{\frac{[420 \text{ J/(s·m·C°)}](7860 \text{ kg/m}^3)}{[79 \text{ J/(s·m·C°)}](10\ 500 \text{ kg/m}^3)}} = \boxed{2.0}$$

b. From Equation (2) we have

$$\frac{\pi r_s^2}{\pi r_i^2} = \frac{\rho_i L_i}{\rho_s L_s} \quad \text{or} \quad \left(\frac{r_s}{r_i}\right)^2 = \frac{\rho_i L_i}{\rho_s L_s}$$

Solving for the ratio of the radii, we have

$$\frac{r_s}{r_i} = \sqrt{\frac{\rho_i}{\rho_s}\left(\frac{L_i}{L_s}\right)} = \sqrt{\frac{7860 \text{ kg/m}^3}{10\ 500 \text{ kg/m}^3}\left(\frac{1}{2.0}\right)} = \boxed{0.61}$$

---

45. **REASONING**   Power is the change in energy per unit time, and according to the Stefan–Boltzmann law, it is $Q/t = e\sigma T^4 A$ (Equation 13.2). We seek the *net* power, which is the power the stove emits minus the power the stove absorbs from its environment. Since the stove has a higher temperature than its environment, the stove emits more power than it absorbs. Thus, we will find that the net radiant power emitted by the stove is not zero. In fact, it is the net power that the stove radiates that has warmed the room to its temperature of 29 °C and sustains that temperature.

SOLUTION   The net power P_{net} generated by the stove is the power $P_{emitted}$ that the stove emits minus the power $P_{absorbed}$ that the stove absorbs from its environment

$$P_{net} = P_{emitted} - P_{absorbed} \quad (1)$$

According to the Stefan–Boltzmann law (Equation 13.2), the heated stove (temperature T) emits a power that is

$$P_{emitted} = \frac{Q_{emitted}}{t} = e\sigma T^4 A$$

This expression can be substituted into Equation 1 above. The radiant power that the stove absorbs from the room is identical to the power that the stove would emit at the constant room temperature of 29 °C (302 K). With T_0 representing the temperature (in kelvins) of the room, the Stefan–Boltzmann law indicates that

$$P_{absorbed} = \frac{Q_{absorbed}}{t} = e\sigma T_0^4 A$$

We can also substitute this result into Equation (1) above.

$$P_{net} = P_{emitted} - P_{absorbed} = e\sigma T^4 A - e\sigma T_0^4 A = e\sigma A(T^4 - T_0^4)$$

Using this expression and the information given in the problem, we can now solve for the net radiant power emitted by the stove.

$$P_{net} = 0.900[5.67 \times 10^{-8} \text{ J(s} \cdot \text{m} \cdot \text{K}^4)](3.50 \text{ m}^2)[(471 \text{ K})^2 - (302 \text{ K})^2] = \boxed{7.30 \times 10^3 \text{ W}}$$

---

47. ***CONCEPTS*** (i) The heat that must be removed is the same in both cases. When water freezes, it changes from the liquid phase to the solid phase. The heat that must be removed to make the water freeze is $Q = mL_f$, where m is the mass of the water and L_f is the latent heat of fusion for water. The mass is the same in both cases and so is L_f, since it a characteristic of the water. (ii) No. The temperature of the water does not change as the freezing process takes place. The heat removed serves only to change the water from the liquid phase to the solid phase. Only after all the water has frozen does the temperature of the ice begin to fall below 0 °C. (iii) The water freezes because it loses more heat to radiation than it gains. The gain occurs because the environment radiates heat and the water absorbs it. However, the temperature T_0 of the environment is less than the temperature T of the water. As a result, the environmental radiation cannot offset completely the loss of heat due to radiation from the water. (iv) It will take longer when the area is smaller. This is because the amount of energy radiated in a given time is proportional to the area from which the radiation occurs. A smaller area means that less energy is radiated per second, so more time will be required to freeze the water by removing heat via radiation.

CALCULATIONS We use Equation 13.3 to take into account that the water both gains and loses heat via radiation. This expression gives the net power lost, the net power being the net heat divided by the time. Thus, we have

$$\frac{Q}{t} = e\sigma A(T^4 - T_0^4) \quad \text{or} \quad t = \frac{Q}{e\sigma A(T^4 - T_0^4)}$$

Since $Q = mL_f$ (Equation 12.5), and taking the latent heat of fusion for water as $L_f = 33.5 \times 10^4$ J/kg, we find

(a) Smaller area:

$$t = \frac{mL_f}{e\sigma A(T^4 - T_0^4)} = \frac{(0.50 \text{ kg})(33.5 \times 10^4 \text{ J/kg})}{0.60[5.67 \times 10^{-8} \text{ J/(s} \cdot \text{m}^2 \cdot \text{K}^4)](0.035 \text{ m}^2)[(273 \text{ K})^4 - (261 \text{ K})^4]}$$

$$= \boxed{1.5 \times 10^5 \text{ s}}$$

(b) Larger area:

$$t = \frac{mL_f}{e\sigma A\left(T^4 - T_0^4\right)} = \frac{(0.50 \text{ kg})\left(33.5 \times 10^4 \text{ J/kg}\right)}{0.60\left[5.67 \times 10^{-8} \text{ J/}\left(\text{s} \cdot \text{m}^2 \cdot \text{K}^4\right)\right]\left(1.5 \text{ m}^2\right)\left[\left(273 \text{ K}\right)^4 - \left(261 \text{ K}\right)^4\right]}$$

$$= \boxed{3.6 \times 10^3 \text{ s}}$$

CHAPTER 14 | *THE IDEAL GAS LAW AND KINETIC THEORY*

1. ***REASONING AND SOLUTION*** Since hemoglobin has a molecular mass of 64 500 u, the mass per mole of hemoglobin is 64 500 g/mol. The number of hemoglobin molecules per mol is Avogadro's number, or 6.022×10^{23} mol^{-1}. Therefore, one molecule of hemoglobin has a mass (in kg) of

$$\left(\frac{64\ 500 \text{ g/mol}}{6.022 \times 10^{23} \text{ mol}^{-1}} \right) \left(\frac{1 \text{ kg}}{1000 \text{ g}} \right) = \boxed{1.07 \times 10^{-22} \text{ kg}}$$

5. ***REASONING*** The mass (in grams) of the active ingredient in the standard dosage is the number of molecules in the dosage times the mass per molecule (in grams per molecule). The mass per molecule can be obtained by dividing the molecular mass (in grams per mole) by Avogadro's number. The molecular mass is the sum of the atomic masses of the molecule's atomic constituents.

SOLUTION Using N to denote the number of molecules in the standard dosage and m_{molecule} to denote the mass of one molecule, the mass (in grams) of the active ingredient in the standard dosage can be written as follows:

$$m = N m_{\text{molecule}}$$

Using M to denote the molecular mass (in grams per mole) and recognizing that $m_{\text{molecule}} = \dfrac{M}{N_A}$, where N_A is Avogadro's number and is the number of molecules per mole, we have

$$m = N m_{\text{molecule}} = N \left(\frac{M}{N_A} \right)$$

M (in grams per mole) is equal to the molecular mass in atomic mass units. We can obtain this quantity by referring to the periodic table on the inside of the back cover of the text to find the molecular masses of the constituent atoms in the active ingredient. Thus, we have

$$\text{Molecular mass} = \underbrace{22(12.011 \text{ u})}_{\text{Carbon}} + \underbrace{23(1.00794 \text{ u})}_{\text{Hydrogen}} + \underbrace{1(35.453 \text{ u})}_{\text{Chlorine}} + \underbrace{2(14.0067 \text{ u})}_{\text{Nitrogen}} + \underbrace{2(15.9994 \text{ u})}_{\text{Oxygen}}$$

$$= 382.89 \text{ u}$$

The mass of the active ingredient in the standard dosage is

$$m = N\left(\frac{M}{N_A}\right) = \left(1.572 \times 10^{19} \text{ molecules}\right)\left(\frac{382.89 \text{ g/mol}}{6.022 \times 10^{23} \text{ molecules/mol}}\right) = \boxed{1.00 \times 10^{-2} \text{ g}}$$

9. ***REASONING*** The initial solution contains N_0 molecules of arsenic trioxide, and this number is reduced by a factor of 100 with each dilution. After one dilution, in other words, the number of arsenic trioxide molecules in the solution is $N_1 = N_0/100$. After 2 dilutions, the number remaining is $N_2 = N_1/100 = N_0/100^2$. Therefore, after d dilutions, the number N of arsenic trioxide molecules that remain in the solution is given by

$$N = \frac{N_0}{100^d} \tag{1}$$

We seek the maximum value of d for which at least one molecule of arsenic trioxide remains. Setting $N = 1$ in Equation (1), we obtain $100^d = N_0$. Taking the common logarithm of both sides yields $\log(100^d) = \log N_0$, which we solve for d:

$$d \log 100 = \log N_0 \quad \text{or} \quad 2d = \log N_0 \quad \text{or} \quad d = \tfrac{1}{2}\log N_0 \tag{2}$$

The initial number N_0 of molecules of arsenic trioxide in the undiluted solution depends upon the number n_0 of moles of the substance present via $N_0 = n_0 N_A$, where N_A is Avogadro's number. Because we know the original mass m of the arsenic trioxide, we can find the number n_0 of moles from the relation $n_0 = \dfrac{m}{\text{Mass per mole}}$. We will use the chemical formula As_2O_3 for arsenic trioxide, and the periodic table found on the inside of the back cover of the textbook to determine the mass per mole of arsenic trioxide.

SOLUTION The mass per mole of arsenic trioxide (As_2O_3) is the sum of the atomic masses of its constituent atoms. There are two atoms of arsenic and three atoms of oxygen per molecule, so we have

$$\text{Molecular mass} = \underbrace{2(74.9216 \text{ u})}_{\substack{\text{Mass of 2} \\ \text{arsenic atoms}}} + \underbrace{3(15.9994 \text{ u})}_{\substack{\text{Mass of 3} \\ \text{oxygen atoms}}} = 197.8414 \text{ u}$$

Therefore, the mass per mole of arsenic trioxide is 197.8414 g/mol. From the relations $N_0 = n_0 N_A$ and $n_0 = \dfrac{m}{\text{Mass per mole}}$, we see that the initial number of arsenic trioxide molecules in the sample is

$$N_0 = n_0 N_{\text{A}} = \left(\frac{m}{\text{Mass per mole}} \right) N_{\text{A}} = \frac{m N_{\text{A}}}{\text{Mass per mole}} \qquad (3)$$

Substituting Equation (3) into Equation (2), we obtain

$$d = \tfrac{1}{2} \log N_0 = \tfrac{1}{2} \log \left(\frac{m N_{\text{A}}}{\text{Mass per mole}} \right) = \tfrac{1}{2} \log \left[\frac{(18.0 \text{ g})(6.022 \times 10^{23} \text{ mol}^{-1})}{197.8414 \text{ g/mol}} \right] = 11.4$$

It is only possible to conduct integral numbers of dilutions, so we conclude from this result that $d = 11$ dilutions result in a solution with at least one molecule of arsenic trioxide remaining, while $d = 12$ dilutions yield a solution in which there may or may not be any molecules of arsenic trioxide present. Thus, the original solution may undergo at most $\boxed{11 \text{ dilutions}}$.

---

11. **REASONING** Both gases fill the balloon to the same pressure P, volume V, and temperature T. Assuming that both gases are ideal, we can apply the ideal gas law $PV = nRT$ to each and conclude that the same number of moles n of each gas is needed to fill the balloon. Furthermore, the number of moles can be calculated from the mass m (in grams) and the mass per mole M (in grams per mole), according to $n = \dfrac{m}{M}$. Using this expression in the equation $n_{\text{Helium}} = n_{\text{Nitrogen}}$ will allow us to obtain the desired mass of nitrogen.

SOLUTION Since the number of moles of helium equals the number of moles of nitrogen, we have

$$\underbrace{\frac{m_{\text{Helium}}}{M_{\text{Helium}}}}_{\substack{\text{Number of moles} \\ \text{of helium}}} = \underbrace{\frac{m_{\text{Nitrogen}}}{M_{\text{Nitrogen}}}}_{\substack{\text{Number of moles} \\ \text{of nitrogen}}}$$

Solving for m_{Nitrogen} and taking the values of mass per mole for helium (He) and nitrogen (N_2) from the periodic table on the inside of the back cover of the text, we find

$$m_{\text{Nitrogen}} = \frac{M_{\text{Nitrogen}} m_{\text{Helium}}}{M_{\text{Helium}}} = \frac{(28.0 \text{ g/mol})(0.16 \text{ g})}{4.00 \text{ g/mol}} = \boxed{1.1 \text{ g}}$$

---

15. **REASONING AND SOLUTION** According to the ideal gas law (Equation 14.1), the total number of moles n of fresh air in a normal breath is

$$n = \frac{PV}{RT} = \frac{(1.0 \times 10^5 \text{ Pa})(5.0 \times 10^{-4} \text{ m}^3)}{[8.31 \text{ J/(mole} \cdot \text{K)}](310 \text{ K})} = 1.94 \times 10^{-2} \text{ mol}$$

The total number of molecules in a normal breath is nN_A, where N_A is Avogadro's number. Since fresh air contains approximately 21% oxygen, the total number of oxygen molecules in a normal breath is $(0.21)nN_A$ or

$$(0.21)(1.94 \times 10^{-2} \text{ mol})(6.022 \times 10^{23} \text{ mol}^{-1}) = \boxed{2.5 \times 10^{21}}$$

---

19. **REASONING** According to Equation 11.1, the mass density ρ of a substance is defined as its mass m divided by its volume V: $\rho = m/V$. The mass of nitrogen is equal to the number n of moles of nitrogen times its mass per mole: $m = n$ (Mass per mole). The number of moles can be obtained from the ideal gas law (see Equation 14.1) as $n = (PV)/(RT)$. The mass per mole (in g/mol) of nitrogen has the same numerical value as its molecular mass (which we know).

SOLUTION Substituting $m = n$ (Mass per mole) into $\rho = m/V$, we obtain

$$\rho = \frac{m}{V} = \frac{n(\text{Mass per mole})}{V} \tag{1}$$

Substituting $n = (PV)/(RT)$ from the ideal gas law into Equation 1 gives the following result:

$$\rho = \frac{n(\text{Mass per mole})}{V} = \frac{\left(\dfrac{P\cancel{V}}{RT}\right)(\text{Mass per mole})}{\cancel{V}} = \frac{P(\text{Mass per mole})}{RT}$$

The pressure is 2.0 atmospheres, or $P = 2(1.013 \times 10^5 \text{ Pa})$. The molecular mass of nitrogen is given as 28 u, which means that its mass per mole is 28 g/mol. Expressed in terms of kilograms per mol, the mass per mole is

$$\text{Mass per mole} = \left(28\frac{\cancel{g}}{\text{mol}}\right)\left(\frac{1 \text{ kg}}{10^3 \cancel{g}}\right)$$

The density of the nitrogen gas is

$$\rho = \frac{P(\text{Mass per mole})}{RT} = \frac{2(1.013 \times 10^5 \text{ Pa})\left(28\dfrac{\cancel{g}}{\text{mol}}\right)\left(\dfrac{1 \text{ kg}}{10^3 \cancel{g}}\right)}{[8.31 \text{ J}/(\text{mol} \cdot \text{K})](310 \text{ K})} = \boxed{2.2 \text{ kg/m}^3}$$

---

27. **REASONING AND SOLUTION** If the pressure at the surface is P_1 and the pressure at a depth h is P_2, we have that $P_2 = P_1 + \rho gh$. We also know that $P_1V_1 = P_2V_2$. Then,

$$\frac{V_1}{V_2} = \frac{P_2}{P_1} = \frac{P_1 + \rho gh}{P_1} = 1 + \frac{\rho gh}{P_1}$$

Therefore,

$$\frac{V_1}{V_2} = 1 + \frac{(1.000 \times 10^3 \text{ kg/m}^3)(9.80 \text{ m/s}^2)(0.200 \text{ m})}{1.01 \times 10^5 \text{ Pa}} = \boxed{1.02}$$

---

31. **REASONING** According to the ideal gas law (Equation 14.1), $PV = nRT$. Since n, the number of moles of the gas, is constant, $n_1 R = n_2 R$. Therefore, $P_1 V_1 / T_1 = P_2 V_2 / T_2$, where $T_1 = 273$ K and T_2 is the temperature we seek. Since the beaker is cylindrical, the volume V of the gas is equal to Ad, where A is the cross-sectional area of the cylindrical volume and d is the height of the region occupied by the gas, as measured from the bottom of the beaker. With this substitution for the volume, the expression obtained from the ideal gas law becomes

$$\frac{P_1 d_1}{T_1} = \frac{P_2 d_2}{T_2} \qquad (1)$$

where the pressures P_1 and P_2 are equal to the sum of the atmospheric pressure and the pressure caused by the mercury in each case. These pressures can be determined using Equation 11.4. Once the pressures are known, Equation (1) can be solved for T_2.

SOLUTION Using Equation 11.4, we obtain the following values for the pressures P_1 and P_2. Note that the initial height of the mercury is $h_1 = \frac{1}{2}(1.520 \text{ m}) = 0.760 \text{ m}$, while the final height of the mercury is $h_2 = \frac{1}{4}(1.520 \text{ m}) = 0.380 \text{ m}$.

$$P_1 = P_0 + \rho gh_1 = (1.01 \times 10^5 \text{ Pa}) + \left[(1.36 \times 10^4 \text{kg/m}^3)(9.80 \text{ m/s}^2)(0.760 \text{ m}) \right] = 2.02 \times 10^5 \text{ Pa}$$

$$P_2 = P_0 + \rho gh_2 = (1.01 \times 10^5 \text{ Pa}) + \left[(1.36 \times 10^4 \text{kg/m}^3)(9.80 \text{ m/s}^2)(0.380 \text{ m}) \right] = 1.52 \times 10^5 \text{ Pa}$$

In these pressure calculations, the density of mercury is $\rho = 1.36 \times 10^4 \text{ kg/m}^3$. In Equation (1) we note that $d_1 = 0.760$ m and $d_2 = 1.14$ m. Solving Equation (1) for T_2 and substituting values, we obtain

$$T_2 = \left(\frac{P_2 d_2}{P_1 d_1} \right) T_1 = \left[\frac{(1.52 \times 10^5 \text{ Pa})(1.14 \text{ m})}{(2.02 \times 10^5 \text{ Pa})(0.760 \text{ m})} \right] (273 \text{ K}) = \boxed{308 \text{ K}}$$

---

33. **REASONING** The smoke particles have the same average translational kinetic energy as the air molecules, namely, $\frac{1}{2} m v_{rms}^2 = \frac{3}{2} kT$, according to Equation 14.6. In this expression m is

the mass of a smoke particle, v_{rms} is the rms speed of a particle, k is Boltzmann's constant, and T is the Kelvin temperature. We can obtain the mass directly from this equation.

SOLUTION Solving Equation 14.6 for the mass m, we find

$$m = \frac{3kT}{v_{rms}^2} = \frac{3\left(1.38\times10^{-23}\ \text{J/K}\right)\left(301\ \text{K}\right)}{\left(2.8\times10^{-3}\ \text{m/s}\right)^2} = \boxed{1.6\times10^{-15}\ \text{kg}}$$

---

37. **REASONING AND SOLUTION** Using the expressions for $\overline{v^2}$ and $\left(\overline{v}\right)^2$ given in the statement of the problem, we obtain:

a. $\quad \overline{v^2} = \frac{1}{3}\left(v_1^2 + v_2^2 + v_3^2\right) = \frac{1}{3}\left[(3.0\ \text{m/s})^2 + (7.0\ \text{m/s})^2 + (9.0\ \text{m/s})^2\right] = \boxed{46.3\ \text{m}^2/\text{s}^2}$

b. $\quad \left(\overline{v}\right)^2 = \left[\frac{1}{3}\left(v_1 + v_2 + v_3\right)\right]^2 = \left[\frac{1}{3}(3.0\ \text{m/s} + 7.0\ \text{m/s} + 9.0\ \text{m/s})\right]^2 = \boxed{40.1\ \text{m}^2/\text{s}^2}$

$\overline{v^2}$ and $\left(\overline{v}\right)^2$ are *not* equal, because they are two different physical quantities.

---

43. **REASONING AND SOLUTION**
a. Assuming that the direction of travel of the bullets is positive, the average change in momentum per second is

$$\Delta p/\Delta t = m\Delta v/\Delta t = (200)(0.0050\ \text{kg})[(0\ \text{m/s}) - 1200\ \text{m/s})]/(10.0\ \text{s}) = \boxed{-120\ \text{N}}$$

b. The average force exerted on the bullets is $\overline{F} = \Delta p/\Delta t$. According to Newton's third law, the average force exerted on the wall is $-\overline{F} = \boxed{120\ \text{N}}$.

c. The pressure P is the magnitude of the force on the wall per unit area, so

$$P = (120\ \text{N})/(3.0\times10^{-4}\ \text{m}^2) = \boxed{4.0\times10^5\ \text{Pa}}$$

49. **REASONING** The mass m of methane that diffuses out of the tank in a time t is given by $m = \dfrac{(DA\,\Delta C)t}{L}$ (Equation 14.8), where D is the diffusion constant for methane, A is the cross-sectional area of the pipe, L is the length of the pipe, and $\Delta C = C_1 - C_2$ is the difference in the concentration of methane at the two ends of the pipe. The higher concentration C_1 at the tank-end of the pipe is given, and the concentration C_2 at the end of the pipe that is open to the atmosphere is zero.

SOLUTION Solving $m = \dfrac{(DA\,\Delta C)t}{L}$ (Equation 14.8) for the cross-sectional area A, we obtain

$$A = \frac{mL}{(D\,\Delta C)t} \tag{1}$$

The elapsed time t must be converted from hours to seconds:

$$t = (12\ \cancel{hr})\left(\frac{3600\ s}{1\ \cancel{hr}}\right) = 4.32\times10^4\ s$$

Therefore, the cross-sectional area A of the pipe is

$$A = \frac{\left(9.00\times10^{-4}\ kg\right)(1.50\ m)}{\left(2.10\times10^{-5}\ m^2/s\right)\left(0.650\ kg/m^3 - 0\ kg/m^3\right)\left(4.32\times10^4\ s\right)} = \boxed{2.29\times10^{-3}\ m^2}$$

51. **REASONING AND SOLUTION**

a. The average concentration is $C_{av} = (1/2)\,(C_1 + C_2) = (1/2)C_2 = m/V = m/(AL)$, so that $C_2 = 2m/(AL)$. Fick's law then becomes $m = DAC_2 t/L = DA(2m/AL)t/L = 2Dmt/L^2$. Solving for t yields

$$\boxed{t = L^2/(2D)}$$

b. Substituting the given data into this expression yields

$$t = (2.5\times10^{-2}\ m)^2/[2(1.0\times10^{-5}\ m^2/s)] = \boxed{31\ s}$$

55. **REASONING** According to the ideal gas law (Equation 14.1), $PV = nRT$. Since n, the number of moles, is constant, $n_1 R = n_2 R$. Thus, according to Equation 14.1, we have

$$\frac{P_1 V_1}{T_1} = \frac{P_2 V_2}{T_2}$$

SOLUTION Solving for T_2, we have

$$T_2 = \left(\frac{P_2}{P_1}\right)\left(\frac{V_2}{V_1}\right)T_1 = \left(\frac{48.5\,P_1}{P_1}\right)\left[\frac{V_1/16}{V_1}\right](305\text{ K}) = \boxed{925\text{ K}}$$

---

57. **REASONING** The behavior of the molecules is described by Equation 14.5: $PV = \frac{2}{3}N(\frac{1}{2}mv_{rms}^2)$. Since the pressure and volume of the gas are kept constant, while the number of molecules is doubled, we can write $P_1 V_2 = P_2 V_2$, where the subscript 1 refers to the initial condition, and the subscript 2 refers to the conditions after the number of molecules is doubled. Thus,

$$\frac{2}{3}N_1\left[\frac{1}{2}m(v_{rms})_1^2\right] = \frac{2}{3}N_2\left[\frac{1}{2}m(v_{rms})_2^2\right] \quad\text{or}\quad N_1(v_{rms})_1^2 = N_2(v_{rms})_2^2$$

The last expression can be solved for $(v_{rms})_2$, the final translational rms speed.

SOLUTION Since the number of molecules is doubled, $N_2 = 2N_1$. Solving the last expression above for $(v_{rms})_2$, we find

$$(v_{rms})_2 = (v_{rms})_1\sqrt{\frac{N_1}{N_2}} = (463\text{ m/s})\sqrt{\frac{N_1}{2N_1}} = \frac{463\text{ m/s}}{\sqrt{2}} = \boxed{327\text{ m/s}}$$

---

59. **REASONING** The ideal gas law specifies that the pressure P, the volume V, the number of moles n, and the Kelvin temperature T are related according to $PV = nRT$ (Equation 14.1), where R is the universal gas constant. We will apply this expression to the situation where the oxygen gas is stored in the tank (pressure $= P_{tank}$, the volume $= V_{tank}$, the number of moles $= n$, and the Kelvin temperature $= T_{tank}$) and to the situation where the oxygen is being administered to a patient (pressure $= P_{patient}$, the volume $= V_{patient}$, the number of moles $= n$, and the Kelvin temperature $= T_{patient}$). The number of moles is the same in both situations.

SOLUTION The ideal gas law gives

$$\frac{P_{patient}V_{patient}}{T_{patient}} = \frac{P_{tank}V_{tank}}{T_{tank}} = nR$$

Solving for $V_{patient}$ shows that

$$V_{patient} = \frac{T_{patient}P_{tank}V_{tank}}{P_{patient}T_{tank}} = \frac{(297 \text{ K})(65.0 \text{ atm})(1.00 \text{ m}^3)}{(1.00 \text{ atm})(288 \text{ K})} = \boxed{67.0 \text{ m}^3}$$

---

65. **REASONING** The same type of molecules (e.g., O_2 molecules) within a gas have different speeds, even though the gas itself has a constant temperature. The rms speed v_{rms} of the molecules is a kind of *average speed* and is related to the kinetic energy (also an average) according to Equation 6.2. Thus, the average translational kinetic energy $\overline{KE}$ of a molecule is $\overline{KE} = \frac{1}{2}mv_{rms}^2$, where m is the mass of a molecule. We know that the average kinetic energy depends on the Kelvin temperature T of the gas through the relation $\overline{KE} = \frac{3}{2}kT$ (Equation 14.6). We can find the rms speed of each type of molecule from a knowledge of its mass and the temperature.

SOLUTION The rms speed v_{rms} of a molecule in a gas is related to the average kinetic energy $\overline{KE}$ by

$$\overline{KE} = \frac{1}{2}mv_{rms}^2$$

where m is its mass. Solving Equation 6.2 above for v_{rms} gives the following:

$$v_{rms} = \sqrt{\frac{2\overline{KE}}{m}} \qquad (1)$$

To use Equation (1) above, we must find an expression for $\overline{KE}$. Since the gas is assumed to be an ideal gas, the average kinetic energy of a molecule is directly proportional to the Kelvin temperature T of the gas according to Equation 14.6:

$$\overline{KE} = \frac{3}{2}kT$$

where k is Boltzmann's constant. By substituting this relation into Equation (1) above, we can obtain an expression for the rms speed in terms of the known variables.

$$v_{rms} = \sqrt{\frac{2\overline{KE}}{m}} = \sqrt{\frac{2(\frac{3}{2}kT)}{m}}$$

In this result, m is the mass of a gas molecule (in kilograms). Since the molecular masses of nitrogen and oxygen are given in atomic mass units (28.0 u and 32.0 u, respectively), we must convert them to kilograms by using the conversion factor $1\,u = 1.6605 \times 10^{-27}$ kg (see Section 14.1). Thus,

$$\textit{Nitrogen}\quad m_{N2} = (28.0\ u)\left(\frac{1.6605 \times 10^{-27}\ \text{kg}}{1\ u}\right) = 4.65 \times 10^{-26}\ \text{kg}$$

$$\textit{Oxygen}\quad m_{O2} = (32.0\ u)\left(\frac{1.6605 \times 10^{-27}\ \text{kg}}{1\ u}\right) = 5.31 \times 10^{-26}\ \text{kg}$$

The rms speed for each type of molecule is then

$$\textit{Nitrogen}\quad v_{\text{rms}} = \sqrt{\frac{2\left(\frac{3}{2}kT\right)}{m_{N2}}} = \sqrt{\frac{2\left[\frac{3}{2}(1.38 \times 10^{-23}\ \text{J/K})(293\ \text{K})\right]}{4.65 \times 10^{-26}\ \text{kg}}} = \boxed{511\ \text{m/s}}$$

$$\textit{Oxygen}\quad v_{\text{rms}} = \sqrt{\frac{2\left(\frac{3}{2}kT\right)}{m_{O2}}} = \sqrt{\frac{2\left[\frac{3}{2}(1.38 \times 10^{-23}\ \text{J/K})(293\ \text{K})\right]}{5.31 \times 10^{-26}\ \text{kg}}} = \boxed{478\ \text{m/s}}$$

Notice that both types of the molecules have the same average kinetic energy, since they have the same temperature. However, the rms speed of the nitrogen molecules is larger due to its smaller mass.

---

67. ***CONCEPTS*** **(i)** Every time an atom collides with a wall and rebounds, the atom exerts a force on the wall. Imagine opening your hand so it is flat and having someone throw a ball straight at it. Your hand is like the wall, and as the ball rebounds, you can feel the force. Intuitively, you would expect the force to become greater as the speed and mass of the ball become greater. This is indeed the case. **(ii)** Yes. Pressure is defined as the magnitude of the force exerted perpendicularly on the wall divided by the area of the wall. Since the atoms exert a force, they also exert a pressure. **(iii)** Yes. The Kelvin temperature is proportional to the average kinetic energy of an atom. The average kinetic energy, in turn, is proportional to the mass of an atom and the square of the rms speed. Since we know both of these quantities, we can determine the temperature of the gas.

CALCULATIONS **(a)** The magnitude F of the force exerted on a wall is given by

$$F = \left(\frac{N}{3}\right)\left(\frac{mv_{\text{rms}}^2}{L}\right)$$

where N is the number of atoms in the box, m is the mass of a single atom, v_{rms} is the rms speed of the atoms, and L is the length of one side of the box. The volume of the cubical box is $\left(2.0\times10^2 \text{ cm}\right)^3 = 8.0\times10^6 \text{ cm}^3$. The number N of atoms is equal to the number of atoms per cubic centimeter times the volume of the box in cubic centimeters:

$$N = \left(\frac{1 \text{ atom}}{\text{cm}^3}\right)\left(8.0\times10^6 \text{ cm}^3\right) = 8.0\times10^6$$

The magnitude of the force acting on one wall is

$$F = \left(\frac{N}{3}\right)\left(\frac{mv_{\text{rms}}^2}{L}\right) = \left(\frac{8.0\times10^6}{3}\right)\left[\frac{\left(1.67\times10^{-27} \text{ kg}\right)\left(260 \text{ m/s}\right)^2}{2.0 \text{ m}}\right]$$

$$= \boxed{1.5\times10^{-16} \text{ N}}$$

(b) The pressure is the magnitude of the force divided by the area A of the wall:

$$P = \frac{F}{A} = \frac{1.5\times10^{-16} \text{ N}}{\left(2.0 \text{ m}\right)^2} = \boxed{3.8\times10^{-17} \text{ Pa}}$$

(c) The Kelvin temperature T of the hydrogen atoms is related to the average kinetic energy of an atom by

$$\tfrac{3}{2}kT = \tfrac{1}{2}mv_{\text{rms}}^2$$

where k is Boltzmann's constant. Solving this equation for the temperature gives

$$T = \frac{mv_{\text{rms}}^2}{3k} = \frac{\left(1.67\times10^{-27} \text{ kg}\right)\left(260 \text{ m/s}\right)^2}{3\left(1.38\times10^{-23} \text{ J/K}\right)} = \boxed{2.7 \text{ K}}$$

This is a frigid 2.7 kelvins above absolute zero.

CHAPTER 15 |*THERMODYNAMICS*

3. ***REASONING*** Energy in the form of work leaves the system, while energy in the form of heat enters. More energy leaves than enters, so we expect the internal energy of the system to decrease, that is, we expect the change ΔU in the internal energy to be negative. The first law of thermodynamics will confirm our expectation. As far as the environment is concerned, we note that when the system loses energy, the environment gains it, and when the system gains energy the environment loses it. Therefore, the change in the internal energy of the environment must be opposite to that of the system.

SOLUTION
a. The system gains heat so Q is positive, according to our convention. The system does work, so W is also positive, according to our convention. Applying the first law of thermodynamics from Equation 15.1, we find for the system that

$$\Delta U = Q - W = (77 \text{ J}) - (164 \text{ J}) = \boxed{-87 \text{ J}}$$

As expected, this value is negative, indicating a decrease.

b. The change in the internal energy of the environment is opposite to that of the system, so that $\boxed{\Delta U_{\text{environment}} = +87 \text{ J}}$.

---

7. ***REASONING*** The change ΔU in the weight lifter's internal energy is given by the first law of thermodynamics as $\Delta U = Q - W$ (Equation 15.1), where Q is the heat and W is the work. The amount of heat is that required to evaporate the water (perspiration) and is mL_v (see Equation 12.5), where m is the mass of the water and L_v is the latent heat of vaporization of perspiration.

SOLUTION
a. The heat required to evaporate the water is energy that leaves the weight lifter's body along with the evaporated water, so this heat Q in the first law of thermodynamics is negative. Therefore, we substitute $Q = -mL_v$ into the first law and obtain

$$\Delta U = Q - W = -mL_v - W$$

$$= -(0.150 \text{ kg})(2.42 \times 10^6 \text{ J/kg}) - (1.40 \times 10^5 \text{ J}) = \boxed{-5.03 \times 10^5 \text{ J}}$$

b. Since 1 nutritional Calorie = 4186 J, the number of nutritional calories is

$$\left(5.03 \times 10^5 \text{ J}\right)\left(\frac{1 \text{ Calorie}}{4186 \text{ J}}\right) = \boxed{1.20 \times 10^2 \text{ nutritional Calories}}$$

9. **REASONING**   According to Equation 15.2, $W = P\Delta V$, the average pressure $\overline{P}$ of the expanding gas is equal to $\overline{P} = W/\Delta V$, where the work W done by the gas on the bullet can be found from the work-energy theorem (Equation 6.3). Assuming that the barrel of the gun is cylindrical with radius r, the volume of the barrel is equal to its length L multiplied by the area (πr^2) of its cross section.   Thus, the change in volume of the expanding gas is $\Delta V = L\pi r^2$.

SOLUTION   The work done by the gas on the bullet is given by Equation 6.3 as

$$W = \tfrac{1}{2}m(v_{final}^2 - v_{initial}^2) = \tfrac{1}{2}(2.6\times10^{-3} \text{ kg})[(370 \text{ m/s})^2 - 0] = 180 \text{ J}$$

The average pressure of the expanding gas is, therefore,

$$\overline{P} = \frac{W}{\Delta V} = \frac{180 \text{ J}}{(0.61\,\text{m})\pi(2.8\times10^{-3}\,\text{m})^2} = \boxed{1.2\times10^7 \text{ Pa}}$$

13. **REASONING AND SOLUTION**
   a. Starting at point A, the work done during the first (vertical) straight-line segment is

$$W_1 = P_1\Delta V_1 = P_1(0 \text{ m}^3) = 0 \text{ J}$$

For the second (horizontal) straight-line segment, the work is

$$W_2 = P_2\Delta V_2 = 10(1.0\times10^4 \text{ Pa})6(2.0\times10^{-3} \text{ m}^3) = 1200 \text{ J}$$

For the third (vertical) straight-line segment the work is

$$W_3 = P_3\Delta V_3 = P_3(0 \text{ m}^3) = 0 \text{ J}$$

For the fourth (horizontal) straight-line segment the work is

$$W_4 = P_4\Delta V_4 = 15(1.0\times10^4 \text{ Pa})6(2.0\times10^{-3} \text{ m}^3) = 1800 \text{ J}$$

The total work done is

$$W = W_1 + W_2 + W_3 + W_4 = \boxed{+3.0\times10^3 \text{ J}}$$

   b. Since the total work is positive, work is done $\boxed{\text{by the system}}$.

17. **REASONING** The pressure P of the gas remains constant while its volume increases by an amount ΔV. Therefore, the work W done by the expanding gas is given by $W = P\Delta V$ (Equation 15.2). ΔV is known, so if we can obtain a value for W, we can use this expression to calculate the pressure. To determine W, we turn to the first law of thermodynamics

$\Delta U = Q - W$ (Equation 15.1), where Q is the heat and ΔU is the change in the internal energy. ΔU is given, so to use the first law to determine W we need information about Q. According to Equation 12.4, the heat needed to raise the temperature of a mass m of material by an amount ΔT is $Q = cm\Delta T$ where c is the material's specific heat capacity.

SOLUTION According to Equation 15.2, the pressure P of the expanding gas can be determined from the work W and the change ΔV in volume of the gas according to

$$P = \frac{W}{\Delta V}$$

Using the first law of thermodynamics, we can write the work as $W = Q - \Delta U$ (Equation 15.1). With this substitution, the expression for the pressure becomes

$$P = \frac{W}{\Delta V} = \frac{Q - \Delta U}{\Delta V} \tag{1}$$

Using Equation 12.4, we can write the heat as $Q = cm\Delta T$, which can then be substituted into Equation (1). Thus,

$$P = \frac{Q - \Delta U}{\Delta V} = \frac{cm\Delta T - \Delta U}{\Delta V}$$

$$= \frac{\left[1080 \text{ J}/\left(\text{kg}\cdot\text{C}°\right)\right]\left(24.0\times10^{-3} \text{ kg}\right)\left(53.0 \text{ C}°\right) - 939 \text{ J}}{1.40\times10^{-3} \text{ m}^3} = \boxed{3.1\times10^5 \text{ Pa}}$$

---

21. **REASONING** Since the gas is expanding adiabatically, the work done is given by Equation 15.4 as $W = \frac{3}{2}nR\left(T_i - T_f\right)$. Once the work is known, we can use the first law of thermodynamics to find the change in the internal energy of the gas.

SOLUTION
a. The work done by the expanding gas is

$$W = \frac{3}{2}nR\left(T_i - T_f\right) = \frac{3}{2}(5.0 \text{ mol})\left[8.31 \text{ J}/\left(\text{mol}\cdot\text{K}\right)\right](370 \text{ K} - 290 \text{ K}) = \boxed{+5.0\times10^3 \text{ J}}$$

b. Since the process is adiabatic, $Q = 0$ J, and the change in the internal energy is

$$\Delta U = Q - W = 0 - 5.0\times10^3 \text{ J} = \boxed{-5.0\times10^3 \text{ J}}$$

---

25. **REASONING** When the expansion is isothermal, the work done can be calculated from Equation (15.3): $W = nRT \ln(V_f / V_i)$. When the expansion is adiabatic, the work done can be calculated from Equation 15.4: $W = \frac{3}{2} nR(T_i - T_f)$.

Since the gas does the same amount of work whether it expands adiabatically or isothermally, we can equate the right hand sides of these two equations. We also note that since the initial temperature is the same for both cases, the temperature T in the isothermal expansion is the same as the initial temperature T_i for the adiabatic expansion. We then have

$$nRT_i \ln\left(\frac{V_f}{V_i}\right) = \frac{3}{2} nR(T_i - T_f) \quad \text{or} \quad \ln\left(\frac{V_f}{V_i}\right) = \frac{\frac{3}{2}(T_i - T_f)}{T_i}$$

SOLUTION Solving for the ratio of the volumes gives

$$\frac{V_f}{V_i} = e^{\frac{3}{2}(T_i - T_f)/T_i} = e^{\frac{3}{2}(405\,\text{K} - 245\,\text{K})/(405\,\text{K})} = \boxed{1.81}$$

---

29. **REASONING AND SOLUTION**
Step A → B
The internal energy of a monatomic ideal gas is $U = (3/2)nRT$. Thus, the change is

$$\Delta U = \frac{3}{2} nR\,\Delta T = \frac{3}{2}(1.00\,\text{mol})\left[8.31\,\text{J}/(\text{mol}\cdot\text{K})\right](800.0\,\text{K} - 400.0\,\text{K}) = \boxed{4990\,\text{J}}$$

The work for this constant pressure step is $W = P\Delta V$. But the ideal gas law applies, so

$$W = P\Delta V = nR\,\Delta T = (1.00\,\text{mol})\left[8.31\,\text{J}/(\text{mol}\cdot\text{K})\right](800.0\,\text{K} - 400.0\,\text{K}) = \boxed{3320\,\text{J}}$$

The first law of thermodynamics indicates that the heat is

$$Q = \Delta U + W = \frac{3}{2} nR\,\Delta T + nR\,\Delta T$$

$$= \frac{5}{2}(1.00\,\text{mol})\left[8.31\,\text{J}/(\text{mol}\cdot\text{K})\right](800.0\,\text{K} - 400.0\,\text{K}) = \boxed{8310\,\text{J}}$$

Step B → C
The internal energy of a monatomic ideal gas is $U = (3/2)nRT$. Thus, the change is

$$\Delta U = \frac{3}{2} nR\,\Delta T = \frac{3}{2}(1.00\,\text{mol})\left[8.31\,\text{J}/(\text{mol}\cdot\text{K})\right](400.0\,\text{K} - 800.0\,\text{K}) = \boxed{-4990\,\text{J}}$$

The volume is constant in this step, so the work done by the gas is $\boxed{W = 0\,\text{J}}$.

The first law of thermodynamics indicates that the heat is

$$Q = \Delta U + W = \Delta U = \boxed{-4990 \text{ J}}$$

Step C → D

The internal energy of a monatomic ideal gas is $U = (3/2)nRT$. Thus, the change is

$$\Delta U = \tfrac{3}{2} nR\,\Delta T = \tfrac{3}{2}(1.00 \text{ mol})\left[8.31 \text{ J/(mol·K)}\right](200.0 \text{ K} - 400.0 \text{ K}) = \boxed{-2490 \text{ J}}$$

The work for this constant pressure step is $W = P\Delta V$. But the ideal gas law applies, so

$$W = P\Delta V = nR\,\Delta T = (1.00 \text{ mol})\left[8.31 \text{ J/(mol·K)}\right](200.0 \text{ K} - 400.0 \text{ K}) = \boxed{-1660 \text{ J}}$$

The first law of thermodynamics indicates that the heat is

$$Q = \Delta U + W = \tfrac{3}{2} nR\,\Delta T + nR\,\Delta T$$

$$= \tfrac{5}{2}(1.00 \text{ mol})\left[8.31 \text{ J/(mol·K)}\right](200.0 \text{ K} - 400.0 \text{ K}) = \boxed{-4150 \text{ J}}$$

Step D → A

The internal energy of a monatomic ideal gas is $U = (3/2)nRT$. Thus, the change is

$$\Delta U = \tfrac{3}{2} nR\,\Delta T = \tfrac{3}{2}(1.00 \text{ mol})\left[8.31 \text{ J/(mol·K)}\right](400.0 \text{ K} - 200.0 \text{ K}) = \boxed{2490 \text{ J}}$$

The volume is constant in this step, so the work done by the gas is $\boxed{W = 0 \text{ J}}$

The first law of thermodynamics indicates that the heat is

$$Q = \Delta U + W = \Delta U = \boxed{2490 \text{ J}}$$

---

33. ***REASONING*** AND ***SOLUTION*** Let the left be side 1 and the right be side 2. Since the partition moves to the right, side 1 does work on side 2, so that the work values involved satisfy the relation $W_1 = -W_2$. Using Equation 15.4 for each work value, we find that

$$\tfrac{3}{2} nR\left(T_{1i} - T_{1f}\right) = -\tfrac{3}{2} nR\left(T_{2i} - T_{2f}\right) \qquad \text{or}$$

$$T_{1f} + T_{2f} = T_{1i} + T_{2i} = 525 \text{ K} + 275 \text{ K} = 8.00 \times 10^2 \text{ K}$$

We now seek a second equation for the two unknowns T_{1f} and T_{2f}. Equation 15.5 for an adiabatic process indicates that $P_{1i}V_{1i}^{\gamma} = P_{1f}V_{1f}^{\gamma}$ and $P_{2i}V_{2i}^{\gamma} = P_{2f}V_{2f}^{\gamma}$. Dividing these two equations and using the facts that $V_{1i} = V_{2i}$ and $P_{1f} = P_{2f}$, gives

$$\frac{P_{1i}V_{1i}^{\gamma}}{P_{2i}V_{2i}^{\gamma}} = \frac{P_{1f}V_{1f}^{\gamma}}{P_{2f}V_{2f}^{\gamma}} \qquad \text{or} \qquad \frac{P_{1i}}{P_{2i}} = \left(\frac{V_{1f}}{V_{2f}}\right)^{\gamma}$$

Using the ideal gas law, we find that

$$\frac{P_{1i}}{P_{2i}} = \left(\frac{V_{1f}}{V_{2f}}\right)^{\gamma} \qquad \text{becomes} \qquad \frac{nRT_{1i}/V_{1i}}{nRT_{2i}/V_{2i}} = \left(\frac{nRT_{1f}/P_{1f}}{nRT_{2f}/P_{2f}}\right)^{\gamma}$$

Since $V_{1i} = V_{2i}$ and $P_{1f} = P_{2f}$, the result above reduces to

$$\frac{T_{1i}}{T_{2i}} = \left(\frac{T_{1f}}{T_{2f}}\right)^{\gamma} \qquad \text{or} \qquad \frac{T_{1f}}{T_{2f}} = \left(\frac{T_{1i}}{T_{2i}}\right)^{1/\gamma} = \left(\frac{525 \text{ K}}{275 \text{ K}}\right)^{1/\gamma} = 1.474$$

Using this expression for the ratio of the final temperatures in $T_{1f} + T_{2f} = 8.00 \times 10^2$ K, we find that

a. $\boxed{T_{1f} = 477 \text{ K}}$ and b. $\boxed{T_{2f} = 323 \text{ K}}$

---

35. ***REASONING AND SOLUTION*** According to the first law of thermodynamics (Equation 15.1), $\Delta U = U_f - U_i = Q - W$. Since the internal energy of this gas is doubled by the addition of heat, the initial and final internal energies are U and $2U$, respectively. Therefore,

$$\Delta U = U_f - U_i = 2U - U = U$$

Equation 15.1 for this situation then becomes $U = Q - W$. Solving for Q gives

$$Q = U + W \qquad\qquad (1)$$

The initial internal energy of the gas can be calculated from Equation 14.7:

$$U = \frac{3}{2}nRT = \frac{3}{2}(2.5 \text{ mol})\left[8.31 \text{ J/(mol·K)}\right](350 \text{ K}) = 1.1 \times 10^4 \text{ J}$$

a. If the process is carried out isochorically (i.e., at constant volume), then W = 0, and the heat required to double the internal energy is

$$Q = U + W = U + 0 = \boxed{1.1 \times 10^4 \text{ J}}$$

b.   If the process is carried out isobarically (i.e., at constant pressure), then $W = P\Delta V$, and Equation (1) above becomes

$$Q = U + W = U + P\Delta V \qquad (2)$$

From the ideal gas law, $PV = nRT$, we have that $P\Delta V = nR\Delta T$, and Equation (2) becomes

$$Q = U + nR\Delta T \qquad (3)$$

The internal energy of an ideal gas is directly proportional to its Kelvin temperature. Since the internal energy of the gas is doubled, the final Kelvin temperature will be twice the initial Kelvin temperature, or $\Delta T = 350$ K.   Substituting values into Equation (3) gives

$$Q = 1.1 \times 10^4 \text{ J} + (2.5 \text{ mol})[8.31 \text{ J/(mol} \cdot \text{K)}](350 \text{ K}) = \boxed{1.8 \times 10^4 \text{ J}}$$

---

37. *REASONING* When the temperature of a gas changes as a result of heat Q being added, the change ΔT in temperature is related to the amount of heat according to $Q = Cn\Delta T$ (Equation 15.6), where C is the molar specific heat capacity, and n is the number of moles. The heat Q_V added under conditions of constant volume is $Q_V = C_V n \Delta T_V$, where C_V is the specific heat capacity at constant volume and is given by $C_V = \frac{3}{2}R$ (Equation 15.8) and R is the universal gas constant. The heat Q_P added under conditions of constant pressure is $Q_P = C_P n \Delta T_P$, where C_P is the specific heat capacity at constant pressure and is given by $C_P = \frac{5}{2}R$ (Equation 15.7). It is given that $Q_V = Q_P$, and this fact will allow us to find the change in temperature of the gas whose pressure remains constant.

SOLUTION Setting $Q_V = Q_P$, gives

$$\underbrace{C_V n \Delta T_V}_{Q_V} = \underbrace{C_P n \Delta T_P}_{Q_P}$$

Algebraically eliminating n and solving for ΔT_P, we obtain

$$\Delta T_P = \left(\frac{C_V}{C_P} \right) \Delta T_V = \left(\frac{\frac{3}{2}R}{\frac{5}{2}R} \right)(75 \text{ K}) = \boxed{45 \text{ K}}$$

45. *REASONING*

a. The efficiency e of a heat engine is given by $e = \dfrac{|W|}{|Q_H|}$ (Equation 15.11), where $|Q_H|$ is the magnitude of the input heat needed to do an amount of work of magnitude $|W|$. The "input energy" used by the athlete is equal to the magnitude $|\Delta U|$ of the athlete's internal energy change, so the efficiency of a "human heat engine" can be expressed as

$$e = \frac{|W|}{|\Delta U|} \qquad \text{or} \qquad |\Delta U| = \frac{|W|}{e} \qquad (1)$$

b. We will use the first law of thermodynamics $\Delta U = Q - W$ (Equation 15.1) to find the magnitude $|Q|$ of the heat that the athlete gives off. We note that the internal energy change ΔU is negative, since the athlete spends this energy in order to do work.

SOLUTION

a. Equation (1) gives the magnitude of the internal energy change:

$$|\Delta U| = \frac{|W|}{e} = \frac{5.1 \times 10^4 \text{ J}}{0.11} = \boxed{4.6 \times 10^5 \text{ J}}$$

b. From part (a), the athlete's internal energy change is $\Delta U = -4.6 \times 10^5$ J. The first law of dynamics $\Delta U = Q - W$ (Equation 15.1), therefore, yields the heat Q given off by the athlete:

$$Q = \Delta U + W = -4.6 \times 10^5 \text{ J} + 5.1 \times 10^4 \text{ J} = -4.1 \times 10^5 \text{ J}$$

The magnitude of the heat given off is, thus, $\boxed{4.1 \times 10^5 \text{ J}}$.

---

49. *REASONING* The efficiency e of an engine can be expressed as (see Equation 15.13) $e = 1 - |Q_C| / |Q_H|$, where $|Q_C|$ is the magnitude of the heat delivered to the cold reservoir and $|Q_H|$ is the magnitude of the heat supplied to the engine from the hot reservoir. Solving this equation for $|Q_C|$ gives $|Q_C| = (1 - e)|Q_H|$. We will use this expression twice, once for the improved engine and once for the original engine. Taking the ratio of these expressions will give us the answer that we seek.

SOLUTION Taking the ratio of the heat rejected to the cold reservoir by the improved engine to that for the original engine gives

$$\frac{|Q_{C, \text{ improved}}|}{|Q_{C, \text{ original}}|} = \frac{(1 - e_{\text{improved}})|Q_{H, \text{ improved}}|}{(1 - e_{\text{original}})|Q_{H, \text{ original}}|}$$

But the input heat to both engines is the same, so $|Q_{H, \text{ improved}}| = |Q_{H, \text{ original}}|$. Thus, the ratio becomes

$$\frac{\left|Q_{C,\text{ improved}}\right|}{\left|Q_{C,\text{ original}}\right|} = \frac{1-e_{\text{improved}}}{1-e_{\text{original}}} = \frac{1-0.42}{1-0.23} = \boxed{0.75}$$

---

53. ***REASONING*** The efficiency e of a Carnot engine is given by Equation 15.15, $e = 1 - (T_C / T_H)$, where, according to Equation 15.14, $\left|Q_C\right| / \left|Q_H\right| = T_C / T_H$. Since the efficiency is given along with T_C and $\left|Q_C\right|$, Equation 15.15 can be used to calculate T_H. Once T_H is known, the ratio T_C / T_H is thus known, and Equation 15.14 can be used to calculate $\left|Q_H\right|$.

SOLUTION
a. Solving Equation 15.15 for T_H gives

$$T_H = \frac{T_C}{1-e} = \frac{378 \text{ K}}{1-0.700} = \boxed{1260 \text{ K}}$$

b. Solving Equation 15.14 for $\left|Q_H\right|$ gives

$$\left|Q_H\right| = \left|Q_C\right| \left(\frac{T_H}{T_C}\right) = (5230 \text{ J}) \left(\frac{1260 \text{ K}}{378 \text{ K}}\right) = \boxed{1.74 \times 10^4 \text{ J}}$$

---

59. ***REASONING*** AND ***SOLUTION*** The temperature of the gasoline engine input is $T_1 = 904$ K, the exhaust temperature is $T_2 = 412$ K, and the air temperature is $T_3 = 300$ K. The efficiency of the engine/exhaust is

$$e_1 = 1 - (T_2/T_1) = 0.544$$

The efficiency of the second engine is

$$e_2 = 1 - (T_3/T_2) = 0.272$$

The magnitude of the work done by each segment is $\left|W_1\right| = e_1\left|Q_{H1}\right|$ and $\left|W_2\right| = e_2\left|Q_{H2}\right| = e_2\left|Q_{C1}\right|$ since

$$\left|Q_{H2}\right| = \left|Q_{C1}\right|$$

Now examine $\left(\left|W_1\right| + \left|W_2\right|\right)/\left|W_1\right|$ to find the ratio of the total work produced by both engines to that produced by the first engine alone.

$$\frac{|W_1| + |W_2|}{|W_1|} = \frac{e_1 |Q_{H1}| + e_2 |Q_{C1}|}{e_1 |Q_{H1}|} = 1 + \left(\frac{e_2}{e_1}\right)\left(\frac{|Q_{C1}|}{|Q_{H1}|}\right)$$

But, $e_1 = 1 - \left(|Q_{C1}|/|Q_{H1}|\right)$, so that $|Q_{C1}|/|Q_{H1}| = 1 - e_1$. Therefore,

$$\frac{|W_1| + |W_2|}{|W_1|} = 1 + \frac{e_2}{e_1}\left(1 - e_1\right) = 1 + \frac{e_2}{e_1} - e_2 = 1 + 0.500 - 0.272 = \boxed{1.23}$$

---

61. **REASONING** The expansion from point a to point b and the compression from point c to point d occur isothermally, and we will apply the first law of thermodynamics to these parts of the cycle in order to obtain expressions for the input and rejected heats, magnitudes $|Q_H|$ and $|Q_C|$, respectively. In order to simplify the resulting expression for $|Q_C|/|Q_H|$, we will then use the fact that the expansion from point b to point c and the compression from point d to point a are adiabatic.

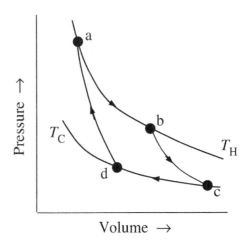

SOLUTION According to the first law of thermodynamics, the change in internal energy ΔU is given by $\Delta U = Q - W$ (Equation 15.1), where Q is the heat and W is the work. Since the internal energy of an ideal gas is proportional to the Kelvin temperature and the temperature is constant for an isothermal process, it follows that $\Delta U = 0$ J for such a case.

The work of isothermal expansion or compression for an ideal gas is $W = nRT \ln\left(V_f / V_i\right)$ (Equation 15.3), where n is the number of moles, R is the universal gas constant, T is the Kelvin temperature, V_f is the final volume of the gas, and V_i is the initial volume. We have, then, that

$$\Delta U = Q - W \quad \text{or} \quad 0 = Q - nRT \ln\left(\frac{V_f}{V_i}\right) \quad \text{or} \quad Q = nRT \ln\left(\frac{V_f}{V_i}\right)$$

Applying this result for Q to the isothermal expansion (temperature $= T_H$) from point a to point b and the isothermal compression (temperature $= T_C$) from point c to point d, we have

$$Q_H = nRT_H \ln\left(\frac{V_b}{V_a}\right) \quad \text{and} \quad Q_C = nRT_C \ln\left(\frac{V_d}{V_c}\right)$$

where $V_f = V_b$ and $V_i = V_a$ for the isotherm at T_H and $V_f = V_d$ and $V_i = V_c$ for the isotherm at T_C. In this problem, we are interested in the magnitude of the heats. For Q_H, this poses no difficulty, since $V_b > V_a$, $\ln\left(V_b / V_a\right)$ is positive, and we have

$$\left|Q_{\mathrm{H}}\right| = nRT_{\mathrm{H}} \, \ln\left(\frac{V_{\mathrm{b}}}{V_{\mathrm{a}}}\right) \tag{1}$$

However, for Q_{C}, we need to be careful, because $V_{\mathrm{c}} > V_{\mathrm{d}}$ and $\ln\left(V_{\mathrm{d}}/V_{\mathrm{c}}\right)$ is negative. Thus, we write for the magnitude of Q_{C} that

$$\left|Q_{\mathrm{C}}\right| = -nRT_{\mathrm{C}} \, \ln\left(\frac{V_{\mathrm{d}}}{V_{\mathrm{c}}}\right) = nRT_{\mathrm{C}} \, \ln\left(\frac{V_{\mathrm{c}}}{V_{\mathrm{d}}}\right) \tag{2}$$

According to Equations (1) and (2), the ratio of the magnitudes of the rejected and input heats is

$$\frac{\left|Q_{\mathrm{C}}\right|}{\left|Q_{\mathrm{H}}\right|} = \frac{nRT_{\mathrm{C}} \, \ln\left(\dfrac{V_{\mathrm{c}}}{V_{\mathrm{d}}}\right)}{nRT_{\mathrm{H}} \, \ln\left(\dfrac{V_{\mathrm{b}}}{V_{\mathrm{a}}}\right)} = \frac{T_{\mathrm{C}} \, \ln\left(\dfrac{V_{\mathrm{c}}}{V_{\mathrm{d}}}\right)}{T_{\mathrm{H}} \, \ln\left(\dfrac{V_{\mathrm{b}}}{V_{\mathrm{a}}}\right)} \tag{3}$$

We now consider the adiabatic parts of the Carnot cycle. For the adiabatic expansion or compression of an ideal gas the initial pressure and volume (P_{i} and V_{i}) are related to the final pressure and volume (P_{f} and V_{f}) according to

$$P_{\mathrm{i}} V_{\mathrm{i}}^{\gamma} = P_{\mathrm{f}} V_{\mathrm{f}}^{\gamma} \tag{15.5}$$

where γ is the ratio of the specific heats at constant pressure and constant volume. It is also true that $P = nRT/V$ (Equation 14.1), according to the ideal gas law. Substituting this expression for the pressure into Equation 15.5 gives

$$\left(\frac{nRT_{\mathrm{i}}}{V_{\mathrm{i}}}\right) V_{\mathrm{i}}^{\gamma} = \left(\frac{nRT_{\mathrm{f}}}{V_{\mathrm{f}}}\right) V_{\mathrm{f}}^{\gamma} \quad \text{or} \quad T_{\mathrm{i}} V_{\mathrm{i}}^{\gamma-1} = T_{\mathrm{f}} V_{\mathrm{f}}^{\gamma-1}$$

Applying this result to the adiabatic expansion from point b to point c and to the adiabatic compression from point d to point a, we obtain

$$T_{\mathrm{H}} V_{\mathrm{b}}^{\gamma-1} = T_{\mathrm{C}} V_{\mathrm{c}}^{\gamma-1} \quad \text{and} \quad T_{\mathrm{C}} V_{\mathrm{d}}^{\gamma-1} = T_{\mathrm{H}} V_{\mathrm{a}}^{\gamma-1}$$

Dividing the first of these equations by the second shows that

$$\frac{T_{\mathrm{H}} V_{\mathrm{b}}^{\gamma-1}}{T_{\mathrm{H}} V_{\mathrm{a}}^{\gamma-1}} = \frac{T_{\mathrm{C}} V_{\mathrm{c}}^{\gamma-1}}{T_{\mathrm{C}} V_{\mathrm{d}}^{\gamma-1}} \quad \text{or} \quad \frac{V_{\mathrm{b}}^{\gamma-1}}{V_{\mathrm{a}}^{\gamma-1}} = \frac{V_{\mathrm{c}}^{\gamma-1}}{V_{\mathrm{d}}^{\gamma-1}} \quad \text{or} \quad \frac{V_{\mathrm{b}}}{V_{\mathrm{a}}} = \frac{V_{\mathrm{c}}}{V_{\mathrm{d}}}$$

With this result, Equation (3) becomes

$$\frac{\left|Q_{\mathrm{C}}\right|}{\left|Q_{\mathrm{H}}\right|} = \frac{T_{\mathrm{C}} \, \ln\left(\dfrac{V_{\mathrm{c}}}{V_{\mathrm{d}}}\right)}{T_{\mathrm{H}} \, \ln\left(\dfrac{V_{\mathrm{b}}}{V_{\mathrm{a}}}\right)} = \boxed{\frac{T_{\mathrm{C}}}{T_{\mathrm{H}}}}$$

63. **REASONING** The coefficient of performance of an air conditioner is $|Q_C|/|W|$, according to Equation 15.16, where $|Q_C|$ is the magnitude of the heat removed from the house and $|W|$ is the magnitude of the work required for the removal. In addition, we know that the first law of thermodynamics (energy conservation) applies, so that $|W|=|Q_H|-|Q_C|$, according to Equation 15.12. In this equation $|Q_H|$ is the magnitude of the heat discarded outside. While we have no direct information about $|Q_C|$ and $|Q_H|$, we do know that the air conditioner is a Carnot device. This means that Equation 15.14 applies: $|Q_C|/|Q_H|=T_C/T_H$. Thus, the given temperatures will allow us to calculate the coefficient of performance.

SOLUTION Using Equation 15.16 for the definition of the coefficient of performance and Equation 15.12 for the fact that $|W|=|Q_H|-|Q_C|$, we have

$$\text{Coefficient of performance} = \frac{|Q_C|}{|W|} = \frac{|Q_C|}{|Q_H|-|Q_C|} = \frac{|Q_C|/|Q_H|}{1-|Q_C|/|Q_H|}$$

Equation 15.14 applies, so that $|Q_C|/|Q_H|=T_C/T_H$. With this substitution, we find

$$\text{Coefficient of performance} = \frac{|Q_C|/|Q_H|}{1-|Q_C|/|Q_H|} = \frac{T_C/T_H}{1-T_C/T_H}$$

$$= \frac{T_C}{T_H-T_C} = \frac{297\ \text{K}}{(311\ \text{K})-(297\ \text{K})} = \boxed{21}$$

71. **REASONING** The net heat added to the room is $|Q_H|-|Q_C|$, where $|Q_H|$ is the magnitude of the heat put into the room by the unit and $|Q_C|$ is the magnitude of the heat removed by the unit. To determine this net heat, we will use the fact that energy conservation applies, so that $|Q_H|=|W|+|Q_C|$ (Equation 15.12). We will also use the fact that the coefficient of performance COP of the air conditioner is $\text{COP}=|Q_C|/|W|$ (Equation 15.16). In Equations 15.12 and 15.16, $|W|$ is the amount of work needed to operate the unit.

SOLUTION Rearranging Equation 15.12 for energy conservation, we have

$$|Q_H|=|W|+|Q_C| \qquad \text{or} \qquad \text{Net heat}=|Q_H|-|Q_C|=|W| \tag{1}$$

Thus, the net heat is just the work $|W|$ needed to operate the unit. We can obtain this work directly from the coefficient of performance as specified by Equation 15.16:

$$\text{COP} = \frac{|Q_C|}{|W|} \qquad \text{or} \qquad |W| = \frac{|Q_C|}{\text{COP}} \qquad (2)$$

Substituting $|W|$ from Equation (2) into Equation (1), we find that

$$\text{Net heat} = |Q_H| - |Q_C| = |W| = \frac{|Q_C|}{\text{COP}} = \frac{7.6 \times 10^4 \text{ J}}{2.0} = \boxed{3.8 \times 10^4 \text{ J}}$$

---

73. **REASONING** Let the coefficient of performance be represented by the symbol COP. Then according to Equation 15.16, $\text{COP} = |Q_C|/|W|$. From the statement of energy conservation for a Carnot refrigerator (Equation 15.12), $|W| = |Q_H| - |Q_C|$. Combining Equations 15.16 and 15.12 leads to

$$\text{COP} = \frac{|Q_C|}{|Q_H| - |Q_C|} = \frac{|Q_C|/|Q_C|}{\left(|Q_H| - |Q_C|\right)/|Q_C|} = \frac{1}{\left(|Q_H|/|Q_C|\right) - 1}$$

Replacing the ratio of the heats with the ratio of the Kelvin temperatures, according to Equation 15.14, leads to

$$\text{COP} = \frac{1}{T_H/T_C - 1} \qquad (1)$$

The heat $|Q_C|$ that must be removed from the refrigerator when the water is cooled can be calculated using Equation 12.4, $|Q_C| = cm\Delta T$; therefore,

$$|W| = \frac{|Q_C|}{\text{COP}} = \frac{cm\Delta T}{\text{COP}} \qquad (2)$$

SOLUTION
a. Substituting values into Equation (1) gives

$$\text{COP} = \frac{1}{\dfrac{T_H}{T_C} - 1} = \frac{1}{\dfrac{(20.0 + 273.15) \text{ K}}{(6.0 + 273.15) \text{ K}} - 1} = \boxed{2.0 \times 10^1}$$

b. Substituting values into Equation (2) gives

$$|W| = \frac{cm\Delta T}{\text{COP}} = \frac{\left[4186 \text{ J}/(\text{kg} \cdot \text{C}^\circ)\right](5.00 \text{ kg})(20.0 \text{ °C} - 6.0 \text{ °C})}{2.0 \times 10^1} = \boxed{1.5 \times 10^4 \text{ J}}$$

77. **REASONING AND SOLUTION** The change in entropy ΔS of a system for a process in which heat Q enters or leaves the system reversibly at a constant temperature T is given by Equation 15.18, $\Delta S = (Q/T)_\text{R}$. For a phase change, $Q = mL$, where L is the latent heat (see Section 12.8).

a. If we imagine a reversible process in which 3.00 kg of ice melts into water at 273 K, the change in entropy of the water molecules is

$$\Delta S = \left(\frac{Q}{T}\right)_\text{R} = \left(\frac{mL_\text{f}}{T}\right)_\text{R} = \frac{(3.00 \text{ kg})\left(3.35 \times 10^5 \text{ J/kg}\right)}{273 \text{ K}} = \boxed{3.68 \times 10^3 \text{ J/K}}$$

b. Similarly, if we imagine a reversible process in which 3.00 kg of water changes into steam at 373 K, the change in entropy of the water molecules is

$$\Delta S = \left(\frac{Q}{T}\right)_\text{R} = \left(\frac{mL_\text{v}}{T}\right)_\text{R} = \frac{(3.00 \text{ kg})\left(2.26 \times 10^6 \text{ J/kg}\right)}{373 \text{ K}} = \boxed{1.82 \times 10^4 \text{ J/K}}$$

c. Since the change in entropy is greater for the vaporization process than for the fusion process, the $\boxed{\text{vaporization process creates more disorder}}$ in the collection of water molecules.

---

83. **REASONING** According to the first law of thermodynamics (Equation 15.1), $\Delta U = Q - W$. For a monatomic ideal gas (Equation 14.7), $U = \frac{3}{2}nRT$. Therefore, for the process in question, the change in the internal energy is $\Delta U = \frac{3}{2}nR\Delta T$. Combining the last expression for ΔU with Equation 15.1 yields

$$\tfrac{3}{2}nR\Delta T = Q - W$$

This expression can be solved for ΔT.

SOLUTION
a. The heat is $Q = +1200 \text{ J}$, since it is absorbed by the system. The work is $W = +2500 \text{ J}$, since it is done *by* the system. Solving the above expression for ΔT and substituting the values for the data given in the problem statement, we have

$$\Delta T = \frac{Q - W}{\tfrac{3}{2}nR} = \frac{1200 \text{ J} - 2500 \text{ J}}{\tfrac{3}{2}(0.50 \text{ mol})[8.31 \text{ J} / (\text{mol} \cdot \text{K})]} = \boxed{-2.1 \times 10^2 \text{ K}}$$

b. Since $\Delta T = T_\text{final} - T_\text{initial}$ is negative, T_initial must be greater than T_final; this change represents a $\boxed{\text{decrease}}$ in temperature.

Alternatively, one could deduce that the temperature decreases from the following physical argument. Since the system loses more energy in doing work than it gains in the form of heat, the internal energy of the system decreases. Since the internal energy of an ideal gas depends only on the temperature, a decrease in the internal energy must correspond to a decrease in the temperature.

---

87. ***REASONING AND SOLUTION***
a. Since the energy that becomes unavailable for doing work is zero for the process, we have from Equation 15.19, $W_{unavailable} = T_0 \Delta S_{universe} = 0$. Therefore, $\Delta S_{universe} = 0$ and according to the discussion in Section 15.11, the process is $\boxed{\text{reversible}}$.

b. Since the process is reversible, we have (see Section 15.11)

$$\Delta S_{universe} = \Delta S_{system} + \Delta S_{surroundings} = 0$$

Therefore,

$$\Delta S_{surroundings} = -\Delta S_{system} = \boxed{-125 \text{ J/K}}$$

---

91. ***REASONING*** For segment AB, there is no work, since the volume is constant. For segment BC the process is isobaric and Equation 15.2 applies. For segment CA, the work can be obtained as the area under the line CA in the graph.

SOLUTION
a. For segment *AB*, the process is isochoric, that is, the volume is constant. For a process in which the volume is constant, no work is done, so $\boxed{W = 0 \text{ J}}$.

b. For segment BC, the process is isobaric, that is, the pressure is constant. Here, the volume is increasing, so the gas is expanding against the outside environment. As a result, the gas does work, which is positive according to our convention. Using Equation 15.2 and the data in the drawing, we obtain

$$W = P\left(V_f - V_i\right)$$

$$= \left(7.0\times10^5 \text{ Pa}\right)\left[\left(5.0\times10^{-3} \text{ m}^3\right) - \left(2.0\times10^{-3} \text{ m}^3\right)\right] = \boxed{+2.1\times10^3 \text{ J}}$$

c. For segment CA, the volume of the gas is decreasing, so the gas is being compressed and work is being done on it. Therefore, the work is negative, according to our convention. The magnitude of the work is the area under the segment CA. We estimate that this area is 15 of the squares in the graphical grid. The area of each square is

$$(1.0\times10^5 \text{ Pa})(1.0\times10^{-3} \text{ m}^3) = 1.0\times10^2 \text{ J}$$

The work, then, is

$$W = -15 (1.0\times10^2 \text{ J}) = \boxed{-1.5\times10^3 \text{ J}}$$

93. **REASONING** The change in the internal energy of the gas can be found using the first law of thermodynamics, since the heat added to the gas is known and the work can be calculated by using Equation 15.2, $W = P \Delta V$. The molar specific heat capacity at constant pressure can be evaluated by using Equation 15.6 and the ideal gas law.

SOLUTION
a. The change in the internal energy is

$$\Delta U = Q - W = Q - P\Delta V$$

$$= 31.4 \text{ J} - \left(1.40 \times 10^4 \text{ Pa}\right)\left(8.00 \times 10^{-4} \text{ m}^3 - 3.00 \times 10^{-4} \text{ m}^3\right) = \boxed{24.4 \text{ J}}$$

b. According to Equation 15.6, the molar specific heat capacity at constant pressure is $C_P = Q/(n \, \Delta T)$. The term $n \, \Delta T$ can be expressed in terms of the pressure and change in volume by using the ideal gas law:

$$P \, \Delta V = n \, R \, \Delta T \quad \text{or} \quad n \, \Delta T = P \, \Delta V / R$$

Substituting this relation for $n \, \Delta T$ into $C_P = Q/(n \, \Delta T)$, we obtain

$$C_P = \frac{Q}{\dfrac{P\Delta V}{R}} = \frac{31.4 \text{ J}}{\dfrac{\left(1.40 \times 10^4 \text{ Pa}\right)\left(5.00 \times 10^{-4} \text{ m}^3\right)}{8.31 \text{ J/(mol} \cdot \text{K)}}} = \boxed{37.3 \text{ J/(mol} \cdot \text{K)}}$$

---

99. **REASONING AND SOLUTION** We wish to find an expression for the overall efficiency e in terms of the efficiencies e_1 and e_2. From the problem statement, the overall efficiency of the two-engine device is

$$e = \frac{|W_1| + |W_2|}{|Q_H|} \tag{1}$$

where $|Q_H|$ is the input heat to engine 1. The efficiency of a heat engine is defined by Equation 15.11, $e = |W|/|Q_H|$, so we can write

$$|W_1| = e_1 |Q_H| \tag{2}$$

and

$$|W_2| = e_2 |Q_{H2}|$$

Since the heat rejected by engine 1 is used as input heat for the second engine, $\left|Q_{\text{H2}}\right| = \left|Q_{\text{C1}}\right|$, and the expression above for $\left|W_2\right|$ can be written as

$$\left|W_2\right| = e_2 \left|Q_{\text{C1}}\right| \tag{3}$$

According to Equation 15.12, we have $\left|Q_{\text{C1}}\right| = \left|Q_{\text{H}}\right| - \left|W_1\right|$, so that Equation (3) becomes

$$\left|W_2\right| = e_2 \left(\left|Q_{\text{H}}\right| - \left|W_1\right|\right) \tag{4}$$

Substituting Equations (2) and (4) into Equation (1) gives

$$e = \frac{e_1 \left|Q_{\text{H}}\right| + e_2 \left(\left|Q_{\text{H}}\right| - \left|W_1\right|\right)}{\left|Q_{\text{H}}\right|} = \frac{e_1 \left|Q_{\text{H}}\right| + e_2 \left(\left|Q_{\text{H}}\right| - e_1 \left|Q_{\text{H}}\right|\right)}{\left|Q_{\text{H}}\right|}$$

Algebraically canceling the $\left|Q_{\text{H}}\right|$'s in the right hand side of the last expression gives the desired result:

$$\boxed{e = e_1 + e_2 - e_1 e_2}$$

---

101. **REASONING**   The conservation of energy dictates that the heat delivered into the house (the hot reservoir) equals the energy from the work done by the heat pump plus the energy in the form of heat taken from the cold outdoors (the cold reservoir). The heat delivered into the house is given, so that we can use energy conservation to determine the work, provided that we can obtain a value for the heat taken from the outdoors. Since we are dealing with an ideal heat pump, we can obtain this value by using Equation 15.14, which relates the ratio of the magnitudes of the heats for the cold and hot reservoirs to the ratio of the reservoir temperatures (in kelvins).

The energy-conservation principle requires that $\left|Q_{\text{H}}\right| = \left|W\right| + \left|Q_{\text{C}}\right|$ (Equation 15.12), where $\left|Q_{\text{H}}\right|$, $\left|W\right|$, and $\left|Q_{\text{C}}\right|$ are, respectively, the magnitudes of the heat delivered into the house (the hot reservoir), the work done by the heat pump, and the heat taken from the cold outdoors (the cold reservoir). Solving for $\left|W\right|$ gives Equation (1) below.

$$\left|W\right| = \left|Q_{\text{H}}\right| - \left|Q_{\text{C}}\right| \qquad (1)$$

In this result, we have a value for $\left|Q_{\text{H}}\right|$ but we need to find an expression for $\left|Q_{\text{C}}\right|$. To obtain a value for $\left|Q_{\text{C}}\right|$, we will use the information given about the temperatures T_{H} for the hot reservoir and T_C for the cold reservoir. According to Equation 15.14, $|Q_C|/|Q_H| = T_C/T_H$, which can be solved for $|Q_C|$ to show that

$$|Q_C| = |Q_H|\left(\frac{T_C}{T_H}\right)$$

We can now substitute this result into Equation (1) above.

$$|W| = |Q_H| - |Q_C| = |Q_H| - |Q_H|\left(\frac{T_C}{T_H}\right) = |Q_H|\left(1 - \frac{T_C}{T_H}\right)$$

It follows that the magnitude of the work for the two given outdoor temperatures is

Outdoor temperature of 273 K $|W| = |Q_H|\left(1 - \frac{T_C}{T_H}\right) = (3350\ \text{J})\left(1 - \frac{273\ \text{K}}{294\ \text{K}}\right) = \boxed{239\ \text{J}}$

Outdoor temperature of 252 K $|W| = |Q_H|\left(1 - \frac{T_C}{T_H}\right) = (3350\ \text{J})\left(1 - \frac{252\ \text{K}}{294\ \text{K}}\right) = \boxed{479\ \text{J}}$

More work must be done when the outdoor temperature is lower, because the heat is pumped up a greater temperature "hill."

---

103. ***CONCEPTS*** **(i)** The change is greater with engine 2. Kinetic energy is $\text{KE} = \frac{1}{2}mv^2$, where m is the mass of the crate and v is its speed. The change in the kinetic energy is the final minus the initial value, or $\text{KE}_f - \text{KE}_0$. Since the crate starts from rest, it has zero initial kinetic energy. Thus the change is equal to the final kinetic energy. Since engine 2 gives the crate the greater final speed, it causes the greater change in kinetic energy. **(ii)** The work-energy theorem states that the net work done on an object equals the change in the object's kinetic energy, or $W = \text{KE}_f - \text{KE}_0$. The net work is the work done by the net force. The surface on which the crate moves is horizontal, and the crate does not leave it. Therefore, the upward normal force that the surface applies to the crate must balance the downward weight of the crate. Furthermore, the surface is frictionless, so there is no friction force. The net force acting on the crate, then, consists of the single force due to the tension in the rope, which arises from the action of the engine. Thus, the work done by the engine is, in fact, the net work done on the crate. But we know that engine 2 causes the crate's kinetic energy to change by the greater amount, so that the engine must do more work. **(iii)** The temperature of the hot reservoir for engine 2 is greater. We know that engine 2 does more work, but each engine receives the same 1450 J of input heat. Therefore, engine 2 derives more work from the input heat. In other words, it is more efficient. But the efficiency of a Carnot engine depends only on the Kelvin temperatures of its hot and cold reservoirs. Since both engines use the same cold reservoir whose temperature is 275 K, only the temperatures of the hot

reservoirs are different. Higher temperatures for the hot reservoir are associated with greater efficiencies, so the temperature of the hot reservoir for engine 2 is greater.

CALCULATIONS The efficiency e of a heat engine is the magnitude of the work $|W|$ divided by the magnitude of the input heat $|Q_H|$, or $e = |W|/|Q_H|$. The efficiency of a Carnot engine is $e_{Carnot} = 1 - T_C/T_H$, where T_C and T_H are, respectively, the temperatures of the cold and hot reservoirs. Combining these two equations, we have

$$1 - \frac{T_C}{T_H} = \frac{|W|}{|Q_H|}$$

But $|W|$ is the magnitude of the net work done on the crate, and it equals the change in the crate's kinetic energy, or $|W| = KE_f - KE_0 = \frac{1}{2}mv^2$.

With this substitution, the efficiency expression becomes

$$1 - \frac{T_C}{T_H} = \frac{\frac{1}{2}mv^2}{|Q_H|}$$

Solving for the temperature T_H, we find

$$T_H = \frac{T_C}{1 - \frac{mv^2}{2|Q_H|}}$$

Using this expression, we can calculate the temperature of the hot reservoir for each engine:

Engine 1 $\quad T_H = \dfrac{275 \text{ K}}{1 - \dfrac{(125 \text{ kg})(2.00 \text{ m/s})^2}{2(1450 \text{ J})}} = \boxed{332 \text{ K}}$

Engine 2 $\quad T_H = \dfrac{275 \text{ K}}{1 - \dfrac{(125 \text{ kg})(3.00 \text{ m/s})^2}{2(1450 \text{ J})}} = \boxed{449 \text{ K}}$

As expected, the value of T_H for engine 2 is greater.

CHAPTER 16 | WAVES AND SOUND

1. **REASONING** Since light behaves as a wave, its speed v, frequency f, and wavelength λ are related to according to $v = f\lambda$ (Equation 16.1). We can solve this equation for the frequency in terms of the speed and the wavelength.

SOLUTION Solving Equation 16.1 for the frequency, we find that

$$f = \frac{v}{\lambda} = \frac{3.00 \times 10^8 \text{ m/s}}{5.45 \times 10^{-7} \text{ m}} = \boxed{5.50 \times 10^{14} \text{ Hz}}$$

---

5. **REASONING** When the end of the Slinky is moved up and down continuously, a transverse wave is produced. The distance between two adjacent crests on the wave, is, by definition, one wavelength. The wavelength λ is related to the speed and frequency of a periodic wave by $\lambda = v/f$ (Equation 16.1). In order to use Equation 16.1, we must first determine the frequency of the wave. The wave on the Slinky will have the same frequency as the simple harmonic motion of the hand. According to the data given in the problem statement, the frequency is $f = (2.00 \text{ cycles})/(1 \text{ s}) = 2.00 \text{ Hz}$.

SOLUTION Substituting the values for λ and f, we find that the distance between crests is

$$\lambda = \frac{v}{f} = \frac{0.50 \text{ m/s}}{2.00 \text{ Hz}} = \boxed{0.25 \text{ m}}$$

---

13. **REASONING** According to Equation 16.2, the linear density of the string is given by $(m/L) = F/v^2$, where the speed v of waves on the middle C string is given by Equation 16.1, $v = f\lambda = \left(\dfrac{1}{T}\right)\lambda$, where T is the period.

SOLUTION Combining Equations 16.2 and 16.1 and using the given data, we obtain

$$m/L = \frac{F}{v^2} = \frac{FT^2}{\lambda^2} = \frac{(944 \text{ N})(3.82 \times 10^{-3} \text{ s})^2}{(1.26 \text{ m})^2} = \boxed{8.68 \times 10^{-3} \text{ kg/m}}$$

---

17. **REASONING** The speed v of a transverse wave on a string is given by $v = \sqrt{F/(m/L)}$ (Equation 16.2), where F is the tension and m/L is the mass per unit length (or linear density) of the string. The strings are identical, so they have the same mass per unit length. However, the tensions are different. In part (a) of the text drawing, the string supports the entire weight of the 26-N block, so the tension in the string is 26 N. In part (b), the block is supported by the part of the string on the left side of the middle pulley and the part of the string on the right

side. Each part supports one-half of the block's weight, or 13 N. Thus, the tension in the string is 13 N.

SOLUTION

a. The speed of the transverse wave in part (a) of the text drawing is

$$v = \sqrt{\frac{F}{m/L}} = \sqrt{\frac{26 \text{ N}}{0.065 \text{ kg/m}}} = \boxed{2.0 \times 10^1 \text{ m/s}}$$

b. The speed of the transverse wave in part (b) of the drawing is

$$v = \sqrt{\frac{F}{m/L}} = \sqrt{\frac{13 \text{ N}}{0.065 \text{ kg/m}}} = \boxed{1.4 \times 10^1 \text{ m/s}}$$

---

27. **REASONING**   Since the wave is traveling in the +*x* direction, its form is given by Equation 16.3 as

$$y = A \sin\left(2\pi ft - \frac{2\pi x}{\lambda}\right)$$

We are given that the amplitude is $A = 0.35$ m. However, we need to evaluate $2\pi f$ and $\frac{2\pi}{\lambda}$.

Although the wavelength λ is not stated directly, it can be obtained from the values for the speed v and the frequency f, since we know that $v = f\lambda$ (Equation 16.1).

SOLUTION Since the frequency is $f = 14$ Hz, we have

$$2\pi f = 2\pi(14 \text{ Hz}) = 88 \text{ rad/s}$$

It follows from Equation 16.1 that

$$\frac{2\pi}{\lambda} = \frac{2\pi f}{v} = \frac{2\pi(14 \text{ Hz})}{5.2 \text{ m/s}} = 17 \text{ m}^{-1}$$

Using these values for $2\pi f$ and $\frac{2\pi}{\lambda}$ in Equation 16.3, we have

$$y = A \sin\left(2\pi ft - \frac{2\pi x}{\lambda}\right)$$

$$\boxed{y = (0.35 \text{ m})\sin\left[(88 \text{ rad/s})t - (17 \text{ m}^{-1})x\right]}$$

29. **REASONING** The speed of a wave on the string is given by Equation 16.2 as $v = \sqrt{\dfrac{F}{m/L}}$, where F is the tension in the string and m/L is the mass per unit length (or linear density) of the string. The wavelength λ is the speed of the wave divided by its frequency f (Equation 16.1).

SOLUTION
a. The speed of the wave on the string is

$$v = \sqrt{\frac{F}{(m/L)}} = \sqrt{\frac{15 \text{ N}}{0.85 \text{ kg/m}}} = \boxed{4.2 \text{ m/s}}$$

b. The wavelength is

$$\lambda = \frac{v}{f} = \frac{4.2 \text{ m/s}}{12 \text{ Hz}} = \boxed{0.35 \text{ m}}$$

c. The amplitude of the wave is $A = 3.6$ cm $= 3.6 \times 10^{-2}$ m. Since the wave is moving along the $-x$ direction, the mathematical expression for the wave is given by Equation 16.4 as

$$y = A \sin\left(2\pi f t + \frac{2\pi x}{\lambda}\right)$$

Substituting in the numbers for A, f, and λ, we have

$$y = A \sin\left(2\pi f t + \frac{2\pi x}{\lambda}\right) = \left(3.6 \times 10^{-2} \text{ m}\right)\sin\left[2\pi\left(12 \text{ Hz}\right)t + \frac{2\pi x}{0.35 \text{ m}}\right]$$

$$= \boxed{\left(3.6 \times 10^{-2} \text{ m}\right)\sin\left[\left(75 \text{ rad/s}\right)t + \left(18 \text{ m}^{-1}\right)x\right]}$$

---

31. **REASONING** The speed v, frequency f, and wavelength λ of the sound are related according to $v = f\lambda$ (Equation 16.1). This expression can be solved for the wavelength in terms of the speed and the frequency. The speed of sound in seawater is 1522 m/s, as given in Table 16.1. While the frequency is not given directly, the period T is known and is related to the frequency according to $f = 1/T$ (Equation 10.5).

SOLUTION Substituting the frequency from Equation 10.5 into Equation 16.1 gives

$$v = f\lambda = \left(\frac{1}{T}\right)\lambda$$

Solving this result for the wavelength yields

$$\lambda = vT = (1522 \text{ m/s})(71 \times 10^{-3} \text{ s}) = \boxed{110 \text{ m}}$$

---

35. ***REASONING AND SOLUTION***   The speed of sound in an ideal gas is given by Equation 16.5, $v = \sqrt{\gamma kT / m}$. The ratio of the speed of sound v_2 in the container (after the temperature change) to the speed v_1 (before the temperature change) is

$$\frac{v_2}{v_1} = \sqrt{\frac{T_2}{T_1}}$$

Thus, the new speed is

$$v_2 = v_1 \sqrt{\frac{T_2}{T_1}} = (1220 \text{ m/s})\sqrt{\frac{405 \text{ K}}{201 \text{ K}}} = \boxed{1730 \text{ m/s}}$$

---

43. ***REASONING***   The sound will spread out uniformly in all directions. For the purposes of counting the echoes, we will consider only the sound that travels in a straight line parallel to the ground and reflects from the vertical walls of the cliff. Let the distance between the hunter and the closer cliff be x_1 and the distance from the hunter to the further cliff be x_2.

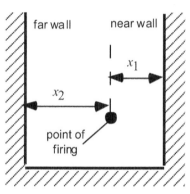

The first echo arrives at the location of the hunter after traveling a total distance $2x_1$ in a time t_1, so that, if v_s is the speed of sound, $t_1 = 2x_1 / v_s$. Similarly, the second echo arrives after reflection from the far wall and in an amount of time t_2 after the firing of the gun.   The quantity t_2 is related to the distance x_2 and the speed of sound v_s according to $t_2 = 2x_2 / v_s$. The time difference between the first and second echo is, therefore

$$\Delta t = t_2 - t_1 = \frac{2}{v_s}(x_2 - x_1) \tag{1}$$

The third echo arrives in a time t_3 after the second echo.   It arises from the sound of the second echo that is reflected from the closer cliff wall. Thus, $t_3 = 2x_1 / v_s$, or, solving for x_1, we have

$$x_1 = \frac{v_s t_3}{2} \tag{2}$$

Combining Equations (1) and (2), we obtain

$$\Delta t = t_2 - t_1 = \frac{2}{v_s}\left(x_2 - \frac{v_s t_3}{2}\right)$$

Solving for x_2, we have

$$x_2 = \frac{v_s}{2}(\Delta t + t_3) \tag{3}$$

The distance between the cliffs can be found from $d = x_1 + x_2$, where x_1 and x_2 can be determined from Equations (2) and (3), respectively.

SOLUTION According to Equation (2), the distance x_1 is

$$x_1 = \frac{(343 \text{ m/s})(1.1 \text{ s})}{2} = 190 \text{ m}$$

According to Equation (3), the distance x_2 is

$$x_2 = \frac{(343 \text{ m/s})}{2}(1.6 \text{ s} + 1.1 \text{ s}) = 460 \text{ m}$$

Therefore, the distance between the cliffs is

$$d = x_1 + x_2 = 190 \text{ m} + 460 \text{ m} = \boxed{650 \text{ m}}$$

---

47. **REASONING** Let v_P represent the speed of the primary wave and v_S the speed of the secondary wave. The travel times for the primary and secondary waves are t_P and t_S, respectively. If x is the distance from the earthquake to the seismograph, then $t_P = x/v_P$ and $t_S = x/v_S$. The difference in the arrival times is

$$t_S - t_P = \frac{x}{v_S} - \frac{x}{v_P} = x\left(\frac{1}{v_S} - \frac{1}{v_P}\right)$$

We can use this equation to find the distance from the earthquake to the seismograph.

SOLUTION Solving the equation above for x gives

$$x = \frac{t_S - t_P}{\dfrac{1}{v_S} - \dfrac{1}{v_P}} = \frac{78 \text{ s}}{\dfrac{1}{4.5 \times 10^3 \text{ m/s}} - \dfrac{1}{8.0 \times 10^3 \text{ m/s}}} = \boxed{8.0 \times 10^5 \text{ m}}$$

---

51. **REASONING** We must determine the time t for the warning to travel the vertical distance $h = 10.0$ m from the prankster to the ears of the man when he is just under the window. The

desired distance above the man's ears is the distance that the balloon would travel in this time and can be found with the aid of the equations of kinematics.

SOLUTION Since sound travels with constant speed v_s, the distance h and the time t are related by $h = v_s t$. Therefore, the time t required for the warning to reach the ground is

$$t = \frac{h}{v_s} = \frac{10.0 \text{ m}}{343 \text{ m/s}} = 0.0292 \text{ s}$$

We now proceed to find the distance that the balloon travels in this time. To this end, we must find the balloon's final speed v_y, after falling from rest for 10.0 m. Since the balloon is dropped from rest, we use Equation 3.6b ($v_y^2 = v_{0y}^2 + 2a_y y$) with $v_{0y} = 0$ m/s:

$$v_y = \sqrt{v_{0y}^2 + 2a_y y} = \sqrt{(0 \text{ m/s})^2 + 2(9.80 \text{ m/s}^2)(10.0 \text{ m})} = 14.0 \text{ m/s}$$

Using this result, we can find the balloon's speed 0.0292 seconds before it hits the man by solving Equation 3.3b ($v_y = v_{0y} + a_y t$) for v_{0y}:

$$v_{0y} = v_y - a_y t = \left[(14.0 \text{ m/s}) - (9.80 \text{ m/s}^2)(0.0292 \text{ s}) \right] = 13.7 \text{ m/s}$$

Finally, we can find the desired distance y above the man's head from Equation 3.5b:

$$y = v_{0y} t + \frac{1}{2} a_y t^2 = (13.7 \text{ m/s})(0.0292 \text{ s}) + \frac{1}{2}(9.80 \text{ m/s}^2)(0.0292 \text{ s})^2 = \boxed{0.404 \text{ m}}$$

---

53. ***REASONING AND SOLUTION*** Since the sound radiates uniformly in all directions, at a distance r from the source, the energy of the sound wave is distributed over the area of a sphere of radius r. Therefore, according to $I = \dfrac{P}{4\pi r^2}$ (Equation 16.9) with $r = 3.8$ m, the power radiated from the source is

$$P = 4\pi I r^2 = 4\pi (3.6 \times 10^{-2} \text{ W/m}^2)(3.8 \text{ m})^2 = \boxed{6.5 \text{ W}}$$

57. **_REASONING AND SOLUTION_** According to Equation 16.8, the power radiated by the speaker is $P = IA = I\pi r^2$, where r is the radius of the circular opening. Thus, the radiated power is

$$P = (17.5 \text{ W/m}^2)(\pi)(0.0950 \text{ m})^2 = 0.496 \text{ W}$$

As a percentage of the electrical power, this is

$$\frac{0.496 \text{ W}}{25.0 \text{ W}} \times 100 \% = \boxed{1.98 \%}$$

59. **_REASONING_** Intensity I is power P divided by the area A, or $I = \dfrac{P}{A}$, according to Equation 16.8. The area is given directly, but the power is not. Therefore, we need to recast this expression in terms of the data given in the problem. Power is the change in energy per unit time, according to Equation 6.10b. In this case the energy is the heat Q that causes the temperature of the lasagna to increase. Thus, the power is $P = \dfrac{Q}{t}$, where t denotes the time. As a result, Equation 16.8 for the intensity becomes

$$I = \frac{P}{A} = \frac{Q}{t\,A} \tag{1}$$

According to Equation 12.4, the heat that must be supplied to increase the temperature of a substance of mass m by an amount ΔT is $Q = cm\Delta T$, where c is the specific heat capacity. Substituting this expression into Equation (1) gives

$$I = \frac{Q}{t\,A} = \frac{cm\Delta T}{t\,A} \tag{2}$$

SOLUTION Equation (2) reveals that the intensity of the microwaves is

$$I = \frac{cm\Delta T}{t\,A} = \frac{\left[3200 \text{ J/}(\text{kg}\cdot\text{C}^\circ)\right](0.35 \text{ kg})(72 \text{ C}^\circ)}{(480 \text{ s})\left(2.2\times10^{-2} \text{ m}^2\right)} = \boxed{7.6\times10^3 \text{ W/m}^2}$$

65. **_REASONING_** This is a situation in which the intensities I_{man} and I_{woman} (in watts per square meter) detected by the man and the woman are compared using the intensity level β, expressed in decibels. This comparison is based on Equation 16.10, which we rewrite as follows:

$$\beta = (10 \text{ dB})\log\left(\frac{I_{\text{man}}}{I_{\text{woman}}}\right)$$

SOLUTION Using Equation 16.10, we have

$$\beta = 7.8 \text{ dB} = (10 \text{ dB}) \log\left(\frac{I_{\text{man}}}{I_{\text{woman}}}\right) \quad \text{or} \quad \log\left(\frac{I_{\text{man}}}{I_{\text{woman}}}\right) = \frac{7.8 \text{ dB}}{10 \text{ dB}} = 0.78$$

Solving for the intensity ratio gives

$$\frac{I_{\text{man}}}{I_{\text{woman}}} = 10^{0.78} = \boxed{6.0}$$

71. **REASONING** If I_1 is the sound intensity produced by a single person, then NI_1 is the sound intensity generated by N people. The sound intensity level generated by N people is given by Equation 16.10 as

$$\beta_N = (10 \text{ dB}) \log\left(\frac{NI_1}{I_0}\right)$$

where I_0 is the threshold of hearing. Solving this equation for N yields

$$N = \left(\frac{I_0}{I_1}\right) 10^{\frac{\beta_N}{10 \text{ dB}}} \tag{1}$$

We also know that the sound intensity level for one person is

$$\beta_1 = (10 \text{ dB}) \log\left(\frac{I_1}{I_0}\right) \quad \text{or} \quad I_1 = I_0 \, 10^{\frac{\beta_1}{10 \text{ dB}}} \tag{2}$$

Equations (1) and (2) are all that we need in order to find the number of people at the football game.

SOLUTION Substituting the expression for I_1 from Equation (2) into Equation (1) gives the desired result.

$$N = \frac{I_0 \, 10^{\frac{\beta_N}{10 \text{ dB}}}}{I_0 \, 10^{\frac{\beta_1}{10 \text{ dB}}}} = \frac{10^{\frac{109 \text{ dB}}{10 \text{ dB}}}}{10^{\frac{60.0 \text{ dB}}{10 \text{ dB}}}} = \boxed{79\ 400}$$

75. ***REASONING AND SOLUTION*** The sound intensity level β in decibels (dB) is related to the sound intensity I according to Equation 16.10, $\beta = (10 \text{ dB}) \log (I / I_0)$, where the quantity I_0 is the reference intensity. According to the problem statement, when the sound intensity level triples, the sound intensity also triples; therefore,

$$3\beta = (10 \text{ dB}) \log \left(\frac{3I}{I_0} \right)$$

Then,

$$3\beta - \beta = (10 \text{ dB}) \log \left(\frac{3I}{I_0} \right) - (10 \text{ dB}) \log \left(\frac{I}{I_0} \right) = (10 \text{ dB}) \log \left(\frac{3I / I_0}{I / I_0} \right)$$

Thus, $2\beta = (10 \text{ dB}) \log 3$ and

$$\beta = (5 \text{ dB}) \log 3 = \boxed{2.39 \text{ dB}}$$

---

77. ***REASONING*** Since you detect a frequency that is smaller than that emitted by the car when the car is stationary, the car must be moving away from you. Therefore, according to Equation 16.12, the frequency f_0 heard by a stationary observer from a source moving away from the observer is given by

$$f_0 = f_s \left(\frac{1}{1 + \dfrac{v_s}{v}} \right)$$

where f_s is the frequency emitted from the source when it is stationary with respect to the observer, v is the speed of sound, and v_s is the speed of the moving source. This expression can be solved for v_s.

SOLUTION We proceed to solve for v_s and substitute the data given in the problem statement. Rearrangement gives

$$\frac{v_s}{v} = \frac{f_s}{f_0} - 1$$

Solving for v_s and noting that $f_0 / f_s = 0.86$ yields

$$v_s = v \left(\frac{f_s}{f_0} - 1 \right) = (343 \text{ m/s}) \left(\frac{1}{0.86} - 1 \right) = \boxed{56 \text{ m/s}}$$

---

83. ***REASONING*** The Doppler shift that occurs here arises because both the source and the observer of the sound are moving. Therefore, the expression for the Doppler-shifted observed frequency f_o is given by Equation 16.15 as

$$f_o = f_s \left(\frac{1 \pm v_o / v}{1 \mp v_s / v} \right)$$

where f_s is the frequency emitted by the source, v_o is the speed of the observer, v_s is the speed of the source, and v is the speed of sound. The observer is moving toward the source, so we use the plus sign in the numerator. The source is moving toward the observer, so we use the minus sign in the denominator. Thus, Equation 16.15 becomes

$$f_o = f_s \left(\frac{1 + v_o / v}{1 - v_s / v} \right)$$

Recognizing that both trucks move at the same speed, we can substitute $v_o = v_s = v_{Truck}$ and solve for v_{Truck}.

SOLUTION Using Equation 16.15 as described in the ***REASONING*** and substituting $v_o = v_s = v_{Truck}$, we have

$$f_o = f_s \left(\frac{1 + v_{Truck} / v}{1 - v_{Truck} / v} \right) \quad \text{or} \quad \frac{f_o}{f_s} - \left(\frac{f_o}{f_s} \right) \left(\frac{v_{Truck}}{v} \right) = 1 + \frac{v_{Truck}}{v}$$

Rearranging, with a view toward solving for v_{Truck}/v, gives

$$\frac{f_o}{f_s} - 1 = \frac{v_{Truck}}{v} + \left(\frac{f_o}{f_s} \right) \left(\frac{v_{Truck}}{v} \right) \quad \text{or} \quad \frac{v_{Truck}}{v} \left(1 + \frac{f_o}{f_s} \right) = \frac{f_o}{f_s} - 1$$

Finally, we obtain

$$\frac{v_{Truck}}{v} = \frac{\frac{f_o}{f_s} - 1}{1 + \frac{f_o}{f_s}} = \frac{1.14 - 1}{1 + 1.14} = \frac{0.14}{2.14} \quad \text{or} \quad v_{Truck} = \left(\frac{0.14}{2.14} \right) (343 \text{ m/s}) = \boxed{22 \text{ m/s}}$$

87. **REASONING**

a. Since the two submarines are approaching each other head on, the frequency f_o detected by the observer (sub B) is related to the frequency f_s emitted by the source (sub A) by

$$f_o = f_s \left(\frac{1 + \dfrac{v_o}{v}}{1 - \dfrac{v_s}{v}} \right) \qquad (16.15)$$

where v_o and v_s are the speed of the observer and source, respectively, and v is the speed of the underwater sound

b. The sound reflected from submarine B has the same frequency that it detects, namely, f_o. Now sub B becomes the source of sound and sub A is the observer. We can still use Equation 16.15 to find the frequency detected by sub A.

SOLUTION

a. The frequency f_o detected by sub B is

$$f_o = f_s \left(\frac{1 + \dfrac{v_o}{v}}{1 - \dfrac{v_s}{v}} \right) = (1550 \text{ Hz}) \left(\frac{1 + \dfrac{8 \text{ m/s}}{1522 \text{ m/s}}}{1 - \dfrac{12 \text{ m/s}}{1522 \text{ m/s}}} \right) = \boxed{1570 \text{ Hz}}$$

b. The sound reflected from submarine B has the same frequency that it detects, namely, 1570 Hz. Now sub B is the source of sound whose frequency is $f_s = 1570$ Hz. The speed of sub B is $v_s = 8$ m/s. The frequency detected by sub A (whose speed is $v_o = 12$ m/s) is

$$f_o = f_s \left(\frac{1 + \dfrac{v_o}{v}}{1 - \dfrac{v_s}{v}} \right) = (1570 \text{ Hz}) \left(\frac{1 + \dfrac{12 \text{ m/s}}{1522 \text{ m/s}}}{1 - \dfrac{8 \text{ m/s}}{1522 \text{ m/s}}} \right) = \boxed{1590 \text{ Hz}}$$

---

91. **REASONING AND SOLUTION** The speed of sound in a liquid is given by Equation 16.6, $v = \sqrt{B_{ad}/\rho}$, where B_{ad} is the adiabatic bulk modulus and ρ is the density of the liquid. Solving for B_{ad}, we obtain $B_{ad} = v^2 \rho$. Values for the speed of sound in fresh water and in ethyl alcohol are given in Table 16.1. The ratio of the adiabatic bulk modulus of fresh water to that of ethyl alcohol at 20°C is, therefore,

$$\frac{\left(B_{ad}\right)_{water}}{\left(B_{ad}\right)_{ethyl\ alcohol}} = \frac{v_{water}^{2}\rho_{water}}{v_{ethyl\ alcohol}^{2}\rho_{ethyl\ alcohol}} = \frac{(1482\ m/s)^{2}(998\ kg/m^{3})}{(1162\ m/s)^{2}(789\ kg/m^{3})} = \boxed{2.06}$$

93. **REASONING** The tension F in the violin string can be found by solving Equation 16.2 for F to obtain $F = mv^{2}/L$, where v is the speed of waves on the string and can be found from Equation 16.1 as $v = f\lambda$.

SOLUTION Combining Equations 16.2 and 16.1 and using the given data, we obtain

$$F = \frac{mv^{2}}{L} = (m/L)f^{2}\lambda^{2} = \left(7.8\times10^{-4}\ kg/m\right)(440\ Hz)^{2}\left(65\times10^{-2}\ m\right)^{2} = \boxed{64\ N}$$

95. **REASONING AND SOLUTION** The intensity level β in decibels (dB) is related to the sound intensity I according to Equation 16.10:

$$\beta = (10\ dB)\log\left(\frac{I}{I_{0}}\right)$$

where the quantity I_{0} is the reference intensity. Therefore, we have

$$\beta_{2} - \beta_{1} = (10\ dB)\log\left(\frac{I_{2}}{I_{0}}\right) - (10\ dB)\log\left(\frac{I_{1}}{I_{0}}\right) = (10\ dB)\log\left(\frac{I_{2}/I_{0}}{I_{1}/I_{0}}\right) = (10\ dB)\log\left(\frac{I_{2}}{I_{1}}\right)$$

Solving for the ratio I_{2}/I_{1}, we find

$$30.0\ dB = (10\ dB)\log\left(\frac{I_{2}}{I_{1}}\right) \qquad or \qquad \frac{I_{2}}{I_{1}} = 10^{3.0} = 1000$$

Thus, we conclude that the sound intensity $\boxed{\text{increases by a factor of } 1000}$.

97. **REASONING** As the transverse wave propagates, the colored dot moves up and down in simple harmonic motion with a frequency of 5.0 Hz. The amplitude (1.3 cm) is the magnitude of the maximum displacement of the dot from its equilibrium position.

SOLUTION The period T of the simple harmonic motion of the dot is $T = 1/f = 1/(5.0\ Hz) = 0.20\ s$. In one period the dot travels through one complete cycle of its motion, and covers a vertical distance of $4\times(1.3\ cm) = 5.2\ cm$. Therefore, in 3.0 s the dot will have traveled a *total vertical distance* of

$$\left(\frac{3.0\ s}{0.20\ s}\right)(5.2\ cm) = \boxed{78\ cm}$$

99. **REASONING** You hear a frequency f_o that is 1.0% lower than the frequency f_s emitted by the source. This means that the frequency you observe is 99.0% of the emitted frequency, so that $f_o = 0.990\,f_s$. You are an observer who is moving away from a stationary source of sound. Therefore, the Doppler-shifted frequency that you observe is specified by Equation 16.14, which can be solved for the bicycle speed v_o.

SOLUTION Equation 16.14, in which v denotes the speed of sound, states that

$$f_o = f_s \left(1 - \frac{v_o}{v} \right)$$

Solving for v_o and using the fact that $f_o = 0.990\,f_s$ reveal that

$$v_o = v \left(1 - \frac{f_o}{f_s} \right) = (343 \text{ m/s}) \left(1 - \frac{0.990\,f_s}{f_s} \right) = \boxed{3.4 \text{ m/s}}$$

---

101. **REASONING** The intensity level β is related to the sound intensity I according to Equation 16.10:

$$\beta = (10 \text{ dB}) \log \left(\frac{I}{I_0} \right)$$

where I_0 is the reference level sound intensity. Solving for I gives

$$I = I_0 10^{\frac{\beta}{10\,\text{dB}}}$$

SOLUTION Taking the ratio of the sound intensity I_{rock} at the rock concert to the intensity I_{jazz} at the jazz fest gives

$$\frac{I_{\text{rock}}}{I_{\text{jazz}}} = \frac{\cancel{I_0}\, 10^{\frac{\beta_{\text{rock}}}{10\,\text{dB}}}}{\cancel{I_0}\, 10^{\frac{\beta_{\text{jazz}}}{10\,\text{dB}}}} = \frac{10^{\frac{115 \text{ dB}}{10\,\text{dB}}}}{10^{\frac{95 \text{ dB}}{10\,\text{dB}}}} = \boxed{1.0 \times 10^2}$$

107. *REASONING* Newton's second law can be used to analyze the motion of the blocks using the methods developed in Chapter 4. We can thus determine an expression that relates the magnitude P of the pulling force to the magnitude F of the tension in the wire. Equation 16.2 $[v = \sqrt{F/(m/L)}]$ can then be used to find the tension in the wire.

SOLUTION The following drawings show a schematic of the situation described in the problem and the free-body diagrams for each block, where $m_1 = 42.0$ kg and $m_2 = 19.0$ kg. The pulling force is **P**, and the tension in the wire gives rise to the forces **F** and **−F**, which act on m_1 and m_2, respectively.

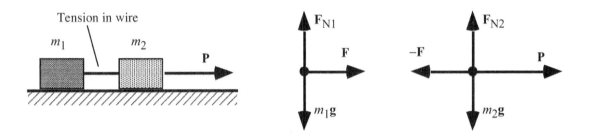

Newton's second law for block 1 is, taking forces that point to the right as positive, $F = m_1 a$, or $a = F/m_1$. For block 2, we obtain $P - F = m_2 a$. Using the expression for a obtained from the equation for block 1, we have

$$P - F = F\left(\frac{m_2}{m_1}\right) \qquad \text{or} \qquad P = F\left(\frac{m_2}{m_1}\right) + F = F\left(\frac{m_2}{m_1} + 1\right)$$

According to Equation 16.2, $F = v^2(m/L)$, where m/L is the mass per unit length of the wire. Combining this expression for F with the expression for P, we have

$$P = v^2(m/L)\left(\frac{m_2}{m_1} + 1\right) = (352 \text{ m/s})^2 (8.50 \times 10^{-4} \text{ kg/m})\left(\frac{19.0 \text{ kg}}{42.0 \text{ kg}} + 1\right) = \boxed{153 \text{ N}}$$

---

111. SSM *REASONING* As the boat moves away from the dock, it is traveling in the same direction as the sound wave (see the figure). Therefore, an observer on the moving boat intercepts fewer condensations and rarefactions per second than someone who is stationary. Consequently, the moving observer hears a frequency f_o that is smaller than the frequency

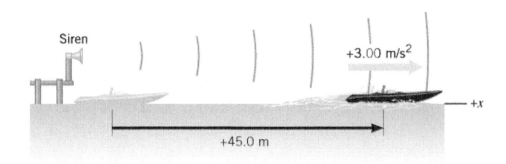

emitted by the siren. The frequency f_o depends on the frequency of the siren and the speed of sound—both of which are known—and on the speed of the boat. The speed of the boat is the magnitude of its velocity, which can be determined by using one of the equations of kinematics.

SOLUTION In this situation we have a stationary source (the siren) and an observer (the person in the boat) who is moving away from the siren. Therefore, the frequency f_0 heard by the moving observer is given by Equation 16.14: $f_o = f_s\left(1 - \dfrac{v_o}{v}\right)$. In this expression, f_s is the frequency of the sound emitted by the stationary siren, v_0 is the speed of the moving observer (the speed of the boat), and v is the speed of sound. The frequency f_s is known, as is the speed of sound ($v = 343$ m/s at 20 °C); see Table 16.1). The speed v_0 of the moving observer will be obtained in the next step.

To determine the velocity of the observer, we note that the initial velocity $v_{0,o}$, acceleration a_x and displacement x are known. Therefore, we turn to Equation 2.9 of the equations of kinematics, which relates these variables to the final velocity v_o of the moving observer by

$$v_o^2 = v_{0,o}^2 + 2a_x x.$$

Taking the square root of each side of this equation gives: $v_o = \sqrt{v_{0,o}^2 + 2a_x x}$. In taking the square root, we have chosen the positive root because the boat is moving in the $+x$ direction. Since the velocity is positive, this expression also gives the magnitude of the velocity, which is the speed of the observer. The expression for v_o can be substituted into Equation 16.14 above. All the variables are known, so the frequency f_o heard by the moving observer can be found.

$$f_o = f_s\left(1 - \frac{v_o}{v}\right) = f_s\left(1 - \frac{\sqrt{v_{0,o}^2 + 2a_x x}}{v}\right)$$

Thus, the frequency of sound heard by the moving observer is:

$$f_o = f_s\left(1 - \frac{\sqrt{v_{0,o}^2 + 2a_x x}}{v}\right) = (755 \text{ Hz})\left[1 - \frac{\sqrt{(0 \text{ m/s})^2 + 2(+3.00 \text{ m/s}^2)(+45.0 \text{ m})}}{343 \text{ m/s}}\right] = \boxed{719 \text{ Hz}}$$

113. [SSM] *CONCEPTS* **(i)** No. The wavelength changes only when the source of the sound is moving. The siren is stationary, so the wavelength does not change. **(ii)** The car and the reflected sound are traveling in opposite directions, the car to the right and the reflected sound to the left. The car intercepts more wave cycles per second than if the car were stationary. Consequently, the driver hears a frequency greater than 2140 Hz. **(iii)** The car and the direct sound are traveling in the same direction. As the direct sound passes the car, the number of wave cycles per second intercepted by the driver is less than if the car were stationary. Thus, the driver hears a frequency that is less than 2140 Hz.

CALCULATIONS **(a)** For the reflected sound, the frequency f_0 that the driver (the "observer") hears is equal to the frequency f_s of the waves emitted by the siren *plus* and additional number of cycles per second, because the car and the reflected sound are moving in opposite directions. The additional number of cycles per second is v_0/λ, where v_0 is the speed of the car and λ is the wavelength of the sound (see the subsection "Moving Observer" in Section 16.9). According to Equation 16.1, the wavelength is equal to the speed of sound v divided by the frequency of the siren, $\lambda = v/\lambda_s$. Thus the frequency heard by the driver can be written as

$$f_0 = f_s + \frac{v_0}{\lambda} = f_s + \frac{v_0}{v/f_s}$$

$$= f_s\left(1 + \frac{v_0}{v}\right) = (2140 \text{ Hz})\left(1 + \frac{27.0 \text{ m/s}}{343 \text{ m/s}}\right) = \boxed{2310 \text{ Hz}}$$

(b) For the direct sound, the frequency f_0 that the driver hears is equal to the frequency f_s of the waves emitted by the siren *minus* v_0/λ, because the car and the direct sound are moving in the same direction:

$$f_0 = f_s - \frac{v_0}{\lambda} = f_s - \frac{v_0}{v/f_s}$$

$$= f_s\left(1 - \frac{v_0}{v}\right) = (2140 \text{ Hz})\left(1 - \frac{27.0 \text{ m/s}}{343 \text{ m/s}}\right) = \boxed{1970 \text{ Hz}}$$

As expected, for the reflected wave, the driver hears a frequency greater than 2140 Hz, while for the direct sound he hears a frequency less than 2140 Hz.

CHAPTER 17 | *THE PRINCIPLE OF LINEAR SUPERPOSITION AND INTERFERENCE PHENOMENA*

3. **REASONING** According to the principle of linear superposition, when two or more waves are present simultaneously at the same place, the resultant wave is the sum of the individual waves. We will use the fact that both pulses move at a speed of 1 cm/s to locate the pulses at the times $t = 1$ s, 2 s, 3 s, and 4 s and, by applying this principle to the places where the pulses overlap, determine the shape of the string.

   SOLUTION The shape of the string at each time is shown in the following drawings:

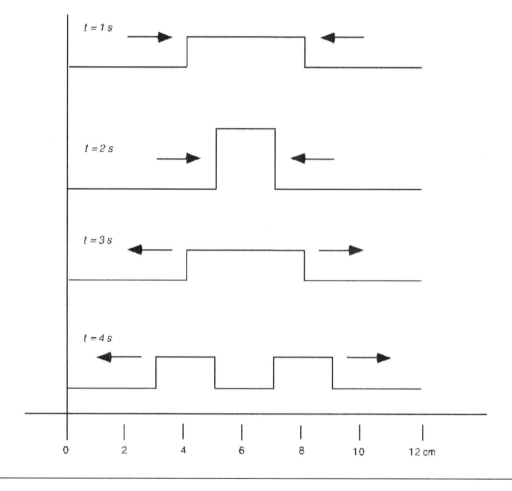

5. **REASONING** According to the principle of linear superposition, the resultant displacement due to two waves is the sum of the displacements due to each wave. In order to find the net displacement at the stated time and positions, then, we will calculate the individual displacements y_1 and y_2 and then find their sum. We note that the phase angles are measured

in radians rather than degrees, so calculators must be set to the radian mode in order to yield valid results.

SOLUTION

a. At $t = 4.00$ s and $x = 2.16$ m, the net displacement y of the string is

$$y = y_1 + y_2$$

$$= (24.0 \text{ mm}) \sin\left[(9.00\pi \text{ rad/s})(4.00 \text{ s}) - (1.25\pi \text{ rad/m})(2.16 \text{ m})\right]$$

$$+ (35.0 \text{ mm}) \sin\left[(2.88\pi \text{ rad/s})(4.00 \text{ s}) + (0.400\pi \text{ rad/m})(2.16 \text{ m})\right]$$

$$= \boxed{+13.3 \text{ mm}}$$

b. The time is still $t = 4.00$ s, but the position is now $x = 2.56$ m. Therefore, the net displacement y is

$$y = y_1 + y_2$$

$$= (24.0 \text{ mm}) \sin\left[(9.00\pi \text{ rad/s})(4.00 \text{ s}) - (1.25\pi \text{ rad/m})(2.56 \text{ m})\right]$$

$$+ (35.0 \text{ mm}) \sin\left[(2.88\pi \text{ rad/s})(4.00 \text{ s}) + (0.400\pi \text{ rad/m})(2.56 \text{ m})\right]$$

$$= \boxed{+48.8 \text{ mm}}$$

7. **REASONING** The geometry of the positions of the loudspeakers and the listener is shown in the following drawing.

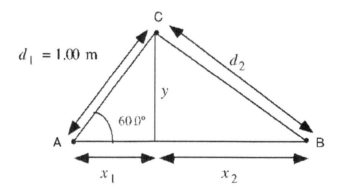

The listener at C will hear either a loud sound or no sound, depending upon whether the interference occurring at C is constructive or destructive. If the listener hears no sound, destructive interference occurs, so

$$d_2 - d_1 = \frac{n\lambda}{2} \qquad n = 1, 3, 5, \text{K} \qquad (1)$$

SOLUTION Since $v = \lambda f$, according to Equation 16.1, the wavelength of the tone is

$$\lambda = \frac{v}{f} = \frac{343 \text{ m/s}}{68.6 \text{ Hz}} = 5.00 \text{ m}$$

Speaker B will be closest to Speaker A when $n = 1$ in Equation (1) above, so

$$d_2 = \frac{n\lambda}{2} + d_1 = \frac{5.00 \text{ m}}{2} + 1.00 \text{ m} = 3.50 \text{ m}$$

From the figure above we have that,

$$x_1 = (1.00 \text{ m}) \cos 60.0^\circ = 0.500 \text{ m}$$

$$y = (1.00 \text{ m}) \sin 60.0^\circ = 0.866 \text{ m}$$

Then

$$x_2^2 + y^2 = d_2^2 = (3.50 \text{ m})^2 \qquad \text{or} \qquad x_2 = \sqrt{(3.50 \text{ m})^2 - (0.866 \text{ m})^2} = 3.39 \text{ m}$$

Therefore, the closest that speaker A can be to speaker B so that the listener hears no sound is $x_1 + x_2 = 0.500 \text{ m} + 3.39 \text{ m} = \boxed{3.89 \text{ m}}$.

---

13. **REASONING** Equation 17.1 specifies the diffraction angle θ according to $\sin \theta = \lambda / D$, where λ is the wavelength of the sound and D is the width of the opening. The wavelength depends on the speed and frequency of the sound. Since the frequency is the same in winter and summer, only the speed changes with the temperature. We can account for the effect of the temperature on the speed by assuming that the air behaves as an ideal gas, for which the speed of sound is proportional to the square root of the Kelvin temperature.

SOLUTION Equation 17.1 indicates that

$$\sin \theta = \frac{\lambda}{D}$$

Into this equation, we substitute $\lambda = v / f$ (Equation 16.1), where v is the speed of sound and f is the frequency:

$$\sin \theta = \frac{\lambda}{D} = \frac{v / f}{D}$$

Assuming that air behaves as an ideal gas, we can use $v = \sqrt{\gamma k T / m}$ (Equation 16.5), where γ is the ratio of the specific heat capacities at constant pressure and constant volume, k is Boltzmann's constant, T is the Kelvin temperature, and m is the average mass of the molecules and atoms of which the air is composed:

$$\sin\theta = \frac{v}{fD} = \frac{1}{fD}\sqrt{\frac{\gamma kT}{m}}$$

Applying this result for each temperature gives

$$\sin\theta_{summer} = \frac{1}{fD}\sqrt{\frac{\gamma kT_{summer}}{m}} \quad \text{and} \quad \sin\theta_{winter} = \frac{1}{fD}\sqrt{\frac{\gamma kT_{winter}}{m}}$$

Dividing the summer-equation by the winter-equation, we find

$$\frac{\sin\theta_{summer}}{\sin\theta_{winter}} = \frac{\dfrac{1}{fD}\sqrt{\dfrac{\gamma kT_{summer}}{m}}}{\dfrac{1}{fD}\sqrt{\dfrac{\gamma kT_{winter}}{m}}} = \sqrt{\frac{T_{summer}}{T_{winter}}}$$

Thus, it follows that

$$\sin\theta_{summer} = \sin\theta_{winter}\sqrt{\frac{T_{summer}}{T_{winter}}} = \sin 15.0°\sqrt{\frac{311\text{ K}}{273\text{ K}}} = 0.276 \quad \text{or} \quad \theta_{summer} = \sin^{-1}(0.276) = \boxed{16.0°}$$

---

19. **REASONING** The beat frequency of two sound waves is the difference between the two sound frequencies. From the graphs, we see that the period of the wave in the upper text figure is 0.020 s, so its frequency is $f_1 = 1/T_1 = 1/(0.020\text{ s}) = 5.0 \times 10^1$ Hz. The frequency of the wave in the lower figure is $f_2 = 1/(0.024\text{ s}) = 4.2 \times 10^1$ Hz.

SOLUTION The beat frequency of the two sound waves is

$$f_{beat} = f_1 - f_2 = 5.0 \times 10^1 \text{ Hz} - 4.2 \times 10^1 \text{ Hz} = \boxed{8\text{ Hz}}$$

---

23. **REASONING** When two frequencies are sounded simultaneously, the beat frequency produced is the difference between the two. Thus, knowing the beat frequency between the tuning fork and one flute tone tells us only the difference between the known frequency and the tuning-fork frequency. It does not tell us whether the tuning-fork frequency is greater or smaller than the known frequency. However, two different beat frequencies and two flute frequencies are given. Consideration of both beat frequencies will enable us to find the tuning-fork frequency.

SOLUTION The fact that a 1-Hz beat frequency is heard when the tuning fork is sounded along with the 262-Hz tone implies that the tuning-fork frequency is either 263 Hz or 261 Hz. We can eliminate one of these values by considering the fact that a 3-Hz beat frequency is heard when the tuning fork is sounded along with the 266-Hz tone. This implies that the tuning-fork frequency is either 269 Hz or 263 Hz. Thus, the tuning-fork frequency must be $\boxed{263\text{ Hz}}$.

29. ***REASONING*** The fundamental frequency f_1 is given by Equation 17.3 with $n = 1$: $f_1 = v/(2L)$. Since values for f_1 and L are given in the problem statement, we can use this expression to find the speed of the waves on the cello string. Once the speed is known, the tension F in the cello string can be found by using Equation 16.2, $v = \sqrt{F/(m/L)}$.

SOLUTION Combining Equations 17.3 and 16.2 yields

$$2Lf_1 = \sqrt{\frac{F}{m/L}}$$

Solving for F, we find that the tension in the cello string is

$$F = 4L^2 f_1^2 (m/L) = 4(0.800 \text{ m})^2 (65.4 \text{ Hz})^2 (1.56 \times 10^{-2} \text{ kg/m}) = \boxed{171 \text{ N}}$$

---

31. ***REASONING*** According to Equation 17.3, the fundamental ($n = 1$) frequency of a string fixed at both ends is related to the wave speed v by $f_1 = v/2L$, where L is the length of the string. Thus, the speed of the wave is $v = 2Lf_1$. Combining this with Equation 16.2, $v = \sqrt{F/(m/L)}$, we find, after some rearranging, that

$$\frac{F}{L^2} = 4f_1^2 (m/L)$$

Since the strings have the same tension and the same lengths between their fixed ends, we have

$$f_{1E}^2 (m/L)_E = f_{1G}^2 (m/L)_G$$

where the symbols "E" and "G" represent the E and G strings on the violin. This equation can be solved for the linear density of the G string.

SOLUTION The linear density of the string is

$$(m/L)_G = \frac{f_{1E}^2}{f_{1G}^2} (m/L)_E = \left(\frac{f_{1E}}{f_{1G}}\right)^2 (m/L)_E$$

$$= \left(\frac{659.3 \text{ Hz}}{196.0 \text{ Hz}}\right)^2 (3.47 \times 10^{-4} \text{ kg/m}) = \boxed{3.93 \times 10^{-3} \text{ kg/m}}$$

37. *REASONING*   We can find the extra length that the D-tuner adds to the E-string by calculating the length of the D-string and then subtracting from it the length of the E string. For standing waves on a string that is fixed at both ends, Equation 17.3 gives the frequencies as $f_n = n(v/2L)$. The ratio of the fundamental frequency of the D-string to that of the E-string is

$$\frac{f_D}{f_E} = \frac{v/(2L_D)}{v/(2L_E)} = \frac{L_E}{L_D}$$

This expression can be solved for the length L_D of the D-string in terms of quantities given in the problem statement.

SOLUTION   The length of the D-string is

$$L_D = L_E \left(\frac{f_E}{f_D} \right) = (0.628 \text{ m}) \left(\frac{41.2 \text{ Hz}}{36.7 \text{ Hz}} \right) = 0.705 \text{ m}$$

The length of the E-string is extended by the D-tuner by an amount

$$L_D - L_E = 0.705 \text{ m} - 0.628 \text{ m} = \boxed{0.077 \text{ m}}$$

---

39. *REASONING*   The beat frequency produced when the piano and the other instrument sound the note (three octaves higher than middle C) is $f_{\text{beat}} = f - f_0$, where f is the frequency of the piano and f_0 is the frequency of the other instrument ($f_0 = 2093 \text{ Hz}$). We can find f by considering the temperature effects and the mechanical effects that occur when the temperature drops from 25.0 °C to 20.0 °C.

SOLUTION   The fundamental frequency f_0 of the wire at 25.0 °C is related to the tension F_0 in the wire by

$$f_0 = \frac{v}{2L_0} = \frac{\sqrt{F_0/(m/L)}}{2L_0} \tag{1}$$

where Equations 17.3 and 16.2 have been combined.

The amount ΔL by which the piano wire attempts to contract is (see Equation 12.2) $\Delta L = \alpha L_0 \Delta T$, where α is the coefficient of linear expansion of the wire, L_0 is its length at 25.0 °C, and ΔT is the amount by which the temperature drops. Since the wire is prevented from contracting, there must be a stretching force exerted at each end of the wire. According to Equation 10.17, the magnitude of this force is

$$\Delta F = Y \left(\frac{\Delta L}{L_0} \right) A$$

where Y is the Young's modulus of the wire, and A is its cross-sectional area. Combining this relation with Equation 12.2, we have

$$\Delta F = Y \left(\frac{\alpha L_0 \Delta T}{L_0} \right) A = \alpha (\Delta T) Y A$$

Thus, the frequency f at the lower temperature is

$$f = \frac{v}{2L_0} = \frac{\sqrt{(F_0 + \Delta F)/(m/L)}}{2L_0} = \frac{\sqrt{\left[F_0 + \alpha (\Delta T) Y A \right]/(m/L)}}{2L_0} \qquad (2)$$

Using Equations (1) and (2), we find that the frequency f is

$$f = f_0 \frac{\sqrt{\left[F_0 + \alpha (\Delta T) Y A \right]/(m/L)}}{\sqrt{F_0/(m/L)}} = f_0 \sqrt{\frac{F_0 + \alpha (\Delta T) Y A}{F_0}}$$

$$f = (2093 \text{ Hz}) \sqrt{\frac{818.0 \text{ N} + (12 \times 10^{-6}/\text{C}^\circ)(5.0 \text{ C}^\circ)(2.0 \times 10^{11} \text{ N/m}^2)(7.85 \times 10^{-7} \text{ m}^2)}{818.0 \text{ N}}}$$

$$= 2105 \text{ Hz}$$

Therefore, the beat frequency is $2105 \text{ Hz} - 2093 \text{ Hz} = \boxed{12 \text{ Hz}}$.

---

41. **REASONING** Equation 17.5 (with $n = 1$) gives the fundamental frequency as $f_1 = v/(4L)$, where L is the length of the auditory canal and v is the speed of sound.

SOLUTION Using Equation 17.5, we obtain

$$f_1 = \frac{v}{4L} = \frac{343 \text{ m/s}}{4(0.029 \text{ m})} = \boxed{3.0 \times 10^3 \text{ Hz}}$$

---

47. **REASONING** The natural frequencies of a tube open at only one end are given by Equation 17.5 as $f_n = n \left(\dfrac{v}{4L} \right)$, where n is any odd integer ($n = 1, 3, 5, \ldots$), v is the speed of sound, and L is the length of the tube. We can use this relation to find the value for n for the 450-Hz sound and to determine the length of the pipe.

SOLUTION

a. The frequency f_n of the 450-Hz sound is given by $450 \text{ Hz} = n\left(\dfrac{v}{4L}\right)$. Likewise, the frequency of the next higher harmonic is $750 \text{ Hz} = (n+2)\left(\dfrac{v}{4L}\right)$, because n is an odd integer and this means that the value of n for the next higher harmonic must be $n + 2$. Taking the ratio of these two relations gives

$$\frac{750 \text{ Hz}}{450 \text{ Hz}} = \frac{(n+2)\left(\dfrac{v}{4L}\right)}{n\left(\dfrac{v}{4L}\right)} = \frac{n+2}{n}$$

Solving this equation for n gives $n = \boxed{3}$.

b. Solving the equation $450 \text{ Hz} = n\left(\dfrac{v}{4L}\right)$ for L and using $n = 3$, we find that the length of the tube is

$$L = n\left(\frac{v}{4f_n}\right) = 3\left[\frac{343 \text{ m/s}}{4(450 \text{ Hz})}\right] = \boxed{0.57 \text{ m}}$$

---

49. **REASONING** The well is open at the top and closed at the bottom, so it can be approximated as a column of air that is open at only one end. According to Equation 17.5, the natural frequencies for such an air column are

$$f_n = n\left(\frac{v}{4L}\right) \quad \text{where} \quad n = 1, 3, 5, \text{K}$$

The depth L of the well can be calculated from the speed of sound, $v = 343$ m/s, and a knowledge of the natural frequencies f_n.

SOLUTION We know that two of the natural frequencies are 42 and 70.0 Hz. The ratio of these two frequencies is

$$\frac{70.0 \text{ Hz}}{42 \text{ Hz}} = \frac{5}{3}$$

Therefore, the value of n for each frequency is $n = 3$ for the 42-Hz sound, and $n = 5$ for the 70.0-Hz sound. Using $n = 3$, for example, the depth of the well is

$$L = \frac{nv}{4f_3} = \frac{3(343 \text{ m/s})}{4(42 \text{ Hz})} = \boxed{6.1 \text{ m}}$$

---

55. **REASONING** When constructive interference occurs again at point C, the path length difference is two wavelengths, or $\Delta s = 2\lambda = 3.20 \text{ m}$. Therefore, we can write the expression for the path length difference as

$$s_{AC} - s_{BC} = \sqrt{s_{AB}^2 + s_{BC}^2} - s_{BC} = 3.20 \text{ m}$$

This expression can be solved for s_{AB}.

SOLUTION Solving for s_{AB}, we find that

$$s_{AB} = \sqrt{(3.20 \text{ m} + 2.40 \text{ m})^2 - (2.40 \text{ m})^2} = \boxed{5.06 \text{ m}}$$

---

59. **REASONING** For standing waves on a string that is clamped at both ends, Equations 17.3 and 16.2 indicate that the standing wave frequencies are

$$f_n = n\left(\frac{v}{2L}\right) \qquad \text{where} \qquad v = \sqrt{\frac{F}{m/L}}$$

Combining these two expressions, we have, with $n = 1$ for the fundamental frequency,

$$f_1 = \frac{1}{2L}\sqrt{\frac{F}{m/L}}$$

This expression can be used to find the ratio of the two fundamental frequencies.

SOLUTION The ratio of the two fundamental frequencies is

$$\frac{f_{\text{old}}}{f_{\text{new}}} = \frac{\dfrac{1}{2L}\sqrt{\dfrac{F_{\text{old}}}{m/L}}}{\dfrac{1}{2L}\sqrt{\dfrac{F_{\text{new}}}{m/L}}} = \sqrt{\frac{F_{\text{old}}}{F_{\text{new}}}}$$

Since $F_{\text{new}} = 4F_{\text{old}}$, we have

$$f_{\text{new}} = f_{\text{old}}\sqrt{\frac{F_{\text{new}}}{F_{\text{old}}}} = f_{\text{old}}\sqrt{\frac{4\,F_{\text{old}}}{F_{\text{old}}}} = f_{\text{old}}\sqrt{4} = (55.0 \text{ Hz})(2) = \boxed{1.10 \times 10^2 \text{ Hz}}$$

---

63. **REASONING** The natural frequencies of the cord are, according to Equation 17.3, $f_n = nv/(2L)$, where $n = 1, 2, 3, ...$ The speed v of the waves on the cord is, according to Equation 16.2, $v = \sqrt{F/(m/L)}$, where F is the tension in the cord. Combining these two expressions, we have

$$f_n = \frac{nv}{2L} = \frac{n}{2L}\sqrt{\frac{F}{m/L}} \qquad \text{or} \qquad \left(\frac{f_n 2L}{n}\right)^2 = \frac{F}{m/L}$$

Applying Newton's second law of motion, $\Sigma F = ma$, to the forces that act on the block and are parallel to the incline gives

$$F - Mg\sin\theta = Ma = 0 \qquad \text{or} \qquad F = Mg\sin\theta$$

where $Mg\sin\theta$ is the component of the block's weight that is parallel to the incline. Substituting this value for the tension into the equation above gives

$$\left(\frac{f_n 2L}{n}\right)^2 = \frac{Mg\sin\theta}{m/L}$$

This expression can be solved for the angle θ and evaluated at the various harmonics. The answer can be chosen from the resulting choices.

SOLUTION Solving this result for $\sin\theta$ shows that

$$\sin\theta = \frac{(m/L)}{Mg}\left(\frac{f_n 2L}{n}\right)^2 = \frac{1.20\times10^{-2}\text{ kg/m}}{(15.0\text{ kg})(9.80\text{ m/s}^2)}\left[\frac{(165\text{ Hz})2(0.600\text{ m})}{n}\right]^2 = \frac{3.20}{n^2}$$

Thus, we have

$$\theta = \sin^{-1}\left(\frac{3.20}{n^2}\right)$$

Evaluating this for the harmonics corresponding to the range of n from $n=2$ to $n=4$, we have

$$\theta = \sin^{-1}\left(\frac{3.20}{2^2}\right) = 53.1° \text{ for } n = 2$$

$$\theta = \sin^{-1}\left(\frac{3.20}{3^2}\right) = 20.8° \text{ for } n = 3$$

$$\theta = \sin^{-1}\left(\frac{3.20}{4^2}\right) = 11.5° \text{ for } n = 4$$

The angles between 15.0° and 90.0° are $\boxed{\theta = 20.8°}$ and $\boxed{\theta = 53.1°}$.

---

65. **CONCEPTS** (i) The diffraction angle in this case is a measure of the angular deviation of the wave from the straight-line path through the aperture. It depends on the diameter D of the aperture and the wavelength λ of the wave, which could be in the form of sound waves or light waves. The relationship between the diffraction angle θ, and D and λ is:

$$\sin\theta = 1.22\frac{\lambda}{D}$$

(ii) The wavelength is given by $\lambda = v/f$, where v is the speed of sound and f is the frequency. (iii) According to the equation $\lambda = v/f$, the wavelength is proportional to the speed v for a given value of the frequency f. Since sound travels at a lower speed in air than in water, the wavelength in air is smaller than that in water, and the frequency does not change in different media. **(iv)** The extent of diffraction is determined by λ/D, the ratio of the wavelength to the diameter of the opening. Smaller ratios lead to a smaller degree of diffraction, or smaller diffraction angles. The wavelength in air is smaller than in water, and the diameter of the opening is the same in both cases. Therefore, the ratio λ/D is smaller in air than in water, and the diffraction angle in air is smaller than the diffraction angle in water.

CALCULATIONS Using $\sin\theta = 1.22\lambda/D$ and $\lambda = v/f$ we have

$$\sin\theta = 1.22\frac{\lambda}{D} = 1.22\frac{v}{fD}$$

Applying this result for air and water, we find

Air: $\qquad \theta = \sin^{-1}\left(1.22\frac{\lambda}{fD}\right) = \sin^{-1}\left[1.22\frac{(343 \text{ m/s})}{(15000 \text{ Hz})(0.20 \text{ m})}\right] = \boxed{8.0°}$

Water: $\qquad \theta = \sin^{-1}\left(1.22\frac{\lambda}{fD}\right) = \sin^{-1}\left[1.22\frac{(1482 \text{ m/s})}{(15000 \text{ Hz})(0.20 \text{ m})}\right] = \boxed{37°}$

As expected, the diffraction angle in air is smaller.

CHAPTER 18 | *ELECTRIC FORCES AND ELECTRIC FIELDS*

1. **REASONING** The charge of a single proton is $+e$, and the charge of a single electron is $-e$, where $e = 1.60 \times 10^{-19}$ C. The net charge of the ionized atom is the sum of the charges of its constituent protons and electrons.

   SOLUTION The ionized atom has 26 protons and 7 electrons, so its net electric charge q is

   $$q = 26(+e) + 7(-e) = +19e = +19(1.60 \times 10^{-19} \text{ C}) = \boxed{+3.04 \times 10^{-18} \text{ C}}$$

5. **REASONING** Identical conducting spheres equalize their charge upon touching. When spheres A and B touch, an amount of charge $+q$, flows from A and instantaneously neutralizes the $-q$ charge on B leaving B momentarily neutral. Then, the remaining amount of charge, equal to $+4q$, is equally split between A and B, leaving A and B each with equal amounts of charge $+2q$. Sphere C is initially neutral, so when A and C touch, the $+2q$ on A splits equally to give $+q$ on A and $+q$ on C. When B and C touch, the $+2q$ on B and the $+q$ on C combine to give a total charge of $+3q$, which is then equally divided between the spheres B and C; thus, B and C are each left with an amount of charge $+1.5q$.

   SOLUTION Taking note of the initial values given in the problem statement, and summarizing the final results determined in the **REASONING** above, we conclude the following:
   a. Sphere C ends up with an amount of charge equal to $\boxed{+1.5q}$.

   b. The charges on the three spheres before they were touched, are, according to the problem statement, $+5q$ on sphere A, $-q$ on sphere B, and zero charge on sphere C. Thus, the total charge on the spheres is $+5q - q + 0 = \boxed{+4q}$.

   c. The charges on the spheres after they are touched are $+q$ on sphere A, $+1.5q$ on sphere B, and $+1.5q$ on sphere C. Thus, the total charge on the spheres is $+q + 1.5q + 1.5q = \boxed{+4q}$.

9. **REASONING** The number N of excess electrons on one of the objects is equal to the charge q on it divided by the charge of an electron $(-e)$, or $N = q/(-e)$. Since the charge on the object is negative, we can write $q = -|q|$, where $|q|$ is the magnitude of the charge. The magnitude of the charge can be found from Coulomb's law (Equation 18.1), which states that the magnitude F of the electrostatic force exerted on each object is given by $F = k|q||q|/r^2$, where r is the distance between them.

SOLUTION The number N of excess electrons on one of the objects is

$$N = \frac{q}{-e} = \frac{-|q|}{-e} = \frac{|q|}{e} \tag{1}$$

To find the magnitude of the charge, we solve Coulomb's law, $F = k|q||q|/r^2$, for $|q|$:

$$|q| = \sqrt{\frac{Fr^2}{k}}$$

Substituting this result into Equation (1) gives

$$N = \frac{|q|}{e} = \frac{\sqrt{\frac{Fr^2}{k}}}{e} = \frac{\sqrt{\frac{(4.55\times10^{-21}\ \text{N})(1.80\times10^{-3}\ \text{m})^2}{8.99\times10^9\ \text{N}\cdot\text{m}^2/\text{C}^2}}}{1.60\times10^{-19}\ \text{C}} = \boxed{8}$$

---

11. **REASONING** Initially, the two spheres are neutral. Since negative charge is removed from the sphere which loses electrons, it then carries a net positive charge. Furthermore, the neutral sphere to which the electrons are added is then negatively charged. Once the charge is transferred, there exists an electrostatic force on each of the two spheres, the magnitude of which is given by Coulomb's law (Equation 18.1), $F = k|q_1||q_2|/r^2$.

SOLUTION

a. Since each electron carries a charge of -1.60×10^{-19} C, the amount of negative charge removed from the first sphere is

$$(3.0\times10^{13}\ \text{electrons})\left(\frac{1.60\times10^{-19}\ \text{C}}{1\ \text{electron}}\right) = 4.8\times10^{-6}\ \text{C}$$

Thus, the first sphere carries a charge $+4.8\times10^{-6}$ C, while the second sphere carries a charge -4.8×10^{-6} C. The magnitude of the electrostatic force that acts on each sphere is, therefore,

$$F = \frac{k|q_1||q_2|}{r^2} = \frac{(8.99\times10^9\ \text{N}\cdot\text{m}^2/\text{C}^2)(4.8\times10^{-6}\ \text{C})^2}{(0.50\ \text{m})^2} = \boxed{0.83\ \text{N}}$$

b. Since the spheres carry charges of opposite sign, the force is $\boxed{\text{attractive}}$.

15. *REASONING AND SOLUTION*

a. Since the gravitational force between the spheres is one of attraction and the electrostatic force must balance it, the electric force must be one of repulsion. Therefore, the charges must have | the same algebraic signs, both positive or both negative |.

b. There are two forces that act on each sphere; they are the gravitational attraction F_G of one sphere for the other, and the repulsive electric force F_E of one sphere on the other. From the problem statement, we know that these two forces balance each other, so that $F_G = F_E$. The magnitude of F_G is given by Newton's law of gravitation (Equation 4.3: $F_G = Gm_1m_2/r^2$), while the magnitude of F_E is given by Coulomb's law (Equation 18.1: $F_E = k|q_1||q_2|/r^2$). Therefore, we have

$$\frac{Gm_1m_2}{r^2} = \frac{k|q_1||q_2|}{r^2} \quad \text{or} \quad Gm^2 = k|q|^2$$

since the spheres have the same mass m and carry charges of the same magnitude $|q|$. Solving for $|q|$, we find

$$|q| = m\sqrt{\frac{G}{k}} = (2.0 \times 10^{-6}\text{ kg})\sqrt{\frac{6.67\times10^{-11}\text{ N}\cdot\text{m}^2/\text{kg}^2}{8.99\times10^9\text{ N}\cdot\text{m}^2/\text{C}^2}} = \boxed{1.7\times10^{-16}\text{ C}}$$

---

17. *REASONING* Each particle will experience an electrostatic force due to the presence of the other charge. According to Coulomb's law (Equation 18.1), the magnitude of the force felt by each particle can be calculated from $F = k|q_1||q_2|/r^2$, where $|q_1|$ and $|q_2|$ are the respective charges on particles 1 and 2 and r is the distance between them. According to Newton's second law, the magnitude of the force experienced by each particle is given by $F = ma$, where a is the acceleration of the particle and we have assumed that the electrostatic force is the only force acting.

SOLUTION

a. Since the two particles have identical positive charges, $|q_1| = |q_2| = |q|$, and we have, using the data for particle 1,

$$\frac{k|q|^2}{r^2} = m_1a_1$$

Solving for $|q|$, we find that

$$|q| = \sqrt{\frac{m_1a_1r^2}{k}} = \sqrt{\frac{(6.00\times10^{-6}\text{ kg})(4.60\times10^3\text{ m/s}^2)(2.60\times10^{-2}\text{ m})^2}{8.99\times10^9\text{ N}\cdot\text{m}^2/\text{C}^2}} = \boxed{4.56\times10^{-8}\text{ C}}$$

b. Since each particle experiences a force of the same magnitude (From Newton's third law), we can write $F_1 = F_2$, or $m_1 a_1 = m_2 a_2$. Solving this expression for the mass m_2 of particle 2, we have

$$m_2 = \frac{m_1 a_1}{a_2} = \frac{(6.00 \times 10^{-6} \text{ kg})(4.60 \times 10^3 \text{ m/s}^2)}{8.50 \times 10^3 \text{ m/s}^2} = \boxed{3.25 \times 10^{-6} \text{ kg}}$$

---

25. **_REASONING_** Consider the drawing at the right. It is given that the charges q_A, q_1, and q_2 are each positive. Therefore, the charges q_1 and q_2 each exert a repulsive force on the charge q_A. As the drawing shows, these forces have magnitudes F_{A1} (vertically downward) and F_{A2} (horizontally to the left). The unknown charge placed at the empty

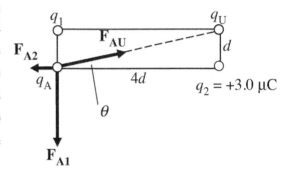

corner of the rectangle is q_U, and it exerts a force on q_A that has a magnitude F_{AU}. In order that the net force acting on q_A point in the vertical direction, the horizontal component of $\mathbf{F}_{AU}$ must cancel out the horizontal force $\mathbf{F}_{A2}$. Therefore, $\mathbf{F}_{AU}$ must point as shown in the drawing, which means that it is an attractive force and q_U must be negative, since q_A is positive.

SOLUTION The basis for our solution is the fact that the horizontal component of $\mathbf{F}_{AU}$ must cancel out the horizontal force $\mathbf{F}_{A2}$. The magnitudes of these forces can be expressed using Coulomb's law $F = k|q||q'|/r^2$, where r is the distance between the charges q and q'. Thus, we have

$$F_{AU} = \frac{k|q_A||q_U|}{(4d)^2 + d^2} \quad \text{and} \quad F_{A2} = \frac{k|q_A||q_2|}{(4d)^2}$$

where we have used the fact that the distance between the charges q_A and q_U is the diagonal of the rectangle, which is $\sqrt{(4d)^2 + d^2}$ according to the Pythagorean theorem, and the fact that the distance between the charges q_A and q_2 is $4d$. The horizontal component of $\mathbf{F}_{AU}$ is $F_{AU} \cos\theta$, which must be equal to F_{A2}, so that we have

$$\frac{k|q_A||q_U|}{(4d)^2 + d^2} \cos\theta = \frac{k|q_A||q_2|}{(4d)^2} \quad \text{or} \quad \frac{|q_U|}{17} \cos\theta = \frac{|q_2|}{16}$$

The drawing in the **REASONING**, reveals that $\cos\theta = (4d)/\sqrt{(4d)^2 + d^2} = 4/\sqrt{17}$. Therefore, we find that

$$\frac{|q_U|}{17}\left(\frac{4}{\sqrt{17}}\right) = \frac{|q_2|}{16} \quad \text{or} \quad |q_U| = \frac{17\sqrt{17}}{64}|q_2| = \frac{17\sqrt{17}}{64}(3.0\times10^{-6}\text{ C}) = \boxed{3.3\times10^{-6}\text{ C}}$$

As discussed in the **REASONING**, the algebraic sign of the charge q_U is $\boxed{\text{negative}}$.

---

27. **REASONING** The charged insulator experiences an electric force due to the presence of the charged sphere shown in the drawing in the text. The forces acting on the insulator are the downward force of gravity (i.e., its weight, $W = mg$), the electrostatic force $F = k|q_1||q_2|/r^2$ (see Coulomb's law, Equation 18.1) pulling to the right, and the tension T in the thread pulling up and to the left at an angle θ with respect to the vertical, as shown in the drawing in the problem statement. We can analyze the forces to determine the desired quantities θ and T.

SOLUTION.
a. We can see from the diagram given with the problem statement that

$$T_x = F \qquad \text{which gives} \qquad T\sin\theta = k|q_1||q_2|/r^2$$

and

$$T_y = W \qquad \text{which gives} \qquad T\cos\theta = mg$$

Dividing the first equation by the second yields

$$\frac{T\sin\theta}{T\cos\theta} = \tan\theta = \frac{k|q_1||q_2|/r^2}{mg}$$

Solving for θ, we find that

$$\theta = \tan^{-1}\left(\frac{k|q_1||q_2|}{mgr^2}\right)$$

$$= \tan^{-1}\left[\frac{(8.99\times10^9\text{ N}\cdot\text{m}^2/\text{C}^2)(0.600\times10^{-6}\text{ C})(0.900\times10^{-6}\text{ C})}{(8.00\times10^{-2}\text{ kg})(9.80\text{ m/s}^2)(0.150\text{ m})^2}\right] = \boxed{15.4°}$$

b. Since $T\cos\theta = mg$, the tension can be obtained as follows:

$$T = \frac{mg}{\cos\theta} = \frac{(8.00\times10^{-2}\text{ kg})(9.80\text{ m/s}^2)}{\cos 15.4°} = \boxed{0.813\text{ N}}$$

29. **REASONING** The electric field created by a point charge is inversely proportional to the square of the distance from the charge, according to Equation 18.3. Therefore, we expect the distance r_2 to be greater than the distance r_1, since the field is smaller at r_2 than it is at r_1. The ratio r_2/r_1, then, should be greater than one.

SOLUTION Applying Equation 18.3 to each position relative to the charge, we have

$$E_1 = \frac{k|q|}{r_1^2} \quad \text{and} \quad E_2 = \frac{k|q|}{r_2^2}$$

Dividing the expression for E_1 by the expression for E_2 gives

$$\frac{E_1}{E_2} = \frac{k|q|/r_1^2}{k|q|/r_2^2} = \frac{r_2^2}{r_1^2}$$

Solving for the ratio r_2/r_1 gives

$$\frac{r_2}{r_1} = \sqrt{\frac{E_1}{E_2}} = \sqrt{\frac{248 \text{ N/C}}{132 \text{ N/C}}} = \boxed{1.37}$$

As expected, this ratio is greater than one.

---

37. **REASONING** The drawing at the right shows the set-up. Here, the electric field **E** points along the +y axis and applies a force of +**F** to the +q charge and a force of –**F** to the –q charge, where $q = 8.0\ \mu\text{C}$ denotes the magnitude of each charge. Each force has the same magnitude of $F = E|q|$, according to Equation 18.2. The torque is measured as discussed in Section 9.1. According to Equation 9.1, the torque produced by each force has a magnitude given by the magnitude of the force times the lever arm, which is the perpendicular distance between the point of application of the force and the axis of rotation. In the drawing the z axis is the axis of rotation and is midway between the ends of the rod. Thus, the lever arm for each force is half the length L of the rod or L/2, and the magnitude of the torque produced by each force is $(E|q|)(L/2)$.

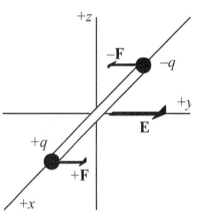

SOLUTION The +**F** and the –**F** force each cause the rod to rotate in the same sense about the z axis. Therefore, the torques from these forces reinforce one another. Using the expression $(E|q|)(L/2)$ for the magnitude of each torque, we find that the magnitude of the net torque is

$$\begin{aligned}\text{Magnitude of} \atop \text{net torque} &= E|q|\left(\frac{L}{2}\right) + E|q|\left(\frac{L}{2}\right) = E|q|L \\ &= \left(5.0\times10^3 \text{ N/C}\right)\left(8.0\times10^{-6} \text{ C}\right)(4.0 \text{ m}) = \boxed{0.16 \text{ N}\cdot\text{m}}\end{aligned}$$

39. **REASONING** Two forces act on the charged ball (charge q); they are the downward force of gravity $m\mathbf{g}$ and the electric force $\mathbf{F}$ due to the presence of the charge q in the electric field $\mathbf{E}$. In order for the ball to float, these two forces must be equal in magnitude and opposite in direction, so that the net force on the ball is zero (Newton's second law). Therefore, $\mathbf{F}$ must point upward, which we will take as the positive direction. According to Equation 18.2, $\mathbf{F} = q\mathbf{E}$. Since the charge q is negative, the electric field $\mathbf{E}$ must point downward, as the product $q\mathbf{E}$ in the expression $\mathbf{F} = q\mathbf{E}$ must be positive, since the force $\mathbf{F}$ points upward. The magnitudes of the two forces must be equal, so that $mg = |q|E$. This expression can be solved for E.

SOLUTION The magnitude of the electric field $\mathbf{E}$ is

$$E = \frac{mg}{|q|} = \frac{(0.012 \text{ kg})(9.80 \text{ m/s}^2)}{18 \times 10^{-6} \text{ C}} = \boxed{6.5 \times 10^3 \text{ N/C}}$$

As discussed in the reasoning, this electric field points $\boxed{\text{downward}}$.

---

45. **REASONING** The two charges lying on the x axis produce no net electric field at the coordinate origin. This is because they have identical charges, are located the same distance from the origin, and produce electric fields that point in opposite directions. The electric field produced by q_3 at the origin points away from the charge, or along the $-y$ direction. The electric field produced by q_4 at the origin points toward the charge, or along the $+y$ direction. The net electric field is, then, $E = -E_3 + E_4$, where E_3 and E_4 can be determined by using Equation 18.3.

SOLUTION The net electric field at the origin is

$$E = -E_3 + E_4 = \frac{-k|q_3|}{r_3^2} + \frac{k|q_4|}{r_4^2}$$

$$= \frac{-\left(8.99 \times 10^9 \text{ N} \cdot \text{m}^2/\text{C}^2\right)\left(3.0 \times 10^{-6} \text{ C}\right)}{\left(5.0 \times 10^{-2} \text{ m}\right)^2} + \frac{\left(8.99 \times 10^9 \text{ N} \cdot \text{m}^2/\text{C}^2\right)\left(8.0 \times 10^{-6} \text{ C}\right)}{\left(7.0 \times 10^{-2} \text{ m}\right)^2}$$

$$= \boxed{+3.9 \times 10^6 \text{ N/C}}$$

The plus sign indicates that $\boxed{\text{the net electric field points along the } +y \text{ direction}}$.

51. ***REASONING AND SOLUTION*** The net electric field at point P in Figure 1 is the vector sum of the fields $\mathbf{E}_+$ and $\mathbf{E}_-$, which are due, respectively, to the charges $+q$ and $-q$. These fields are shown in Figure 2.

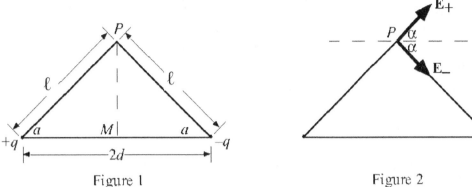

Figure 1   Figure 2

According to Equation 18.3, the magnitudes of the fields E_+ and E_- are the same, since the triangle is an isosceles triangle with equal sides of length ℓ. Therefore, $E_+ = E_- = k|q|/1^2$. The vertical components of these two fields cancel, while the horizontal components reinforce, leading to a total field at point P that is horizontal and has a magnitude of

$$E_P = E_+ \cos \alpha + E_- \cos \alpha = 2\left(\frac{k|q|}{1^2}\right)\cos \alpha$$

At point M in Figure 1, both $\mathbf{E}_+$ and $\mathbf{E}_-$ are horizontal and point to the right. Again using Equation 18.3, we find

$$E_M = E_+ + E_- = \frac{k|q|}{d^2} + \frac{k|q|}{d^2} = \frac{2k|q|}{d^2}$$

Since $E_M/E_P = 9.0$, we have

$$\frac{E_M}{E_P} = \frac{2k|q|/d^2}{2k|q|(\cos \alpha)/1^2} = \frac{1}{(\cos \alpha)d^2/1^2} = 9.0$$

But from Figure 1, we can see that $d/\ell = \cos \alpha$. Thus, it follows that

$$\frac{1}{\cos^3\alpha} = 9.0 \qquad \text{or} \qquad \cos \alpha = \sqrt[3]{1/9.0} = 0.48$$

The value for α is, then, $\alpha = \cos^{-1}(0.48) = \boxed{61°}$.

55.   ***REASONING*** As discussed in Section 18.9, the magnitude of the electric flux Φ_E through a surface is equal to the magnitude of the component of the electric field that is normal to the surface multiplied by the area of the surface, $\Phi_E = E_\perp A$, where $E_\perp$ is the magnitude of the component of **E** that is normal to the surface of area A. We can use this expression and the figure in the text to determine the desired quantities.

SOLUTION
a. The magnitude of the flux through surface 1 is

$$\left(\Phi_E\right)_1 = (E \cos 35°)A_1 = (250 \text{ N/C})(\cos 35°)(1.7 \text{ m}^2) = \boxed{350 \text{ N} \cdot \text{m}^2/\text{C}}$$

b. Similarly, the magnitude of the flux through surface 2 is

$$\left(\Phi_E\right)_2 = (E \cos 55°)A_2 = (250 \text{ N/C})(\cos 55°)(3.2 \text{ m}^2) = \boxed{460 \text{ N} \cdot \text{m}^2/\text{C}}$$

---

61.   ***REASONING*** The electric flux through each face of the cube is given by $\Phi_E = (E \cos\phi)A$ (see Section 18.9) where E is the magnitude of the electric field at the face, A is the area of the face, and ϕ is the angle between the electric field and the outward normal of that face. We can use this expression to calculate the electric flux Φ_E through each of the six faces of the cube.

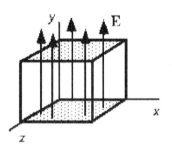

SOLUTION
a.   On the bottom face of the cube, the outward normal points parallel to the $-y$ axis, in the opposite direction to the electric field, and $\phi = 180°$. Therefore,

$$\left(\Phi_E\right)_{\text{bottom}} = (1500 \text{ N/C})(\cos 180°)(0.20 \text{ m})^2 = \boxed{-6.0 \times 10^1 \text{ N} \text{gm}^2/\text{C}}$$

On the top face of the cube, the outward normal points parallel to the $+y$ axis, and $\phi = 0.0°$. The electric flux is, therefore,

$$\left(\Phi_E\right)_{\text{top}} = (1500 \text{ N/C})(\cos 0.0°)(0.20 \text{ m})^2 = \boxed{+6.0 \times 10^1 \text{ N} \text{gm}^2/\text{C}}$$

On each of the other four faces, the outward normals are perpendicular to the direction of the electric field, so $\phi = 90°$. So for each of the four side faces,

$$\left(\Phi_E\right)_{\text{sides}} = (1500 \text{ N/C})(\cos 90°)(0.20 \text{ m})^2 = \boxed{0 \text{ N} \text{gm}^2/\text{C}}$$

b.   The total flux through the cube is

$$\left(\Phi_E\right)_{total} = \left(\Phi_E\right)_{top} + \left(\Phi_E\right)_{bottom} + \left(\Phi_E\right)_{side\ 1} + \left(\Phi_E\right)_{side\ 2} + \left(\Phi_E\right)_{side\ 3} + \left(\Phi_E\right)_{side\ 4}$$

Therefore,

$$\left(\Phi_E\right)_{total} = \left(+6.0\times10^1\ N\text{·}m^2/C\right) + \left(-6.0\times10^1\ N\text{·}m^2/C\right) + 0 + 0 + 0 + 0 = \boxed{0\ N\text{·}m^2/C}$$

---

67. **REASONING**
a. The drawing shows the two point charges q_1 and q_2. Point A is located at $x = 0$ cm, and point B is at $x = +6.0$ cm.

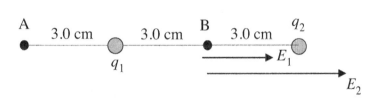

Since q_1 is positive, the electric field points away from it. At point A, the electric field E_1 points to the left, in the $-x$ direction. Since q_2 is negative, the electric field points toward it. At point A, the electric field E_2 points to the right, in the $+x$ direction. The net electric field is $E = -E_1 + E_2$. We can use Equation 18.3, $E = k|q|/r^2$, to find the magnitude of the electric field due to each point charge.

b. The drawing shows the electric fields produced by the charges q_1 and q_2 at point B, which is located at $x = +6.0$ cm.

Since q_1 is positive, the electric field points away from it. At point B, the electric field points to the right, in the $+x$ direction. Since q_2 is negative, the electric field points toward it. At point B, the electric field points to the right, in the $+x$ direction. The net electric field is $E = +E_1 + E_2$.

SOLUTION

a. The net electric field at the origin (point A) is $E = -E_1 + E_2$:

$$E = -E_1 + E_2 = \frac{-k|q_1|}{r_1^2} + \frac{k|q_2|}{r_2^2}$$

$$= \frac{-\left(8.99 \times 10^9 \text{ N·m}^2/\text{C}^2\right)\left(8.5 \times 10^{-6} \text{ C}\right)}{\left(3.0 \times 10^{-2} \text{ m}\right)^2} + \frac{\left(8.99 \times 10^9 \text{ N·m}^2/\text{C}^2\right)\left(21 \times 10^{-6} \text{ C}\right)}{\left(9.0 \times 10^{-2} \text{ m}\right)^2}$$

$$= \boxed{-6.2 \times 10^7 \text{ N/C}}$$

The minus sign tells us that the net electric field points along the $-x$ axis.

b. The net electric field at $x = +6.0$ cm (point B) is $E = E_1 + E_2$:

$$E = E_1 + E_2 = \frac{k|q_1|}{r_1^2} + \frac{k|q_2|}{r_2^2}$$

$$= \frac{\left(8.99 \times 10^9 \text{ N·m}^2/\text{C}^2\right)\left(8.5 \times 10^{-6} \text{ C}\right)}{\left(3.0 \times 10^{-2} \text{ m}\right)^2} + \frac{\left(8.99 \times 10^9 \text{ N·m}^2/\text{C}^2\right)\left(21 \times 10^{-6} \text{ C}\right)}{\left(3.0 \times 10^{-2} \text{ m}\right)^2}$$

$$= \boxed{+2.9 \times 10^8 \text{ N/C}}$$

The plus sign tells us that the net electric field points along the $+x$ axis.

---

69. **REASONING** The electrons transferred increase the magnitudes of the positive and negative charges from $2.00 \ \mu\text{C}$ to a greater value. We can calculate the number N of electrons by dividing the change in the magnitude of the charges by the magnitude e of the charge on an electron. The greater charge that exists after the transfer can be obtained from Coulomb's law and the value given for the magnitude of the electrostatic force.

SOLUTION The number N of electrons transferred is

$$N = \frac{|q_{\text{after}}| - |q_{\text{before}}|}{e}$$

where $\left|q_{\text{after}}\right|$ and $\left|q_{\text{before}}\right|$ are the magnitudes of the charges after and before the transfer of electrons occurs. To obtain $\left|q_{\text{after}}\right|$, we apply Coulomb's law with a value of 68.0 N for the electrostatic force:

$$F = k\frac{\left|q_{\text{after}}\right|^2}{r^2} \qquad \text{or} \qquad \left|q_{\text{after}}\right| = \sqrt{\frac{Fr^2}{k}}$$

Using this result in the expression for N, we find that

$$N = \frac{\sqrt{\dfrac{Fr^2}{k}} - \left|q_{\text{before}}\right|}{e} = \frac{\sqrt{\dfrac{(68.0\ \text{N})(0.0300\ \text{m})^2}{8.99\times10^9\ \text{N}\cdot\text{m}^2/\text{C}^2}} - 2.00\times10^{-6}\ \text{C}}{1.60\times10^{-19}\ \text{C}} = \boxed{3.8\times10^{12}}$$

---

71. **REASONING** The drawing shows the arrangement of the three charges. Let $\mathbf{E}_q$ represent the electric field at the empty corner due to the $-q$ charge. Furthermore, let $\mathbf{E}_1$ and $\mathbf{E}_2$ be the electric fields at the empty corner due to charges $+q_1$ and $+q_2$, respectively.

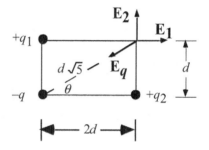

According to the Pythagorean theorem, the distance from the charge $-q$ to the empty corner along the diagonal is given by $\sqrt{(2d)^2 + d^2} = \sqrt{5d^2} = d\sqrt{5}$. The magnitude of each electric field is given by Equation 18.3, $E = k|q|/r^2$. Thus, the magnitudes of each of the electric fields at the empty corner are given as follows:

$$E_q = \frac{k|q|}{r^2} = \frac{k|q|}{\left(d\sqrt{5}\right)^2} = \frac{k|q|}{5d^2}$$

$$E_1 = \frac{k|q_1|}{(2d)^2} = \frac{k|q_1|}{4d^2} \qquad \text{and} \qquad E_2 = \frac{k|q_2|}{d^2}$$

The angle θ that the diagonal makes with the horizontal is $\theta = \tan^{-1}(d/2d) = 26.57°$. Since the net electric field E_{net} at the empty corner is zero, the horizontal component of the net field must be zero, and we have

$$E_1 - E_q \cos 26.57° = 0 \qquad \text{or} \qquad \frac{k|q_1|}{4d^2} - \frac{k|q|\cos 26.57°}{5d^2} = 0$$

Similarly, the vertical component of the net field must be zero, and we have

$$E_2 - E_q \sin 26.57° = 0 \qquad \text{or} \qquad \frac{k|q_2|}{d^2} - \frac{k|q|\sin 26.57°}{5d^2} = 0$$

These last two expressions can be solved for the charge magnitudes $|q_1|$ and $|q_2|$.

SOLUTION Solving the last two expressions for $|q_1|$ and $|q_2|$, we find that

$$|q_1| = \frac{4}{5}q \ \cos 26.57° = \boxed{0.716 \ q}$$

$$|q_2| = \frac{1}{5}q \ \sin 26.57° = \boxed{0.0895 \ q}$$

---

77. **REASONING** Since we know the initial velocity and displacement of the proton, we can determine its final velocity from an equation of kinematics, provided the proton's acceleration can be found. The acceleration is given by Newton's second law as the net force acting on the proton divided by its mass. The net force is the electrostatic force, since the proton is moving in an electric field. The electrostatic force depends on the proton's charge and the electric field, both of which are known.

SOLUTION To obtain the final velocity v_x of the proton we employ Equation 3.6a from the equations of kinematics: $v_x^2 = v_{0x}^2 + 2a_x x$. We have chosen this equation because two of the variables, the initial velocity v_{0x} and the displacement x, are known. Taking the square root of each side of this relation and choosing the + sign, since the proton is moving in the +x direction (see figure below), we arrive at Equation 1: $v_x = +\sqrt{v_{0x}^2 + 2a_x x}$ (1). Although the acceleration a_x is not known, we will obtain an expression for it below.

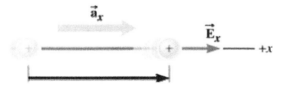

Newton's second law, as given in Equation 4.2a, states that the acceleration a_x of the proton is equal to the net force $\sum F_x$ acting on it divided by the proton's mass m: $a_x = \sum F_x / m$. Only the electrostatic force F_x acts on the proton, so it is the net force. Setting $\sum F_x = F_x$ in Newton's second law gives

$$a_x = \frac{F_x}{m}$$

This expression can be substituted into Equation 1. The electrostatic force is not known, so we proceed to the next step to evaluate it using the concept of the electric field.

Since the proton is moving in a uniform electric field E_x, it experiences an electrostatic force given by $F_x = q_0 E_x$ (Equation 18.2), where q_0 is the charge. Setting $q_0 = e$ for the proton, we have

$$F_x = e E_x$$

All the variables on the right side of this equation are known, so we substitute it into our expression above:

$$v_x = +\sqrt{v_{0x}^2 + 2a_x x} = +\sqrt{v_{0x}^2 + 2\left(\frac{F_x}{m}\right)x} = +\sqrt{v_{0x}^2 + 2\left(\frac{eE_x}{m}\right)x}$$

$$v_x = +\sqrt{v_{0x}^2 + 2\left(\frac{eE_x}{m}\right)x}$$

$$= +\sqrt{(2.5 \times 10^4 \text{ m/s})^2 + 2\left[\frac{(1.60 \times 10^{-19} \text{ C})(2.3 \times 10^3 \text{ N/C})}{1.67 \times 10^{-27} \text{ kg}}\right](2.0 \times 10^{-3} \text{ m})}$$

$$= \boxed{+3.9 \times 10^4 \text{ m/s}}$$

The + sign denotes that the final velocity points along the +x axis.

---

79. **CONCEPTS** (i) Yes. A net electric field is produced by the charges q_1 and q_2, and it exists throughout the entire region that surrounds them. If a test charge were placed at this point, it would experience a force due to this net field. The force would be the product of the charge and the net electric field. (ii) The electric field created by a charge always points away from a positive charge and towards a negative charge. Since q_1 is negative, the electric field $\vec{E}_1$ at P points towards q_1 (see part b of the drawing). (iii) No. The fields have different directions. We must add them as vectors to obtain the net field. Only then can we determine its magnitude.

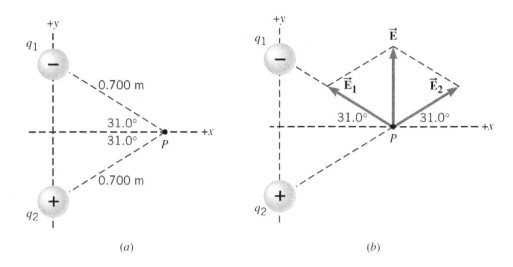

(a)   (b)

CALCULATIONS (a) The magnitudes of the electric fields that q_1 and q_2 produce at P are given by

$$E_1 = \frac{k|q_1|}{r_1^2} = \frac{\left(8.99 \times 10^9 \text{ N} \cdot \text{m}^2/\text{C}^2\right)\left(4.00 \times 10^{-6} \text{ C}\right)}{\left(0.700 \text{ m}\right)^2} = 7.34 \times 10^4 \text{ N/C}$$

$$E_2 = \frac{k|q_2|}{r_2^2} = \frac{\left(8.99 \times 10^9 \text{ N} \cdot \text{m}^2/\text{C}^2\right)\left(4.00 \times 10^{-6} \text{ C}\right)}{\left(0.700 \text{ m}\right)^2} = 7.34 \times 10^4 \text{ N/C}$$

The x and y components of these fields and the total field $\vec{E}$ are given in the following table:

Electric field	x component	y component
$\vec{E}_1$	$-E_1 \cos 31.0° = -6.29 \times 10^4$ N/C	$+E_1 \sin 31.0° = +3.78 \times 10^4$ N/C
$\vec{E}_2$	$+E_2 \cos 31.0° = +6.29 \times 10^4$ N/C	$+E_2 \sin 31.0° = +3.78 \times 10^4$ N/C
$\vec{E}$	$\vec{E}_x = 0$ N/C	$\vec{E}_y = +7.56 \times 10^4$ N/C

The net electric field $\vec{E}$ has only a component along the $+y$ axis, so

$$\boxed{\vec{E} = 7.56 \times 10^4 \text{ N/C, directed along the} + y \text{ axis.}}$$

(b) According to Newton's second law, the acceleration $\vec{a}$ of an object placed at point P is equal to the net force acting on the object divided by the mass of the object. The net force $\vec{F}$

is the product of the charge and the net electric field $\left(\vec{\mathbf{F}} = q_0 \vec{\mathbf{E}}\right)$ (we are ignoring gravitational forces). Therefore, the acceleration is

$$\vec{\mathbf{a}} = \frac{\vec{\mathbf{F}}}{m} = \frac{q_0 \vec{\mathbf{E}}}{m} = \frac{\left(8.00 \times 10^{-6} \ \text{C}\right)\left(7.56 \times 10^4 \ \text{N/C}\right)}{1.20 \times 10^{-3} \ \text{kg}}$$

$$= \boxed{5.04 \times 10^2 \ \text{m/s}^2 \text{, along the} + y \text{ axis}}$$

CHAPTER 19 | *ELECTRIC POTENTIAL ENERGY AND THE ELECTRIC POTENTIAL*

3. **REASONING AND SOLUTION** Combining Equations 19.1 and 19.3, we have

$$W_{AB} = \text{EPE}_A - \text{EPE}_B = q_0(V_A - V_B) = (+1.6 \times 10^{-19}\,\text{C})(0.070\,\text{V}) = \boxed{1.1 \times 10^{-20}\,\text{J}}$$

7. **REASONING** The translational speed of the particle is related to the particle's translational kinetic energy, which forms one part of the total mechanical energy that the particle has. The total mechanical energy is conserved, because only the gravitational force and an electrostatic force, both of which are conservative forces, act on the particle (see Section 6.5). Thus, we will determine the speed at point A by utilizing the principle of conservation of mechanical energy.

SOLUTION The particle's total mechanical energy E is

$$E = \underbrace{\tfrac{1}{2}mv^2}_{\substack{\text{Translational}\\\text{kinetic}\\\text{energy}}} + \underbrace{\tfrac{1}{2}I\omega^2}_{\substack{\text{Rotational}\\\text{kinetic}\\\text{energy}}} + \underbrace{mgh}_{\substack{\text{Gravitational}\\\text{potential}\\\text{energy}}} + \underbrace{\tfrac{1}{2}kx^2}_{\substack{\text{Elastic}\\\text{potential}\\\text{energy}}} + \underbrace{\text{EPE}}_{\substack{\text{Electric}\\\text{potential}\\\text{energy}}}$$

Since the particle does not rotate the angular speed ω is always zero, and since there is no elastic force we may omit the terms $\tfrac{1}{2}I\omega^2$ and $\tfrac{1}{2}kx^2$ from this expression. With this in mind, we express the fact that $E_B = E_A$ (energy is conserved) as follows:

$$\tfrac{1}{2}mv_B^2 + mgh_B + \text{EPE}_B = \tfrac{1}{2}mv_A^2 + mgh_A + \text{EPE}_A$$

This equation can be simplified further, since the particle travels horizontally, so that $h_B = h_A$, with the result that

$$\tfrac{1}{2}mv_B^2 + \text{EPE}_B = \tfrac{1}{2}mv_A^2 + \text{EPE}_A$$

Solving for v_A gives

$$v_A = \sqrt{v_B^2 + \frac{2\left(\text{EPE}_B - \text{EPE}_A\right)}{m}}$$

According to Equation 19.4, the difference in electric potential energies $\text{EPE}_B - \text{EPE}_A$ is related to the electric potential difference $V_B - V_A$:

$$\text{EPE}_B - \text{EPE}_A = q_0 \left(V_B - V_A \right)$$

Substituting this expression into the expression for v_A gives

$$v_A = \sqrt{v_B^2 + \frac{2q_0 \left(V_B - V_A \right)}{m}} = \sqrt{(0 \text{ m/s})^2 + \frac{2\left(-2.0 \times 10^{-5} \text{ C}\right)\left(-36 \text{ V}\right)}{4.0 \times 10^{-6} \text{ kg}}} = \boxed{19 \text{ m/s}}$$

---

9. **REASONING** The only force acting on the moving charge is the conservative electric force. Therefore, the total energy of the charge remains constant. Applying the principle of conservation of energy between locations A and B, we obtain

$$\tfrac{1}{2} m v_A^2 + \text{EPE}_A = \tfrac{1}{2} m v_B^2 + \text{EPE}_B$$

Since the charged particle starts from rest, $v_A = 0$. The difference in potential energies is related to the difference in potentials by Equation 19.4, $\text{EPE}_B - \text{EPE}_A = q(V_B - V_A)$. Thus, we have

$$q(V_A - V_B) = \tfrac{1}{2} m v_B^2 \tag{1}$$

Similarly, applying the conservation of energy between locations C and B gives

$$q(V_C - V_B) = \tfrac{1}{2} m (2v_B)^2 \tag{2}$$

Dividing Equation (1) by Equation (2) yields

$$\frac{V_A - V_B}{V_C - V_B} = \frac{1}{4}$$

This expression can be solved for V_B.

SOLUTION Solving for V_B, we find that

$$V_B = \frac{4V_A - V_C}{3} = \frac{4(452 \text{ V}) - 791 \text{ V}}{3} = \boxed{339 \text{ V}}$$

---

15. **REASONING** The potential of each charge q at a distance r away is given by Equation 19.6 as $V = kq/r$. By applying this expression to each charge, we will be able to find the desired ratio, because the distances are given for each charge.

SOLUTION According to Equation 19.6, the potentials of each charge are

$$V_A = \frac{kq_A}{r_A} \quad \text{and} \quad V_B = \frac{kq_B}{r_B}$$

Since we know that $V_A = V_B$, it follows that

$$\frac{kq_A}{r_A} = \frac{kq_B}{r_B} \quad \text{or} \quad \frac{q_B}{q_A} = \frac{r_B}{r_A} = \frac{0.43 \text{ m}}{0.18 \text{ m}} = \boxed{2.4}$$

---

17. **REASONING** The electric potential at a distance r from a point charge q is given by Equation 19.6 as $V = kq/r$. The total electric potential at location P due to the four point charges is the algebraic sum of the individual potentials.

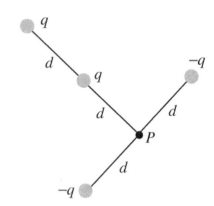

SOLUTION The total electric potential at P is (see the drawing)

$$V = \frac{k(-q)}{d} + \frac{k(+q)}{2d} + \frac{k(+q)}{d} + \frac{k(-q)}{d} = \frac{-kq}{2d}$$

Substituting in the numbers gives

$$V = \frac{-kq}{2d} = \frac{-\left(8.99 \times 10^9 \ \frac{\text{N} \cdot \text{m}^2}{\text{C}^2}\right)\left(2.0 \times 10^{-6} \text{ C}\right)}{2(0.96 \text{ m})} = \boxed{-9.4 \times 10^3 \text{ V}}$$

---

23. **REASONING** Initially, the three charges are infinitely far apart. We will proceed as in Example 8 by adding charges to the triangle, one at a time, and determining the electric potential energy at each step. According to Equation 19.3, the electric potential energy EPE is the product of the charge q and the electric potential V at the spot where the charge is placed, EPE = qV. The total electric potential energy of the group is the sum of the energies of each step in assembling the group.

SOLUTION Let the corners of the triangle be numbered clockwise as 1, 2 and 3, starting with the top corner. When the first charge ($q_1 = 8.00 \ \mu\text{C}$) is placed at a corner 1, the charge has no electric potential energy, $\text{EPE}_1 = 0$. This is because the electric potential V_1 produced by the other two charges at corner 1 is zero, since they are infinitely far away.

Once the 8.00-μC charge is in place, the electric potential V_2 that it creates at corner 2 is

$$V_2 = \frac{kq_1}{r_{21}}$$

where $r_{21} = 5.00$ m is the distance between corners 1 and 2, and $q_1 = 8.00$ μC. When the 20.0-μC charge is placed at corner 2, its electric potential energy EPE_2 is

$$EPE_2 = q_2 V_2 = q_2 \left(\frac{kq_1}{r_{21}} \right)$$

$$= \left(20.0 \times 10^{-6} \text{ C}\right) \left[\frac{\left(8.99 \times 10^9 \text{ N} \cdot \text{m}^2/\text{C}^2\right)\left(8.00 \times 10^{-6} \text{ C}\right)}{5.00 \text{ m}} \right] = 0.288 \text{ J}$$

The electric potential V_3 at the remaining empty corner is the sum of the potentials due to the two charges that are already in place on corners 1 and 2:

$$V_3 = \frac{kq_1}{r_{31}} + \frac{kq_2}{r_{32}}$$

where $q_1 = 8.00$ μC, $r_{31} = 3.00$ m, $q_2 = 20.0$ μC, and $r_{32} = 4.00$ m. When the third charge ($q_3 = -15.0$ μC) is placed at corner 3, its electric potential energy EPE_3 is

$$EPE_3 = q_3 V_3 = q_3 \left(\frac{kq_1}{r_{31}} + \frac{kq_2}{r_{32}} \right) = q_3 k \left(\frac{q_1}{r_{31}} + \frac{q_2}{r_{32}} \right)$$

$$= \left(-15.0 \times 10^{-6} \text{ C}\right)\left(8.99 \times 10^9 \text{ N} \cdot \text{m}^2/\text{C}^2\right) \left(\frac{8.00 \times 10^{-6} \text{ C}}{3.00 \text{ m}} + \frac{20.0 \times 10^{-6} \text{ C}}{4.00 \text{ m}} \right) = -1.034 \text{ J}$$

The electric potential energy of the entire array is given by

$$EPE = EPE_1 + EPE_2 + EPE_3 = 0 + 0.288 \text{ J} + (-1.034 \text{ J}) = \boxed{-0.746 \text{ J}}$$

---

27. ***REASONING*** The only force acting on the moving charge is the conservative electric force. Therefore, the sum of the kinetic energy KE and the electric potential energy EPE is the same at points A and B:

$$\tfrac{1}{2}mv_A^2 + EPE_A = \tfrac{1}{2}mv_B^2 + EPE_B$$

Since the particle comes to rest at B, $v_B = 0$. Combining Equations 19.3 and 19.6, we have

$$\text{EPE}_{A} = qV_{A} = q\left(\frac{kq_{1}}{d}\right)$$

and

$$\text{EPE}_{B} = qV_{B} = q\left(\frac{kq_{1}}{r}\right)$$

where d is the initial distance between the fixed charge and the moving charged particle, and r is the distance between the charged particles after the moving charge has stopped. Therefore, the expression for the conservation of energy becomes

$$\tfrac{1}{2}mv_{A}^{2} + \frac{kqq_{1}}{d} = \frac{kqq_{1}}{r}$$

This expression can be solved for r. Once r is known, the distance that the charged particle moves can be determined.

SOLUTION Solving the expression above for r gives

$$r = \frac{kqq_{1}}{\tfrac{1}{2}mv_{A}^{2} + \frac{kqq_{1}}{d}}$$

$$= \frac{(8.99\times10^{9}\,\text{N}\cdot\text{m}^{2}/\text{C}^{2})(-8.00\times10^{-6}\,\text{C})(-3.00\times10^{-6}\,\text{C})}{\tfrac{1}{2}(7.20\times10^{-3}\,\text{kg})(65.0\,\text{m/s})^{2} + \dfrac{(8.99\times10^{9}\,\text{N}\cdot\text{m}^{2}/\text{C}^{2})(-8.00\times10^{-6}\,\text{C})(-3.00\times10^{-6}\,\text{C})}{0.0450\,\text{m}}}$$

$$= 0.0108\,\text{m}$$

Therefore, the charge moves a distance of $0.0450\,\text{m} - 0.0108\,\text{m} = \boxed{0.0342\,\text{m}}$.

---

33. **REASONING AND SOLUTION** From Equation 19.7a we know that $E = -\dfrac{\Delta V}{\Delta s}$, where ΔV is the potential difference between the two surfaces of the membrane, and Δs is the distance between them. If A is a point on the positive surface and B is a point on the negative surface, then $\Delta V = V_{A} - V_{B} = 0.070\,\text{V}$. The electric field between the surfaces is

$$E = -\frac{\Delta V}{\Delta s} = -\frac{V_{B} - V_{A}}{\Delta s} = \frac{V_{A} - V_{B}}{\Delta s} = \frac{0.070\,\text{V}}{8.0\times10^{-9}\,\text{m}} = \boxed{8.8\times10^{6}\,\text{V/m}}$$

---

35. **REASONING** The magnitude E of the electric field is given by Equation 19.7a (without the minus sign) as $E = \dfrac{\Delta V}{\Delta s}$, where ΔV is the potential difference between the two metal conductors of the spark plug, and Δs is the distance between the two conductors. We can use this relation to find ΔV.

SOLUTION The potential difference between the conductors is

$$\Delta V = E\Delta s = \left(4.7\times10^7 \text{ V/m}\right)\left(0.75\times10^{-3}\text{ m}\right)= \boxed{3.5\times10^4 \text{ V}}$$

37. **REASONING** The drawing shows the electric field **E** and the three points, A, B, and C, in the vicinity of point P, which we take as the origin. We choose the upward direction as being positive. Thus, $E = -4.0 \times 10^3$ V/m, since the electric field points straight down. The electric potential at points A and B can be determined from Equation 19.7a as $\Delta V = -E\,\Delta s$, since E and Δs are known. Since the path from P to C is perpendicular to the electric field, no work is done in moving a charge along such a path. Thus, the potential difference between these two points is zero.

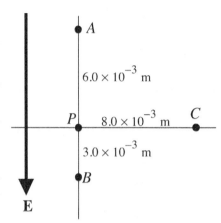

SOLUTION

a. The potential difference between points P and A is $V_A - V_P = -E\,\Delta s$. The potential at A is

$$V_A = V_P - E\,\Delta s = 155 \text{ V } -\left(-4.0\times10^3 \text{ V/m}\right)\left(6.0\times10^{-3}\text{ m}\right)= \boxed{179 \text{ V}}$$

b. The potential difference between points P and B is $V_B - V_P = -E\,\Delta s$. The potential at B is

$$V_B = V_P - E\,\Delta s = 155 \text{ V } -\left(-4.0\times10^3 \text{ V/m}\right)\left(-3.0\times10^{-3}\text{ m}\right)= \boxed{143 \text{ V}}$$

c. Since the path from P to C is perpendicular to the electric field and no work is done in moving a charge along such a path, it follows that $\Delta V = 0$ V. Therefore, $V_C = V_P = \boxed{155 \text{ V}}$.

43. **REASONING** According to Equation 19.11b, the energy stored in a capacitor with capacitance C and potential V across its plates is $\text{Energy} = \frac{1}{2}CV^2$.

SOLUTION Therefore, solving Equation 19.11b for V, we have

$$V = \sqrt{\frac{2(\text{Energy})}{C}} = \sqrt{\frac{2(73 \text{ J})}{120\times10^{-6}\text{F}}} = \boxed{1.1\times10^3 \text{ V}}$$

47. **REASONING AND SOLUTION** Equation 19.10 gives the capacitance for a parallel plate capacitor filled with a dielectric of constant κ: $C = \kappa\varepsilon_0 A/d$. Solving for κ, we have

$$\kappa = \frac{Cd}{\varepsilon_0 A} = \frac{(7.0 \times 10^{-6}\,\text{F})(1.0 \times 10^{-5}\,\text{m})}{(8.85 \times 10^{-12}\,\text{F/m})(1.5\,\text{m}^2)} = \boxed{5.3}$$

51. **REASONING**   According to Equation 19.11b, the energy stored in a capacitor with a capacitance C and potential V across its plates is $\text{Energy} = \frac{1}{2}CV^2$. Once we determine how much energy is required to operate a 75-W light bulb for one minute, we can then use the expression for the energy to solve for V.

SOLUTION   The energy stored in the capacitor, which is equal to the energy required to operate a 75-W bulb for one minute (= 60 s), is

$$\text{Energy} = Pt = (75\,\text{W})(60\,\text{s}) = 4500\,\text{J}$$

Therefore, solving Equation 19.11b for V, we have

$$V = \sqrt{\frac{2(\text{Energy})}{C}} = \sqrt{\frac{2(4500\,\text{J})}{3.3\,\text{F}}} = \boxed{52\,\text{V}}$$

55. **REASONING**   If we assume that the motion of the proton and the electron is horizontal in the $+x$ direction, the motion of the proton is determined by Equation 2.8, $x = v_0 t + \frac{1}{2}a_p t^2$, where x is the distance traveled by the proton, v_0 is its initial speed, and a_p is its acceleration. If the distance between the capacitor places is d, then this relation becomes $\frac{1}{2}d = v_0 t + \frac{1}{2}a_p t^2$, or

$$d = 2v_0 t + a_p t^2 \qquad (1)$$

We can solve Equation (1) for the initial speed v_0 of the proton, but, first, we must determine the time t and the acceleration a_p of the proton . Since the proton strikes the negative plate at the same instant the electron strikes the positive plate, we can use the motion of the electron to determine the time t.

For the electron, $\frac{1}{2}d = \frac{1}{2}a_e t^2$, where we have taken into account the fact that the electron is released from rest. Solving this expression for t we have $t = \sqrt{d/a_e}$. Substituting this expression into Equation (1), we have

$$d = 2v_0 \sqrt{\frac{d}{a_e}} + \left(\frac{a_p}{a_e}\right)d \qquad (2)$$

The accelerations can be found by noting that the magnitudes of the forces on the electron and proton are equal, since these particles have the same magnitude of charge. The force on the electron is $F = eE = eV/d$, and the acceleration of the electron is, therefore,

$$a_e = \frac{F}{m_e} = \frac{eV}{m_e d} \tag{3}$$

Newton's second law requires that $m_e a_e = m_p a_p$, so that

$$\frac{a_p}{a_e} = \frac{m_e}{m_p} \tag{4}$$

Combining Equations (2), (3) and (4) leads to the following expression for v_0, the initial speed of the proton:

$$v_0 = \frac{1}{2}\left(1 - \frac{m_e}{m_p}\right)\sqrt{\frac{eV}{m_e}}$$

SOLUTION Substituting values into the expression above, we find

$$v_0 = \frac{1}{2}\left(1 - \frac{9.11\times10^{-31}\,\text{kg}}{1.67\times10^{-27}\,\text{kg}}\right)\sqrt{\frac{(1.60\times10^{-19}\text{C})(175\text{ V})}{9.11\times10^{-31}\,\text{kg}}} = \boxed{2.77\times10^6\,\text{m/s}}$$

---

57. **REASONING AND SOLUTION** The capacitance is given by

$$C = \frac{\kappa\varepsilon_0 A}{d} = \frac{5\left(8.85\times10^{-12}\,\text{F/m}\right)\left(5\times10^{-6}\,\text{m}^2\right)}{1\times10^{-8}\,\text{m}} = \boxed{2\times10^{-8}\,\text{F}}$$

---

61. **REASONING** The work W_{AB} done by an electric force in moving a charge q_0 from point A to point B and the electric potential difference $V_B - V_A$ are related according to

$V_B - V_A = \dfrac{-W_{AB}}{q_0}$ (Equation 19.4). This expression can be solved directly for q_0.

SOLUTION Solving Equation 19.4 for the charge q_0 gives

$$q_0 = \frac{-W_{AB}}{V_B - V_A} = \frac{W_{AB}}{V_A - V_B} = \frac{2.70\times10^{-3}\text{ J}}{50.0\text{ V}} = \boxed{5.40\times10^{-5}\text{ C}}$$

---

67. **REASONING** According to Equation 19.10, the capacitance of a parallel plate capacitor filled with a dielectric is $C = \kappa\varepsilon_0 A / d$, where κ is the dielectric constant, A is the area of one plate, and d is the distance between the plates.

From the definition of capacitance (Equation 19.8), $q = CV$. Thus, using Equation 19.10, we see that the charge q on a parallel plate capacitor that contains a dielectric is given by $q = (\kappa \varepsilon_0 A / d)V$. Since each dielectric occupies one-half of the volume between the plates, the area of each plate in contact with each material is $A/2$. Thus,

$$q_1 = \frac{\kappa_1 \varepsilon_0 (A/2)}{d} V = \frac{\kappa_1 \varepsilon_0 A}{2d} V \qquad \text{and} \qquad q_2 = \frac{\kappa_2 \varepsilon_0 (A/2)}{d} V = \frac{\kappa_2 \varepsilon_0 A}{2d} V$$

According to the problem statement, the total charge stored by the capacitor is

$$q_1 + q_2 = CV \qquad (1)$$

where q_1 and q_2 are the charges on the plates in contact with dielectrics 1 and 2, respectively.

Using the expressions for q_1 and q_2 above, Equation (1) becomes

$$CV = \frac{\kappa_1 \varepsilon_0 A}{2d} V + \frac{\kappa_2 \varepsilon_0 A}{2d} V = \frac{\kappa_1 \varepsilon_0 A + \kappa_2 \varepsilon_0 A}{2d} V = \frac{(\kappa_1 + \kappa_2)\varepsilon_0 A}{2d} V$$

This expression can be solved for C.

SOLUTION   Solving for C, we obtain $\boxed{C = \dfrac{\varepsilon_0 A(\kappa_1 + \kappa_2)}{2d}}$

---

69. **REASONING**   The charge stored by a capacitor is $q = CV$, according to Equation 19.8. The battery maintains a constant potential difference of $V = 12$ V between the plates of the capacitor while the dielectric is inserted. Inserting the dielectric causes the capacitance C to increase, so that with V held constant, the charge q must increase. Thus, additional charge flows onto the plates. To find the dielectric constant, we will apply Equation 19.8 to the capacitor filled with the dielectric material and then to the empty capacitor.

SOLUTION   According to Equations 19.8 and 19.10, the charge stored by the dielectric-filled capacitor is $q = \kappa C_0 V$, where κ is the dielectric constant, κC_0 is the capacitance of the capacitor with the dielectric inserted, C_0 is the capacitance of the empty capacitor, and V is the voltage provided by the battery. Solving for the dielectric constant gives Equation 1:

$$\kappa = \frac{q}{C_0 V} .$$

In this expression, C_0 and V are known, and we will deal with the unknown quantity q next.

The charge q stored by the dielectric-filled capacitor is the charge q_0 stored by the empty capacitor plus the additional charge Δq that arises when the dielectric is inserted. Therefore, we have

$$q = q_0 + \Delta q \,.$$

This result can be substituted into Equation 1 above. Δq is given, and a value for q_0 will be obtained in the next step.

According to Equation 19.8, the charge stored by the empty capacitor is $q_0 = C_0 V$.

Combining the results of each of these steps algebraically, we find that:

$$\kappa = \frac{q}{C_0 V} = \frac{q_0 + \Delta q}{C_0 V} = \frac{C_0 V + \Delta q}{C_0 V}$$

$$\kappa = \frac{C_0 V + \Delta q}{C_0 V} = 1 + \frac{\Delta q}{C_0 V} = 1 + \frac{2.6 \times 10^{-5}\ \text{C}}{(1.2 \times 10^{-6}\ \text{F})(12\ \text{V})} = \boxed{2.8}$$

---

71. **CONCEPTS** (i) Since both charges are positive, the individual electric-field contributions from q_A and q_B point away from each charge. Thus, at the midpoint they point in opposite directions, the contribution from q_A pointing to the right and the contribution from q_B pointing to the left. (ii) The magnitude of the net electric field is zero, because the electric field from q_A cancels the electric field from q_B. These charges have the same magnitude $|q|$, and the midpoint is the same distance r from each of them. According to the equation $E = k|q|/r^2$, then, each charge produces a field of the same magnitude. Since the individual electric-field contributions from q_A and q_B point in opposite directions at the midpoint, their vector sum is zero. (iii) The total electric potential at the midpoint is the algebraic sum of the individual contributions from each charge. According to $V = kq/r$, each contribution is positive, since each charge is positive. Thus, the total electric potential is also positive. (iv) No. The electric potential is a scalar quantity, not a vector quantity. Therefore, it has no direction associated with it.

CALCULATIONS As discussed in conceptual question (ii), the total electric field at the midpoint is:

$$\boxed{E = 0\ \text{N/C}}$$

Using $V = kq/r$, we find the total potential at the midpoint is

$$V_{\text{Total}} = \frac{kq_A}{r_A} + \frac{kq_B}{r_B}$$

Since $q_A = q_B = +2.4 \times 10^{-9}$ C and $r_A = r_B = 0.25$ m , we have

$$V_{\text{Total}} = \frac{2kq_A}{r_A} = \frac{2\left(8.99 \times 10^9 \text{ N} \cdot \text{m}^2/\text{C}^2\right)\left(+2.4 \times 10^{-9} \text{ C}\right)}{0.25 \text{ m}} = \boxed{170 \text{ V}}$$

CHAPTER 20 | *ELECTRIC CIRCUITS*

3. ***REASONING AND SOLUTION*** First determine the total charge delivered to the battery using Equation 20.1:

$$\Delta q = I\,\Delta t = (6.0 \text{ A})(5.0 \text{ h})\left(\frac{3600 \text{ s}}{1 \text{ h}}\right) = 1.1 \times 10^5 \text{ C}$$

To find the energy delivered to the battery, multiply this charge by the energy per unit charge (i.e., the voltage) to get

$$\text{Energy} = (\Delta q)V = (1.1 \times 10^5 \text{ C})(12 \text{ V}) = \boxed{1.3 \times 10^6 \text{ J}}$$

9. ***REASONING*** The number N of protons that strike the target is equal to the amount of electric charge Δq striking the target divided by the charge e of a proton, $N = (\Delta q)/e$. From Equation 20.1, the amount of charge is equal to the product of the current I and the time Δt. We can combine these two relations to find the number of protons that strike the target in 15 seconds.

   The heat Q that must be supplied to change the temperature of the aluminum sample of mass m by an amount ΔT is given by Equation 12.4 as $Q = cm\Delta T$, where c is the specific heat capacity of aluminum. The heat is provided by the kinetic energy of the protons and is equal to the number of protons that strike the target times the kinetic energy per proton. Using this reasoning, we can find the change in temperature of the block for the 15 second-time interval.

SOLUTION
a. The number N of protons that strike the target is

$$N = \frac{\Delta q}{e} = \frac{I\,\Delta t}{e} = \frac{(0.50 \times 10^{-6} \text{ A})(15 \text{ s})}{1.6 \times 10^{-19} \text{ C}} = \boxed{4.7 \times 10^{13}}$$

b. The amount of heat Q provided by the kinetic energy of the protons is

$$Q = (4.7 \times 10^{13} \text{ protons})(4.9 \times 10^{-12} \text{ J/proton}) = 230 \text{ J}$$

Since $Q = cm\Delta T$ and since Table 12.2 gives the specific heat of aluminum as $c = 9.00 \times 10^2$ J/(kg·C°), the change in temperature of the block is

$$\Delta T = \frac{Q}{cm} = \frac{230 \text{ J}}{\left(9.00 \times 10^2\,\dfrac{\text{J}}{\text{kg} \cdot \text{C}^\circ}\right)(15 \times 10^{-3} \text{ kg})} = \boxed{17 \text{ C}^\circ}$$

15. ***REASONING*** The resistance of a metal wire of length L, cross-sectional area A and resistivity ρ is given by Equation 20.3: $R = \rho L / A$. Solving for A, we have $A = \rho L / R$. We can use this expression to find the ratio of the cross-sectional area of the aluminum wire to that of the copper wire.

SOLUTION Forming the ratio of the areas and using resistivity values from Table 20.1, we have

$$\frac{A_{\text{aluminum}}}{A_{\text{copper}}} = \frac{\rho_{\text{aluminum}} L / R}{\rho_{\text{copper}} L / R} = \frac{\rho_{\text{aluminum}}}{\rho_{\text{copper}}} = \frac{2.82 \times 10^{-8}\ \Omega \cdot \text{m}}{1.72 \times 10^{-8}\ \Omega \cdot \text{m}} = \boxed{1.64}$$

---

19. ***REASONING*** We will ignore any changes in length due to thermal expansion. Although the resistance of each section changes with temperature, the total resistance of the composite does not change with temperature. Therefore,

$$\underbrace{\left(R_{\text{tungsten}}\right)_0 + \left(R_{\text{carbon}}\right)_0}_{\text{At room temperature}} = \underbrace{R_{\text{tungsten}} + R_{\text{carbon}}}_{\text{At temperature } T}$$

From Equation 20.5, we know that the temperature dependence of the resistance for a wire of resistance R_0 at temperature T_0 is given by $R = R_0[1 + \alpha(T - T_0)]$, where α is the temperature coefficient of resistivity. Thus,

$$\left(R_{\text{tungsten}}\right)_0 + \left(R_{\text{carbon}}\right)_0 = \left(R_{\text{tungsten}}\right)_0 (1 + \alpha_{\text{tungsten}} \Delta T) + \left(R_{\text{carbon}}\right)_0 (1 + \alpha_{\text{carbon}} \Delta T)$$

Since ΔT is the same for each wire, this simplifies to

$$\left(R_{\text{tungsten}}\right)_0 \alpha_{\text{tungsten}} = -\left(R_{\text{carbon}}\right)_0 \alpha_{\text{carbon}} \tag{1}$$

This expression can be used to find the ratio of the resistances. Once this ratio is known, we can find the ratio of the lengths of the sections with the aid of Equation 20.3 ($L = RA/\rho$).

SOLUTION From Equation (1), the ratio of the resistances of the two sections of the wire is

$$\frac{\left(R_{\text{tungsten}}\right)_0}{\left(R_{\text{carbon}}\right)_0} = -\frac{\alpha_{\text{carbon}}}{\alpha_{\text{tungsten}}} = -\frac{-0.0005\ [(\text{C}°)^{-1}]}{0.0045\ [(\text{C}°)^{-1}]} = \frac{1}{9}$$

Thus, using Equation 20.3, we find the ratio of the tungsten and carbon lengths to be

$$\frac{L_{\text{tungsten}}}{L_{\text{carbon}}} = \frac{\left(R_0 A / \rho\right)_{\text{tungsten}}}{\left(R_0 A / \rho\right)_{\text{carbon}}} = \frac{\left(R_{\text{tungsten}}\right)_0}{\left(R_{\text{carbon}}\right)_0}\left(\frac{\rho_{\text{carbon}}}{\rho_{\text{tungsten}}}\right) = \left(\frac{1}{9}\right)\left(\frac{3.5 \times 10^{-5}\ \Omega \cdot \text{m}}{5.6 \times 10^{-8}\ \Omega \cdot \text{m}}\right) = \boxed{70}$$

where we have used resistivity values from Table 20.1 and the fact that the two sections have the same cross-sectional areas.

27. **REASONING** According to Equation 6.10b, the energy used is Energy $= Pt$, where P is the power and t is the time. According to Equation 20.6a, the power is $P = IV$, where I is the current and V is the voltage. Thus, Energy $= IVt$, and we apply this result first to the dryer and then to the computer.

SOLUTION The energy used by the dryer is

$$\text{Energy} = Pt = IVt = (16 \text{ A})(240 \text{ V})(45 \text{ min}) \underbrace{\left(\frac{60 \text{ s}}{1.00 \text{ min}} \right)}_{\substack{\text{Converts minutes} \\ \text{to seconds}}} = 1.04 \times 10^7 \text{ J}$$

For the computer, we have

$$\text{Energy} = 1.04 \times 10^7 \text{ J} = IVt = (2.7 \text{ A})(120 \text{ V})t$$

Solving for t we find

$$t = \frac{1.04 \times 10^7 \text{ J}}{(2.7 \text{ A})(120 \text{ V})} = 3.21 \times 10^4 \text{ s} = \left(3.21 \times 10^4 \text{ s}\right)\left(\frac{1.00 \text{ h}}{3600 \text{ s}} \right) = \boxed{8.9 \text{ h}}$$

---

31. **REASONING AND SOLUTION** As a function of temperature, the resistance of the wire is given by Equation 20.5: $R = R_0 \left[1 + \alpha(T - T_0)\right]$, where α is the temperature coefficient of resistivity. From Equation 20.6c, we have $P = V^2 / R$. Combining these two equations, we have

$$P = \frac{V^2}{R_0 \left[1 + \alpha\left(T - T_0\right)\right]} = \frac{P_0}{1 + \alpha\left(T - T_0\right)}$$

where $P_0 = V^2/R_0$, since the voltage is constant. But $P = \frac{1}{2}P_0$, so we find

$$\frac{P_0}{2} = \frac{P_0}{1 + \alpha\left(T - T_0\right)} \qquad \text{or} \qquad 2 = 1 + \alpha\left(T - T_0\right)$$

Solving for T, we find

$$T = \frac{1}{\alpha} + T_0 = \frac{1}{0.0045 \text{ (C}^\circ)^{-1}} + 28^\circ = \boxed{250 \text{ °C}}$$

---

37. **REASONING** The average power is given by Equation 20.15c as $\overline{P} = V_{rms}^2 / R$. In this expression the rms voltage V_{rms} appears. However, we seek the peak voltage V_0. The relation between the two types of voltage is given by Equation 20.13 as $V_{rms} = V_0 / \sqrt{2}$, so we can obtain the peak voltage by using Equation 20.13 to substitute into Equation 20.15c.

SOLUTION Substituting V_{rms} from Equation 20.13 into Equation 20.15c gives

$$\bar{P} = \frac{V_{rms}^2}{R} = \frac{\left(V_0/\sqrt{2}\right)^2}{R} = \frac{V_0^2}{2R}$$

Solving for the peak voltage V_0 gives

$$V_0 = \sqrt{2R\bar{P}} = \sqrt{2(4.0\ \Omega)(55\ W)} = \boxed{21\ V}$$

39. **REASONING**

a. We can obtain the frequency of the alternating current by comparing this specific expression for the current with the more general one in Equation 20.8.

b. The resistance of the light bulb is, according to Equation 20.14, equal to the rms-voltage divided by the rms-current. The rms-voltage is given, and we can obtain the rms-current by dividing the peak current by $\sqrt{2}$, as expressed by Equation 20.12.

c. The average power is given by Equation 20.15a as the product of the rms-current and the rms-voltage.

SOLUTION

a. By comparing $I = (0.707\ A)\sin\left[(314\ Hz)\,t\right]$ with the general expression (see Equation 20.8) for the current in an ac circuit, $I = I_0 \sin 2\pi f t$, we see that

$$2\pi f t = (314\ Hz)\,t \quad \text{or} \quad f = \frac{314\ Hz}{2\pi} = \boxed{50.0\ Hz}$$

b. The resistance is equal to V_{rms}/I_{rms}, where the rms-current is related to the peak current I_0 by $I_{rms} = I_0/\sqrt{2}$. Thus, the resistance of the light bulb is

$$R = \frac{V_{rms}}{I_{rms}} = \frac{V_{rms}}{\dfrac{I_0}{\sqrt{2}}} = \frac{\sqrt{2}(120.0\ V)}{0.707\ A} = \boxed{2.40\times10^2\ \Omega} \tag{20.14}$$

c. The average power is the product of the rms-current and rms-voltage:

$$\bar{P} = I_{rms}V_{rms} = \left(\frac{I_0}{\sqrt{2}}\right)V_{rms} = \left(\frac{0.707\ A}{\sqrt{2}}\right)(120.0\ V) = \boxed{60.0\ W} \tag{20.15a}$$

41. **REASONING** The equivalent series resistance R_s is the sum of the resistances of the three resistors. The potential difference V can be determined from Ohm's law according to $V = IR_s$.

SOLUTION

a. The equivalent resistance is

$$R_s = 25\ \Omega + 45\ \Omega + 75\ \Omega = \boxed{145\ \Omega}$$

b. The potential difference across the three resistors is

$$V = IR_s = (0.51\ \text{A})(145\ \Omega) = \boxed{74\ \text{V}}$$

---

43. **REASONING** Using Ohm's law (Equation 20.2) we can write an expression for the voltage across the original circuit as $V = I_0 R_0$. When the additional resistor R is inserted in series, assuming that the battery remains the same, the voltage across the new combination is given by $V = I(R + R_0)$. Since V is the same in both cases, we can write $I_0 R_0 = I(R + R_0)$. This expression can be solved for R_0.

SOLUTION Solving for R_0, we have

$$I_0 R_0 - IR_0 = IR \quad \text{or} \quad R_0(I_0 - I) = IR$$

Therefore, we find that

$$R_0 = \frac{IR}{I_0 - I} = \frac{(12.0\ \text{A})(8.00\ \Omega)}{15.0\ \text{A} - 12.0\ \text{A}} = \boxed{32\ \Omega}$$

---

47. **REASONING**

a. The greatest voltage for the battery is the voltage that generates the maximum current I that the circuit can tolerate. Once this maximum current is known, the voltage can be calculated according to Ohm's law, as the current times the equivalent circuit resistance for the three resistors in series. To determine the maximum current we note that the power P dissipated in each resistance R is $P = I^2 R$ according to Equation 20.6b. Since the power rating and resistance are known for each resistor, the maximum current that can be tolerated by a resistor is $I = \sqrt{P/R}$. By examining this maximum current for each resistor, we will be able to identify the maximum current that the circuit can tolerate.

b. The battery delivers power to the circuit that is given by the battery voltage times the current, according to Equation 20.6a.

SOLUTION

a. Solving Equation 20.6b for the current, we find that the maximum current for each resistor is as follows:

$$\underbrace{I = \sqrt{\frac{P}{R}} = \sqrt{\frac{4.0\ \text{W}}{2.0\ \Omega}} = 1.4\ \text{A}}_{2.0\text{-}\Omega\ \text{resistor}} \qquad \underbrace{I = \sqrt{\frac{10.0\ \text{W}}{12.0\ \Omega}} = 0.913\ \text{A}}_{12.0\text{-}\Omega\ \text{resistor}} \qquad \underbrace{I = \sqrt{\frac{5.0\ \text{W}}{3.0\ \Omega}} = 1.3\ \text{A}}_{3.0\text{-}\Omega\ \text{resistor}}$$

The smallest of these three values is 0.913 A and is the maximum current that the circuit can tolerate. Since the resistors are connected in series, the equivalent resistance of the circuit is

$$R_{\text{S}} = 2.0\ \Omega + 12.0\ \Omega + 3.0\ \Omega = 17.0\ \Omega$$

Using Ohm's law with this equivalent resistance and the maximum current of 0.913 A reveals that the maximum battery voltage is

$$V = IR_{\text{S}} = (0.913\ \text{A})(17.0\ \Omega) = \boxed{15.5\ \text{V}}$$

b. The power delivered by the battery in part (a) is given by Equation 20.6a as

$$P = IV = (0.913\ \text{A})(15.5\ \text{V}) = \boxed{14.2\ \text{W}}$$

---

53. **REASONING** When the switch is open, no current goes to the resistor R_2. Current exists only in R_1, so it is the equivalent resistance. When the switch is closed, current is sent to both resistors. Since they are wired in parallel, we can use Equation 20.17 to find the equivalent resistance. Whether the switch is open or closed, the power P delivered to the circuit can be found from the relation $P = V^2 / R$ (Equation 20.6c), where V is the battery voltage and R is the equivalent resistance.

SOLUTION

a. When the switch is open, there is current only in resistor R_1. Thus, the equivalent resistance is $R_1 = \boxed{65.0\ \Omega}$.

b. When the switch is closed, there is current in both resistors and, furthermore, they are wired in parallel. The equivalent resistance is

$$\frac{1}{R_{\text{P}}} = \frac{1}{R_1} + \frac{1}{R_2} = \frac{1}{65.0\ \Omega} + \frac{1}{96.0\ \Omega} \qquad \text{or} \qquad R_{\text{P}} = \boxed{38.8\ \Omega} \tag{20.17}$$

c. When the switch is open, the power delivered to the circuit by the battery is given by $P = V^2 / R_1$, since the only resistance in the circuit is R_1. Thus, the power is

$$P = \frac{V^2}{R_1} = \frac{(9.00\ \text{V})^2}{65.0\ \Omega} = \boxed{1.25\ \text{W}} \tag{20.6}$$

d. When the switch is closed, the power delivered to the circuit is $P = V^2 / R_{\text{P}}$, where R_{P} is the equivalent resistance of the two resistors wired in parallel:

$$P = \frac{V^2}{R_{\text{P}}} = \frac{(9.00\ \text{V})^2}{38.8\ \Omega} = \boxed{2.09\ \text{W}} \tag{20.6}$$

55. **REASONING** Since the resistors are connected in parallel, the voltage across each one is the same and can be calculated from Ohm's Law (Equation 20.2: $V = IR$). Once the voltage across each resistor is known, Ohm's law can again be used to find the current in the second resistor. The total power consumed by the parallel combination can be found calculating the power consumed by each resistor from Equation 20.6b: $P = I^2 R$. Then, the total power consumed is the sum of the power consumed by each resistor.

SOLUTION Using data for the second resistor, the voltage across the resistors is equal to

$$V = IR = (3.00 \text{ A})(64.0 \text{ } \Omega) = 192 \text{ V}$$

a. The current through the 42.0-Ω resistor is

$$I = \frac{V}{R} = \frac{192 \text{ V}}{42.0 \text{ } \Omega} = \boxed{4.57 \text{ A}}$$

b. The power consumed by the 42.0-Ω resistor is

$$P = I^2 R = (4.57 \text{ A})^2 (42.0 \text{ } \Omega) = 877 \text{ W}$$

while the power consumed by the 64.0-Ω resistor is

$$P = I^2 R = (3.00 \text{ A})^2 (64.0 \text{ } \Omega) = 576 \text{ W}$$

Therefore the total power consumed by the two resistors is $877 \text{ W} + 576 \text{ W} = \boxed{1450 \text{ W}}$.

---

59. **REASONING AND SOLUTION** The aluminum and copper portions may be viewed a being connected in parallel since the same voltage appears across them. Using a and b to denote the inner and outer radii, respectively, and using Equation 20.3 to express the resistance for each portion, we find for the equivalent resistance that

$$\frac{1}{R_{\text{p}}} = \frac{1}{R_{\text{Al}}} + \frac{1}{R_{\text{Cu}}} = \frac{A_{\text{Cu}}}{\rho_{\text{Cu}}L} + \frac{A_{\text{Al}}}{\rho_{\text{Al}}L} = \frac{\pi a^2}{\rho_{\text{Cu}}L} + \frac{\pi \left(b^2 - a^2 \right)}{\rho_{\text{Al}}L}$$

$$= \frac{\pi \left(2.00 \times 10^{-3} \text{ m} \right)^2}{\left(1.72 \times 10^{-8} \text{ } \Omega \cdot \text{m} \right)(1.50 \text{ m})} + \frac{\pi \left[\left(3.00 \times 10^{-3} \text{ m} \right)^2 - \left(2.00 \times 10^{-3} \text{ m} \right)^2 \right]}{\left(2.82 \times 10^{-8} \text{ } \Omega \cdot \text{m} \right)(1.50 \text{ m})} = \boxed{0.00116 \text{ } \Omega}$$

We have taken resistivity values for copper and aluminum from Table 20.1

---

63. **REASONING** To find the current, we will use Ohm's law, together with the proper equivalent resistance. The coffee maker and frying pan are in series, so their equivalent resistance is given by Equation 20.16 as $R_{\text{coffee}} + R_{\text{pan}}$. This total resistance is in parallel with the resistance of the bread maker, so the equivalent resistance of the parallel combination can be obtained from Equation 20.17 as $R_{\text{p}}^{-1} = (R_{\text{coffee}} + R_{\text{pan}})^{-1} + R_{\text{bread}}^{-1}$.

SOLUTION Using Ohm's law and the expression developed above for R_p^{-1}, we find

$$I = \frac{V}{R_p} = V\left(\frac{1}{R_{coffee} + R_{pan}} + \frac{1}{R_{bread}}\right) = (120 \text{ V})\left(\frac{1}{14\,\Omega + 16\,\Omega} + \frac{1}{23\,\Omega}\right) = \boxed{9.2 \text{ A}}$$

65. **REASONING** When two or more resistors are in series, the equivalent resistance is given by Equation 20.16: $R_s = R_1 + R_2 + R_3 + \dots$. Likewise, when resistors are in parallel, the expression to be solved to find the equivalent resistance is given by Equation 20.17: $\frac{1}{R_p} = \frac{1}{R_1} + \frac{1}{R_2} + \frac{1}{R_3} + \dots$. We will successively apply these to the individual resistors in the figure in the text beginning with the resistors on the right side of the figure.

SOLUTION Since the 4.0-Ω and the 6.0-Ω resistors are in series, the equivalent resistance of the combination of those two resistors is 10.0 Ω. The 9.0-Ω and 8.0-Ω resistors are in parallel; their equivalent resistance is 4.24 Ω. The equivalent resistances of the parallel combination (9.0 Ω and 8.0 Ω) and the series combination (4.0 Ω and the 6.0 Ω) are in parallel; therefore, their equivalent resistance is 2.98 Ω. The 2.98-Ω combination is in series with the 3.0-Ω resistor, so that equivalent resistance is 5.98 Ω. Finally, the 5.98-Ω combination and the 20.0-Ω resistor are in parallel, so the equivalent resistance between the points A and B is $\boxed{4.6\ \Omega}$.

71. **REASONING** The power P delivered to the circuit is, according to Equation 20.6c, $P = V^2 / R_{12345}$, where V is the voltage of the battery and R_{12345} is the equivalent resistance of the five-resistor circuit. The voltage and power are known, so that the equivalent resistance can be calculated. We will use our knowledge of resistors wired in series and parallel to evaluate R_{12345} in terms of the resistance R of each resistor. In this manner we will find the value for R.

SOLUTION First we note that all the resistors are equal, so $R_1 = R_2 = R_3 = R_4 = R_5 = R$. We can find the equivalent resistance R_{12345} as follows. The resistors R_3 and R_4 are in series, so the equivalent resistance R_{34} of these two is $R_{34} = R_3 + R_4 = 2R$. The resistors R_2, R_{34}, and R_5 are in parallel, and the reciprocal of the equivalent resistance R_{2345} is

$$\frac{1}{R_{2345}} = \frac{1}{R_2} + \frac{1}{R_{34}} + \frac{1}{R_5} = \frac{1}{R} + \frac{1}{2R} + \frac{1}{R} = \frac{5}{2R}$$

so $R_{2345} = 2R/5$. The resistor R_1 is in series with R_{2345}, and the equivalent resistance of this combination is the equivalent resistance of the circuit. Thus, we have

$$R_{12345} = R_1 + R_{2345} = R + \frac{2R}{5} = \frac{7R}{5}$$

The power delivered to the circuit is

$$P = \frac{V^2}{R_{12345}} = \frac{V^2}{\left(\dfrac{7R}{5}\right)}$$

Solving for the resistance R, we find that

$$R = \frac{5V^2}{7P} = \frac{5(45 \text{ V})^2}{7(58 \text{ W})} = \boxed{25 \text{ }\Omega}$$

---

73. **REASONING** The terminal voltage of the battery is given by $V_{\text{terminal}} = \text{Emf} - Ir$, where r is the internal resistance of the battery. Since the terminal voltage is observed to be one-half of the emf of the battery, we have $V_{\text{terminal}} = \text{Emf}/2$ and $I = \text{Emf}/(2r)$. From Ohm's law, the equivalent resistance of the circuit is $R = \text{emf}/I = 2r$. We can also find the equivalent resistance of the circuit by considering that the identical bulbs are in parallel across the battery terminals, so that the equivalent resistance of the N bulbs is found from

$$\frac{1}{R_{\text{p}}} = \frac{N}{R_{\text{bulb}}} \qquad \text{or} \qquad R_{\text{p}} = \frac{R_{\text{bulb}}}{N}$$

This equivalent resistance is in series with the battery, so we find that the equivalent resistance of the circuit is

$$R = 2r = \frac{R_{\text{bulb}}}{N} + r$$

This expression can be solved for N.

SOLUTION Solving the above expression for N, we have

$$N = \frac{R_{\text{bulb}}}{2r - r} = \frac{R_{\text{bulb}}}{r} = \frac{15 \text{ }\Omega}{0.50 \text{ }\Omega} = \boxed{30}$$

---

79. **REASONING** The current I can be found by using Kirchhoff's loop rule. Once the current is known, the voltage between points A and B can be determined.

SOLUTION
a. We assume that the current is directed clockwise around the circuit. Starting at the upper-left corner and going clockwise around the circuit, we set the potential drops equal to the potential rises:

$$\underbrace{(5.0 \text{ }\Omega)I + (27 \text{ }\Omega)I + 10.0 \text{ V} + (12 \text{ }\Omega)I + (8.0 \text{ }\Omega)I}_{\text{Potential drops}} = \underbrace{30.0 \text{ V}}_{\text{Potential rises}}$$

Solving for the current gives $\boxed{I = 0.38 \text{ A}}$.

b. The voltage between points A and B is

$$V_{AB} = 30.0 \text{ V} - (0.38 \text{ A})(27 \text{ }\Omega) = \boxed{2.0 \times 10^1 \text{ V}}$$

c. $\boxed{\text{Point B}}$ is at the higher potential.

---

85. **REASONING** In preparation for applying Kirchhoff's rules, we now choose the currents in each resistor. The directions of the currents are arbitrary, and should they be incorrect, the currents will turn out to be negative quantities. Having chosen the currents, we also mark the ends of the resistors with the plus and minus signs that indicate that the currents are directed from higher (+) toward lower (−) potential. These plus and minus signs will guide us when we apply Kirchhoff's loop rule.

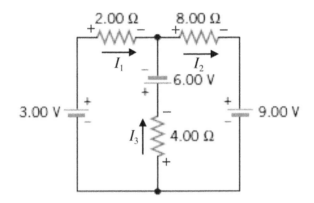

SOLUTION Applying the junction rule to junction B, we find

$$\underbrace{I_1 + I_3}_{\text{Into junction}} = \underbrace{I_2}_{\text{Out of junction}} \tag{1}$$

Applying the loop rule to loop ABCD (going clockwise around the loop), we obtain

$$\underbrace{I_1(2.00 \text{ }\Omega)}_{\text{Potential drops}} = \underbrace{6.00 \text{ V} + I_3(4.00 \text{ }\Omega) + 3.00 \text{ V}}_{\text{Potential rises}} \tag{2}$$

Applying the loop rule to loop BEFC (going clockwise around the loop), we obtain

$$\underbrace{I_2(8.00 \text{ }\Omega) + 9.00 \text{ V} + I_3(4.00 \text{ }\Omega) + 6.00 \text{ V}}_{\text{Potential drops}} = \underbrace{0}_{\text{Potential rises}} \tag{3}$$

Substituting I_2 from Equation (1) into Equation (3) gives

$$\left(I_1 + I_3\right)\left(8.00 \ \Omega\right) + 9.00 \ \text{V} + I_3\left(4.00 \ \Omega\right) + 6.00 \ \text{V} = 0$$

$$I_1\left(8.00 \ \Omega\right) + I_3\left(12.00 \ \Omega\right) + 15.00 \ \text{V} = 0 \qquad (4)$$

Solving Equation (2) for I_1 gives

$$I_1 = 4.50 \ \text{A} + I_3\left(2.00\right)$$

This result may be substituted into Equation (4) to show that

$$\left[4.50 \ \text{A} + I_3\left(2.00\right)\right]\left(8.00 \ \Omega\right) + I_3\left(12.00 \ \Omega\right) + 15.00 \ \text{V} = 0$$

$$I_3\left(28.00 \ \Omega\right) + 51.00 \ \text{V} = 0 \quad \text{or} \quad I_3 = \frac{-51.00 \ \text{V}}{28.00 \ \Omega} = \boxed{-1.82 \ \text{A}}$$

The minus sign indicates that $\boxed{\text{the current in the 4.00-}\Omega \text{ resistor is directed downward}}$, rather than upward as selected arbitrarily in the drawing.

---

87. **REASONING** As discussed in Section 20.11, some of the current (6.20 mA) goes directly through the galvanometer and the remainder I goes through the shunt resistor. Since the resistance of the coil R_C and the shunt resistor R are in parallel, the voltage across each is the same. We will use this fact to determine how much current goes through the shunt resistor. This value, plus the 6.20 mA that goes through the galvanometer, is the maximum current that this ammeter can read.

SOLUTION The voltage across the coil resistance is equal to the voltage across the shunt resistor, so

$$\underbrace{\left(6.20 \times 10^{-3} \text{A}\right)\left(20.0 \ \Omega\right)}_{\substack{\text{Voltage across} \\ \text{coil resistance}}} = \underbrace{\left(I\right)\left(24.8 \times 10^{-3} \ \Omega\right)}_{\substack{\text{Voltage across} \\ \text{shunt resistor}}}$$

So $I = 5.00$ A. The maximum current is 5.00 A + 6.20 mA = $\boxed{5.01 \ \text{A}}$.

---

91. **REASONING AND SOLUTION** For the 20.0 V scale

$$V_1 = I(R_1 + R_c)$$

For the 30.0 V scale

$$V_2 = I(R_2 + R_c)$$

Subtracting and rearranging yields

$$I = \frac{V_2 - V_1}{R_2 - R_1} = \frac{30.0 \text{ V} - 20.0 \text{ V}}{2930 \ \Omega - 1680 \ \Omega} = \boxed{8.00 \times 10^{-3} \text{ A}}$$

Substituting this value into either of the equations for V_1 or V_2 gives $R_c = \boxed{820 \ \Omega}$.

---

95. **REASONING** The equivalent capacitance C_S of a set of three capacitors connected in series is given by $\dfrac{1}{C_S} = \dfrac{1}{C_1} + \dfrac{1}{C_2} + \dfrac{1}{C_3}$ (Equation 20.19). In this case, we know that the equivalent capacitance is $C_S = 3.00 \ \mu\text{F}$, and the capacitances of two of the individual capacitors in this series combination are $C_1 = 6.00 \ \mu\text{F}$ and $C_2 = 9.00 \ \mu\text{F}$. We will use Equation 20.19 to determine the remaining capacitance C_3.

SOLUTION Solving Equation 20.19 for C_3, we obtain

$$\frac{1}{C_3} = \frac{1}{C_S} - \frac{1}{C_1} - \frac{1}{C_2} \qquad \text{or} \qquad C_3 = \frac{1}{\dfrac{1}{C_S} - \dfrac{1}{C_1} - \dfrac{1}{C_2}}$$

Therefore, the third capacitance is

$$C_3 = \frac{1}{\dfrac{1}{3.00 \ \mu\text{F}} - \dfrac{1}{6.00 \ \mu\text{F}} - \dfrac{1}{9.00 \ \mu\text{F}}} = \boxed{18 \ \mu\text{F}}$$

---

101. **REASONING** When two or more capacitors are in series, the equivalent capacitance of the combination can be obtained from Equation 20.19, $\dfrac{1}{C_S} = \dfrac{1}{C_1} + \dfrac{1}{C_2} + \dfrac{1}{C_3}....$ Equation 20.18 gives the equivalent capacitance for two or more capacitors in parallel: $C_p = C_1 + C_2 + C_3 + ...$. The energy stored in a capacitor is given by $\frac{1}{2}CV^2$, according to Equation 19.11. Thus, the energy stored in the series combination is $\frac{1}{2}C_S V_S^2$, where

$$\frac{1}{C_S} = \frac{1}{7.0 \ \mu\text{F}} + \frac{1}{3.0 \ \mu\text{F}} = 0.476 \left(\mu\text{F}\right)^{-1} \qquad \text{or} \qquad C_S = \frac{1}{0.476 \left(\mu\text{F}\right)^{-1}} = 2.10 \ \mu\text{F}$$

Similarly, the energy stored in the parallel combination is $\frac{1}{2}C_p V_p^2$ where

$$C_p = 7.0 \ \mu\text{F} + 3.0 \ \mu\text{F} = 10.0 \ \mu\text{F}$$

The voltage required to charge the parallel combination of the two capacitors to the same total energy as the series combination can be found by equating the two energy expressions and solving for V_p.

SOLUTION Equating the two expressions for the energy, we have

$$\tfrac{1}{2}C_s V_s^2 = \tfrac{1}{2}C_p V_p^2$$

Solving for V_p, we obtain the result

$$V_p = V_s \sqrt{\frac{C_s}{C_p}} = (24\text{ V})\sqrt{\frac{2.10\ \mu\text{F}}{10.0\ \mu\text{F}}} = \boxed{11\text{V}}$$

103.**REASONING** The charge q on a discharging capacitor in a RC circuit is given by Equation 20.22: $q = q_0 e^{-t/RC}$, where q_0 is the original charge at time $t = 0$ s. Once t (time for one pulse) and the ratio q/q_0 are known, this expression can be solved for C.

SOLUTION Since the pacemaker delivers 81 pulses per minute, the time for one pulse is

$$\frac{1\text{ min}}{81\text{ pulses}} \times \frac{60.0\text{ s}}{1.00\text{ min)}} = 0.74\text{ s/pulse}$$

Since one pulse is delivered every time the fully-charged capacitor loses 63.2% of its original charge, the charge remaining is 36.8% of the original charge. Thus, we have $q = (0.368)q_0$, or $q/q_0 = 0.368$.

From Equation 20.22, we have

$$\frac{q}{q_0} = e^{-t/RC}$$

Taking the natural logarithm of both sides, we have,

$$\ln\left(\frac{q}{q_0}\right) = -\frac{t}{RC}$$

Solving for C, we find

$$C = \frac{-t}{R\,\ln(q/q_0)} = \frac{-(0.74\text{ s})}{(1.8\times10^6\ \Omega)\,\ln(0.368)} = \boxed{4.1\times10^{-7}\text{ F}}$$

109.**REASONING** To find the equivalent capacitance of the three capacitors, we must first, following $C_p = C_1 + C_2 + C_3 + \text{L}$ (Equation 20.18), add the capacitances of the two parallel capacitors together. We must then combine the result of Equation 20.18 with the remaining capacitance in accordance with $\dfrac{1}{C_S} = \dfrac{1}{C_1} + \dfrac{1}{C_2} + \dfrac{1}{C_3} + L$ (Equation 20.19). As Equation 20.19 shows, combining capacitors in series *decreases* the overall capacitance, and the resulting equivalent capacitance C_S is *smaller* than any of the capacitances being added. Therefore, the way to maximize the overall equivalent capacitance is to choose the largest capacitance (C_1) to be connected in series with the parallel combination C_{23} of the smaller capacitances.

SOLUTION When C_2 and C_3 are connected in parallel, their equivalent capacitance C_{23} is, from $C_P = C_1 + C_2 + C_3 + L$ (Equation 20.18),

$$C_{23} = C_2 + C_3 \tag{1}$$

When the equivalent capacitance C_{23} is connected in series with C_1, the resulting equivalent capacitance C_S is, according to Equation 20.19 and Equation (1),

$$C_S = \left(\frac{1}{C_S} \right)^{-1} = \left(\frac{1}{C_1} + \frac{1}{C_{23}} \right)^{-1} = \left(\frac{1}{C_1} + \frac{1}{C_2 + C_3} \right)^{-1} = \left(\frac{1}{67~\mu F} + \frac{1}{45~\mu F + 33~\mu F} \right)^{-1} = \boxed{36~\mu F}$$

---

111. ***REASONING AND SOLUTION*** Ohm's law (Equation 20.2), $V = IR$, gives the result directly:

$$R = \frac{V}{I} = \frac{9.0~\text{V}}{0.11~\text{A}} = \boxed{82~\Omega}$$

---

117. ***REASONING*** Since we know that the current in the 8.00-Ω resistor is 0.500 A, we can use Ohm's law ($V = IR$) to find the voltage across the 8.00-Ω resistor. The 8.00-Ω resistor and the 16.0-Ω resistor are in parallel; therefore, the voltages across them are equal. Thus, we can also use Ohm's law to find the current through the 16.0-Ω resistor. The currents that flow through the 8.00-Ω and the 16.0-Ω resistors combine to give the total current that flows through the 20.0-Ω resistor. Similar reasoning can be used to find the current through the 9.00-Ω resistor.

SOLUTION
a. The voltage across the 8.00-Ω resistor is $V_8 = (0.500~\text{A})(8.00~\Omega) = 4.00~\text{V}$. Since this is also the voltage that is across the 16.0-Ω resistor, we find that the current through the 16.0-Ω resistor is $I_{16} = (4.00~\text{V})/(16.0~\Omega) = 0.250~\text{A}$. Therefore, the total current that flows through the 20.0-Ω resistor is

$$I_{20} = 0.500~\text{A} + 0.250~\text{A} = \boxed{0.750~\text{A}}$$

b. The 8.00-Ω and the 16.0-Ω resistors are in parallel, so their equivalent resistance can be obtained from Equation 20.17, $\dfrac{1}{R_{\text{p}}} = \dfrac{1}{R_1} + \dfrac{1}{R_2} + \dfrac{1}{R_3} + ...$, and is equal to 5.33 Ω. Therefore, the equivalent resistance of the upper branch of the circuit is $R_{\text{upper}} = 5.33\ \Omega + 20.0\ \Omega = 25.3\ \Omega$, since the 5.33-$\Omega$ resistance is in series with the 20.0-Ω resistance. Using Ohm's law, we find that the voltage across the upper branch must be $V = (0.750\ \text{A})(25.3\ \Omega) = 19.0\ \text{V}$. Since the lower branch is in parallel with the upper branch, the voltage across both branches must be the same. Therefore, the current through the 9.00-Ω resistor is, from Ohm's law,

$$I_9 = \frac{V_{\text{lower}}}{R_9} = \frac{19.0\ \text{V}}{9.00\ \Omega} = \boxed{2.11\ \text{A}}$$

---

119. **REASONING** The resistance of one of the wires in the extension cord is given by

Equation 20.3: $R = \rho L / A$, where the resistivity of copper is $\rho = 1.72 \times 10^{-8}\ \Omega \cdot \text{m}$, according to Table 20.1. Since the two wires in the cord are in series with each other, their total resistance is $R_{\text{cord}} = R_{\text{wire 1}} + R_{\text{wire 2}} = 2\rho L / A$. Once we find the equivalent resistance of the entire circuit (extension cord + trimmer), Ohm's law can be used to find the voltage applied to the trimmer.

SOLUTION
a. The resistance of the extension cord is

$$R_{\text{cord}} = \frac{2\rho L}{A} = \frac{2(1.72 \times 10^{-8}\ \Omega \cdot \text{m})(46\ \text{m})}{1.3 \times 10^{-6}\ \text{m}^2} = \boxed{1.2\ \Omega}$$

b. The total resistance of the circuit (cord + trimmer) is, since the two are in series,

$$R_{\text{s}} = 1.2\ \Omega + 15.0\ \Omega = 16.2\ \Omega$$

Therefore from Ohm's law (Equation 20.2: $V = IR$), the current in the circuit is

$$I = \frac{V}{R_{\text{s}}} = \frac{120\ \text{V}}{16.2\ \Omega} = 7.4\ \text{A}$$

Finally, the voltage applied to the trimmer alone is (again using Ohm's law),

$$V_{\text{trimmer}} = (7.4\ \text{A})(15.0\ \Omega) = \boxed{110\ \text{V}}$$

---

121. **REASONING** The resistance of a metal wire of length L, cross-sectional area A and resistivity ρ is given by Equation 20.3: $R = \rho L / A$. The volume V_2 of the new wire will be the same as the original volume V_1 of the wire, where volume is the product of length and

cross-sectional area. Thus, $V_1 = V_2$ or $A_1L_1 = A_2L_2$. Since the new wire is three times longer than the first wire, we can write

$$A_1L_1 = A_2L_2 = A_2(3L_1) \quad \text{or} \quad A_2 = A_1/3$$

We can form the ratio of the resistances, use this expression for the area A_2, and find the new resistance.

SOLUTION The resistance of the new wire is determined as follows:

$$\frac{R_2}{R_1} = \frac{\rho L_2/A_2}{\rho L_1/A_1} = \frac{L_2 A_1}{L_1 A_2} = \frac{(3L_1)A_1}{L_1(A_1/3)} = 9$$

Solving for R_2, we find that

$$R_2 = 9R_1 = 9(21.0 \ \Omega) = \boxed{189 \ \Omega}$$

---

123. **REASONING** The foil effectively converts the capacitor into two capacitors in series. Equation 19.10 gives the expression for the capacitance of a capacitor of plate area A and plate separation d (no dielectric): $C_0 = \varepsilon_0 A/d$. We can use this expression to determine the capacitance of the individual capacitors created by the presence of the foil. Then using the fact that the "two capacitors" are in series, we can use Equation 20.19 to find the equivalent capacitance of the system.

SOLUTION Since the foil is placed one-third of the way from one plate of the original capacitor to the other, we have $d_1 = (2/3)d$, and $d_2 = (1/3)d$. Then

$$C_1 = \frac{\varepsilon_0 A}{(2/3)d} = \frac{3\varepsilon_0 A}{2d}$$

and

$$C_2 = \frac{\varepsilon_0 A}{(1/3)d} = \frac{3\varepsilon_0 A}{d}$$

Since these two capacitors are effectively in series, it follows that

$$\frac{1}{C_s} = \frac{1}{C_1} + \frac{1}{C_2} = \frac{1}{3\varepsilon_0 A/(2d)} + \frac{1}{3\varepsilon_0 A/d} = \frac{3d}{3\varepsilon_0 A} = \frac{d}{\varepsilon_0 A}$$

But $C_0 = \varepsilon_0 A/d$, so that $d/(\varepsilon_0 A) = 1/C_0$, and we have

$$\frac{1}{C_s} = \frac{1}{C_0} \qquad \text{or} \qquad \boxed{C_s = C_0}$$

125. **CONCEPTS** (i) The power is given by Equation 20.6c: $P = V^2/R$. (ii) Since there is only one bulb in the circuit, V is the voltage of the battery: $V = 48$ V. (iii) Since the bulbs are identical and are wired in series, each receives one-half the battery voltage V. The power delivered to each bulb is:

$$P = \frac{\left(\frac{1}{2}V\right)^2}{R} = \frac{1}{4}\left(\frac{V^2}{R}\right)$$

This result shows that the power delivered to each bulb in the series circuit is only one-fourth the power delivered to the single-bulb circuit. Thus, the brightness of each bulb decreases. (iv) Since the bulbs are wired in parallel, each receives the full battery voltage V. Thus the power delivered to each bulb remains the same as when only one bulb is in the circuit, so the brightness of each bulb does not change.

CALCULATIONS (a) When only one bulb is in the circuit, the power that the bulb receives is

$$P = \frac{V^2}{R} = \frac{(48 \text{ V})^2}{240 \text{ } \Omega} = \boxed{9.6 \text{ W}}$$

(b) When the two (identical) light bulbs are wired in series, each receives one-half the battery voltage V. The power delivered to each bulb of resistance R is, then,

$$P = \frac{\left(\frac{1}{2}V\right)^2}{R} = \frac{\left[\frac{1}{2}(48 \text{ V})\right]^2}{240 \text{ } \Omega} = \boxed{2.4 \text{ W}}$$

As expected the power delivered to each bulb is only one-fourth the power delivered when there is only one bulb in the circuit.

(c) When the two bulbs are wired in parallel, the voltage across each is the same as the voltage of the battery. Therefore, the power delivered to each bulb is given by

$$P = \frac{V^2}{R} = \frac{(48 \text{ V})^2}{240 \text{ } \Omega} = \boxed{9.6 \text{ W}}$$

As expected, the power delivered, and hence the brightness, does not change relative to that in the single-bulb circuit.

CHAPTER 21 | *MAGNETIC FORCES AND MAGNETIC FIELDS*

1. **REASONING** The electron's acceleration is related to the net force ΣF acting on it by Newton's second law: $a = \Sigma F/m$ (Equation 4.1), where m is the electron's mass. Since we are ignoring the gravitational force, the net force is that caused by the magnetic force, whose magnitude is expressed by Equation 21.1 as $F = |q_0|vB \sin \theta$. Thus, the magnitude of the electron's acceleration can be written as $a = (|q_0|vB \sin \theta)/m$.

   SOLUTION We note that $\theta = 90.0°$, since the velocity of the electron is perpendicular to the magnetic field. The magnitude of the electron's charge is 1.60×10^{-19} C, and the electron's mass is 9.11×10^{-31} kg (see the inside of the front cover), so

$$a = \frac{|q_0|vB \sin \theta}{m}$$

$$= \frac{(1.60 \times 10^{-19} \text{ C})(2.1 \times 10^6 \text{ m/s})(1.6 \times 10^{-5} \text{ T}) \sin 90.0°}{9.11 \times 10^{-31} \text{ kg}} = \boxed{5.9 \times 10^{12} \text{ m/s}^2}$$

3. **REASONING** According to Equation 21.1, the magnitude of the magnetic force on a moving charge is $F = |q_0|vB \sin \theta$. Since the magnetic field points due north and the proton moves eastward, $\theta = 90.0°$. Furthermore, since the magnetic force on the moving proton balances its weight, we have $mg = |q_0|vB \sin \theta$, where m is the mass of the proton. This expression can be solved for the speed v.

   SOLUTION Solving for the speed v, we have

$$v = \frac{mg}{|q_0|B \sin \theta} = \frac{(1.67 \times 10^{-27} \text{ kg})(9.80 \text{ m/s}^2)}{(1.6 \times 10^{-19} \text{ C})(2.5 \times 10^{-5} \text{ T}) \sin 90.0°} = \boxed{4.1 \times 10^{-3} \text{ m/s}}$$

11. **REASONING** The direction in which the electrons are deflected can be determined using Right-Hand Rule No. 1 and reversing the direction of the force (RHR-1 applies to positive charges, and electrons are negatively charged).

Each electron experiences an acceleration a given by Newton's second law of motion, $a = F/m$, where F is the net force and m is the mass of the electron. The only force acting on the electron is the magnetic force, $F = |q_0|vB \sin \theta$, so it is the net force. The speed v of the

264

electron is related to its kinetic energy KE by the relation $KE = \frac{1}{2}mv^2$. Thus, we have enough information to find the acceleration.

SOLUTION

a. According to RHR-1, if you extend your right hand so that your fingers point along the direction of the magnetic field **B** and your thumb points in the direction of the velocity **v** of a *positive* charge, your palm will face in the direction of the force **F** on the positive charge.

For the electron in question, the fingers of the right hand should be oriented downward (direction of **B**) with the thumb pointing to the east (direction of **v**). The palm of the right hand points due north (the direction of **F** on a positive charge). Since the electron is negatively charged, it will be deflected $\boxed{\text{due south}}$.

b. The acceleration of an electron is given by Newton's second law, where the net force is the magnetic force. Thus,

$$a = \frac{F}{m} = \frac{|q_0|vB\sin\theta}{m}$$

Since the kinetic energy is $KE = \frac{1}{2}mv^2$, the speed of the electron is $v = \sqrt{2(KE)/m}$. Thus, the acceleration of the electron is

$$a = \frac{|q_0|vB\sin\theta}{m} = \frac{|q_0|\sqrt{\dfrac{2(KE)}{m}}\,B\sin\theta}{m}$$

$$= \frac{\left(1.60\times10^{-19}\text{ C}\right)\sqrt{\dfrac{2\left(2.40\times10^{-15}\text{ J}\right)}{9.11\times10^{-31}\text{ kg}}}\left(2.00\times10^{-5}\text{ T}\right)\sin 90.0^\circ}{9.11\times10^{-31}\text{ kg}} = \boxed{2.55\times10^{14}\text{ m/s}^2}$$

---

13. **REASONING** The radius $\underline{r}$ of the circular path is given by $r = \dfrac{mv}{|q|B}$ (Equation 21.2), where m and v are the mass and speed of the particle, respectively, $|q|$ is the magnitude of the charge, and B is the magnitude of the magnetic field. This expression can be solved directly for B, since r, m, and v are given and $q = +e$, where $e = 1.60\times10^{-19}$ C.

SOLUTION Solving Equation 21.2 for B gives

$$B = \frac{mv}{|q|r} = \frac{\left(3.06\times10^{-25}\text{ kg}\right)\left(7.2\times10^3\text{ m/s}\right)}{\left|+1.60\times10^{-19}\text{ C}\right|\left(0.10\text{ m}\right)} = \boxed{0.14\text{ T}}$$

17. **REASONING** As discussed in Section 21.4, the mass m of a singly-ionized particle that has been accelerated through a potential difference V and injected into a magnetic field of magnitude B is given by

$$m = \left(\frac{er^2}{2V}\right)B^2 \qquad (1)$$

where $e = 1.60 \times 10^{-19}$ C is the magnitude of the charge of an electron and r is the radius of the particle's path. If the beryllium-10 ions reach the same position in the detector as the beryllium-7 ions, both types of ions must have the same path radius r. Additionally, the accelerating potential difference V is kept constant, so we see that the quantity $\left(\frac{er^2}{2V}\right)$ in Equation (1) is the same for both types of ions.

SOLUTION All that differs between the two situations are the masses (m_7, m_{10}) of the ions and the magnitudes of the magnetic fields (B_7, B_{10}). Solving Equation (1) for the constant quantity $\left(\frac{er^2}{2V}\right)$, we obtain

$$\underbrace{\left(\frac{er^2}{2V}\right)}_{\substack{\text{Same for} \\ \text{both ions}}} = \frac{m_{10}}{B_{10}^2} = \frac{m_7}{B_7^2} \qquad (2)$$

Solving Equation (2) for B_{10}^2, we find that

$$B_{10}^2 = B_7^2 \frac{m_{10}}{m_7} \qquad \text{or} \qquad B_{10} = B_7 \sqrt{\frac{m_{10}}{m_7}} = (0.283 \text{ T})\sqrt{\frac{16.63 \times 10^{-27} \text{ kg}}{11.65 \times 10^{-27} \text{ kg}}} = \boxed{0.338 \text{ T}}$$

---

23. **REASONING** When the proton moves in the magnetic field, its trajectory is a circular path. The proton will just miss the opposite plate if the distance between the plates is equal to the radius of the path. The radius is given by Equation 21.2 as $r = mv/(|q|B)$. This relation can be used to find the magnitude B of the magnetic field, since values for all the other variables are known.

SOLUTION Solving the relation $r = mv/(|q|B)$ for the magnitude of the magnetic field, and realizing that the radius is equal to the plate separation, we find that

$$B = \frac{mv}{|q|r} = \frac{\left(1.67 \times 10^{-27} \text{ kg}\right)\left(3.5 \times 10^6 \text{ m/s}\right)}{\left(1.60 \times 10^{-19} \text{ C}\right)(0.23 \text{ m})} = \boxed{0.16 \text{ T}}$$

The values for the mass and the magnitude of the charge (which is the same as that of the electron) have been taken from the inside of the front cover.

---

25. **REASONING** The particle travels in a semicircular path of radius r, where r is given by Equation 21.2 $\left(r = \dfrac{mv}{|q|B} \right)$. The time spent by the particle in the magnetic field is given by $t = s/v$, where s is the distance traveled by the particle and v is its speed. The distance s is equal to one-half the circumference of a circle ($s = \pi r$).

SOLUTION We find that

$$t = \frac{s}{v} = \frac{\pi r}{v} = \frac{\pi}{v}\left(\frac{mv}{|q|B} \right) = \frac{\pi m}{|q|B} = \frac{\pi(6.0 \times 10^{-8} \text{ kg})}{(7.2 \times 10^{-6} \text{ C})(3.0 \text{ T})} = \boxed{8.7 \times 10^{-3} \text{ s}}$$

---

31. **REASONING** The magnitude F of the magnetic force experienced by the wire is given by $F = ILB \sin \theta$ (Equation 21.3), where I is the current, L is the length of the wire, B is the magnitude of the earth's magnetic field, and θ is the angle between the direction of the current and the magnetic field. Since all the variables are known except B, we can use this relation to find its value.

SOLUTION Solving $F = ILB \sin \theta$ for the magnitude of the magnetic field, we have

$$B = \frac{F}{IL \sin \theta} = \frac{0.15 \text{ N}}{(75 \text{ A})(45 \text{ m}) \sin 60.0^\circ} = \boxed{5.1 \times 10^{-5} \text{ T}}$$

---

35. **REASONING** According to Equation 21.3, the magnetic force has a magnitude of $F = ILB \sin \theta$, where I is the current, B is the magnitude of the magnetic field, L is the length of the wire, and $\theta = 90^\circ$ is the angle of the wire with respect to the field.

SOLUTION Using Equation 21.3, we find that

$$L = \frac{F}{IB \sin \theta} = \frac{7.1 \times 10^{-5} \text{ N}}{(0.66 \text{ A})\left(4.7 \times 10^{-5} \text{ T}\right) \sin 58^\circ} = \boxed{2.7 \text{ m}}$$

39. ***REASONING*** Since the rod does not rotate about the axis at *P*, the net torque relative to that axis must be zero; $\Sigma\tau = 0$ (Equation 9.2). There are two torques that must be considered, one due to the magnetic force and another due to the weight of the rod. We consider both of these to act at the rod's center of gravity, which is at the geometrical center of the rod (length = *L*), because the rod is uniform. According to Right-Hand Rule No. 1, the magnetic force acts perpendicular to the rod and is directed up and to the left in the drawing. Therefore, the magnetic torque is a counterclockwise (positive) torque. Equation 21.3 gives the magnitude *F* of the magnetic force as $F = ILB \sin 90.0°$, since the current is perpendicular to the magnetic field. The weight is *mg* and acts downward, producing a clockwise (negative) torque. The magnitude of each torque is the magnitude of the force times the lever arm (Equation 9.1). Thus, we have for the torques:

$$\tau_{\text{magnetic}} = +\underbrace{\left(ILB\right)}_{\text{force}}\underbrace{\left(L/2\right)}_{\text{lever arm}} \quad \text{and} \quad \tau_{\text{weight}} = -\underbrace{\left(mg\right)}_{\text{force}}\underbrace{\left[\left(L/2\right)\cos\theta\right]}_{\text{lever arm}}$$

Setting the sum of these torques equal to zero will enable us to find the angle θ that the rod makes with the ground.

SOLUTION Setting the sum of the torques equal to zero gives $\Sigma\tau = \tau_{\text{magnetic}} + \tau_{\text{weight}} = 0$, and we have

$$+\left(ILB\right)\left(L/2\right) - \left(mg\right)\left[\left(L/2\right)\cos\theta\right] = 0 \quad \text{or} \quad \cos\theta = \frac{ILB}{mg}$$

$$\theta = \cos^{-1}\left[\frac{\left(4.1\,\text{A}\right)\left(0.45\,\text{m}\right)\left(0.36\,\text{T}\right)}{\left(0.094\,\text{kg}\right)\left(9.80\,\text{m/s}^2\right)}\right] = \boxed{44°}$$

49. ***REASONING*** The torque on the loop is given by Equation 21.4, $\tau = NIAB\sin\phi$. From the drawing in the text, we see that the angle ϕ between the normal to the plane of the loop and the magnetic field is $90° - 35° = 55°$. The area of the loop is $0.70\,\text{m} \times 0.50\,\text{m} = 0.35\,\text{m}^2$.

SOLUTION
a. The magnitude of the net torque exerted on the loop is

$$\tau = NIAB\sin\phi = (75)(4.4\,\text{A})(0.35\,\text{m}^2)(1.8\,\text{T})\sin 55° = \boxed{170\,\text{N}\cdot\text{m}}$$

b. As discussed in the text, when a current-carrying loop is placed in a magnetic field, the loop tends to rotate such that its normal becomes aligned with the magnetic field. The normal to the loop makes an angle of 55° with respect to the magnetic field. Since this angle decreases as the loop rotates, the $\boxed{35° \text{ angle increases}}$.

53. ***REASONING*** AND ***SOLUTION***
a. In Figure 21.27a the magnetic field that exists at the location of each wire points upward. Since the current in each wire is the same, the fields at the locations of the wires also have the same magnitudes. Therefore, a single external field that points ⟨ downward ⟩ will cancel the mutual repulsion of the wires, if this external field has a magnitude that equals that of the field produced by either wire.

b. Equation 21.5 gives the magnitude of the field produced by a long straight wire. The external field must have this magnitude:

$$B = \frac{\mu_0 I}{2\pi r} = \frac{\left(4\pi \times 10^{-7} \text{ T} \cdot \text{m/A}\right)(25 \text{ A})}{2\pi (0.016 \text{ m})} = \boxed{3.1 \times 10^{-4} \text{ T}}$$

---

55. ***REASONING*** The magnitude B of the magnetic field in the interior of a solenoid that has a length much greater than its diameter is given by $B = \mu_0 n I$ (Equation 21.7), where $\mu_0 = 4\pi \times 10^{-7} \text{ T} \cdot \text{m/A}$ is the permeability of free space, n is the number of turns per meter of the solenoid's length, and I is the current in the wire of the solenoid. Since B and I are given, we can solve Equation 21.7 for n.

SOLUTION Solving Equation 21.7 for n, we find that the number of turns per meter of length is

$$n = \frac{B}{\mu_0 I} = \frac{7.0 \text{ T}}{\left(4\pi \times 10^{-7} \text{ T} \cdot \text{m/A}\right)\left(2.0 \times 10^2 \text{ A}\right)} = \boxed{2.8 \times 10^4 \text{ turns/m}}$$

---

57. ***REASONING*** The magnitude of the magnetic field at the center of a circular loop of current is given by Equation 21.6 as $B = N\mu_0 I/(2R)$, where N is the number of turns, μ_0 is the permeability of free space, I is the current, and R is the radius of the loop. The field is perpendicular to the plane of the loop. Magnetic fields are vectors, and here we have two fields, each perpendicular to the plane of the loop producing it. Therefore, the two field vectors are perpendicular, and we must add them as vectors to get the net field. Since they are perpendicular, we can use the Pythagorean theorem to calculate the magnitude of the net field.

SOLUTION Using Equation 21.6 and the Pythagorean theorem, we find that the magnitude of the net magnetic field at the common center of the two loops is

$$B_{\text{net}} = \sqrt{\left(\frac{N\mu_0 I}{2R}\right)^2 + \left(\frac{N\mu_0 I}{2R}\right)^2} = \sqrt{2}\left(\frac{N\mu_0 I}{2R}\right)$$

$$= \frac{\sqrt{2}(1)\left(4\pi \times 10^{-7} \text{ T} \cdot \text{m/A}\right)(1.7 \text{ A})}{2(0.040 \text{ m})} = \boxed{3.8 \times 10^{-5} \text{ T}}$$

65. **REASONING** According to Equation 21.6 the magnetic field at the center of a circular, current-carrying loop of N turns and radius r is $B = N\mu_0 I/(2r)$. The number of turns N in the coil can be found by dividing the total length L of the wire by the circumference after it has been wound into a circle. The current in the wire can be found by using Ohm's law, $I = V/R$.

SOLUTION The number of turns in the wire is

$$N = \frac{L}{2\pi r}$$

The current in the wire is

$$I = \frac{V}{R} = \frac{12.0 \text{ V}}{(5.90\times10^{-3}\ \Omega/\text{m})L} = \frac{2.03\times10^{3}}{L} \text{ A}$$

Therefore, the magnetic field at the center of the coil is

$$B = N\left(\frac{\mu_0 I}{2r}\right) = \left(\frac{L}{2\pi r}\right)\left(\frac{\mu_0 I}{2r}\right) = \frac{\mu_0 L I}{4\pi r^2}$$

$$= \frac{\left(4\pi\times10^{-7}\,\text{T}\cdot\text{m/A}\right)L\left(\dfrac{2.03\times10^{3}}{L}\text{ A}\right)}{4\pi(0.140\text{ m})^2} = \boxed{1.04\times10^{-2}\text{ T}}$$

71. **REASONING** AND **SOLUTION** The drawing at the right shows an end-on view of the solid cylinder. The dots represent the current in the cylinder coming out of the paper toward you. The dashed circle of radius r is the closed path used in Ampère's law and is centered on the axis of the cylinder. Equation 21.8 gives Ampère's law as $\Sigma B_\parallel \Delta l = \mu_0 I$. Because of the symmetry of the arrangement in the drawing, we have $B_\parallel = B$ for all $\Delta \ell$ on the circular path, so that Ampère's law becomes

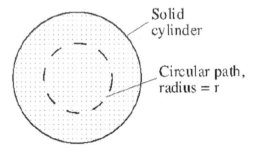

Solid cylinder

Circular path, radius = r

$$\Sigma B_\parallel \Delta l = B(\Sigma\Delta l) = \mu_0 I$$

In this result, $\Sigma\Delta\ell = 2\pi r$, the circumference of the circle. The current I is the part of the total current that comes through the area πr^2 bounded by the dashed path. We can calculate this current by using the current per unit cross-sectional area of the solid cylinder. This current per unit area is called the current density. The current I is the current density times the area πr^2:

$$I = \underbrace{\left(\frac{I_0}{\pi R^2}\right)}_{\substack{\text{current} \\ \text{density}}}\left(\pi r^2\right) = \frac{I_0 r^2}{R^2}$$

Thus, Ampère's law becomes

$$B\left(\Sigma \Delta l\right) = \mu_0 I \quad \text{or} \quad B\left(2\pi r\right) = \mu_0\left(\frac{I_0 r^2}{R^2}\right) \quad \text{or} \quad B = \boxed{\frac{\mu_0 I_0 r}{2\pi R^2}}$$

---

73. **REASONING** The angle θ between the electron's velocity and the magnetic field can be found from Equation 21.1,

$$\sin \theta = \frac{F}{|q| vB}$$

According to Newton's second law, the magnitude F of the force is equal to the product of the electron's mass m and the magnitude a of its acceleration, $F = ma$.

SOLUTION The angle θ is

$$\theta = \sin^{-1}\left(\frac{ma}{|q| vB}\right) = \sin^{-1}\left[\frac{\left(9.11\times10^{-31}\ \text{kg}\right)\left(3.50\times10^{14}\ \text{m/s}^2\right)}{\left(1.60\times10^{-19}\ \text{C}\right)\left(6.80\times10^6\ \text{m/s}\right)\left(8.70\times10^{-4}\ \text{T}\right)}\right] = \boxed{19.7^\circ}$$

---

75. **REASONING** According to Equation 21.4, the maximum torque is $\tau_{\text{max}} = NIAB$, where N is the number of turns in the coil, I is the current, $A = \pi r^2$ is the area of the circular coil, and B is the magnitude of the magnetic field. Since the coil contains only one turn, the length L of the wire is the circumference of the circle, so that $L = 2\pi r$ or $r = L/(2\pi)$. Since N, I, and B are known we can solve for L.

SOLUTION According to Equation 21.4 and the fact that $r = L/(2\pi)$, we have

$$\tau_{\text{max}} = NI\pi r^2 B = NI\pi\left(\frac{L}{2\pi}\right)^2 B$$

Solving this result for L gives

$$L = \sqrt{\frac{4\pi\,\tau_{\text{max}}}{NIB}} = \sqrt{\frac{4\pi\left(8.4\times10^{-4}\ \text{N}\cdot\text{m}\right)}{(1)(3.7\ \text{A})(0.75\ \text{T})}} = \boxed{0.062\ \text{m}}$$

81. **REASONING**   From the discussion in Section 21.3, we know that when a charged particle moves perpendicular to a magnetic field, the trajectory of the particle is a circle. The drawing at the right shows a particle moving in the plane of the paper (the magnetic field is perpendicular to the paper). If the particle is moving initially through the coordinate origin and to the right (along the +x axis), the subsequent circular path of the particle will intersect the y axis at the greatest possible value, which is equal to twice the radius r of the circle.

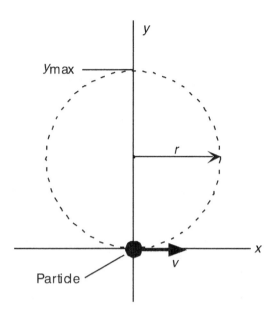

SOLUTION
a. From the drawing above, it can be seen that the largest value of y is equal to the diameter ($2r$) of the circle. When the particle passes through the coordinate origin its velocity must be parallel to the +x axis. Thus, the angle is $\boxed{\theta = 0°}$.

b. The maximum value of y is twice the radius r of the circle. According to Equation 21.2, the radius of the circular path is $r = mv / (|q|B)$. The maximum value y_{max} is, therefore,

$$y_{max} = 2r = 2\left(\frac{mv}{|q|B}\right) = 2\left[\frac{(3.8\times10^{-8} \text{ kg})(44 \text{ m/s})}{(7.3\times10^{-6} \text{ C})(1.6 \text{ T})}\right] = \boxed{0.29 \text{ m}}$$

---

85. **REASONING**   The magnetic moment of the rotating charge can be found from the expression *Magnetic moment* $= NIA$, as discussed in Section 21.6. For this situation, $N = 1$. Thus, we need to find the current and the area for the rotating charge. This can be done by resorting to first principles.

SOLUTION   The current for the rotating charge is, by definition (see Equation 20.1), $I = \Delta q / \Delta t$, where Δq is the amount of charge that passes by a given point during a time interval Δt. Since the charge passes by once per revolution, we can find the current by dividing the total rotating charge by the period T of revolution.

$$I = \frac{\Delta q}{T} = \frac{\Delta q}{2\pi/\omega} = \frac{\omega \Delta q}{2\pi} = \frac{(150 \text{ rad/s})(4.0\times10^{-6}\text{C})}{2\pi} = 9.5\times10^{-5}\text{A}$$

The area of the rotating charge is $A = \pi r^2 = \pi (0.20 \text{ m})^2 = 0.13 \text{ m}^2$. Therefore, the magnetic moment is

$$Magnetic\ moment = NIA = (1)(9.5 \times 10^{-5}\,A)(0.13\,m^2) = \boxed{1.2 \times 10^{-5}\ A \cdot m^2}$$

87. **CONCEPTS** (i) The net force is the vector sum of all the forces acting on the particle, which in this case are the magnetic force and the electric force. (ii) First, remember that the direction for either a positive or negative charge is perpendicular to the plane formed by the magnetic field vector $\vec{B}$ and the velocity vector $\vec{v}$. Then, apply Right-Hand Rule No. 1 (RHR-1) to find the direction of the force as if the charge were positive. Finally, reverse the direction indicated by RHR-1 to determine the direction of the force on the negative charge. (iii) The direction for either a positive or negative charge is along the line of the electric field. The electric force on a positive charge is in the same direction as the electric field, while the force on a negative charge is in the direction opposite the electric field. (iv) The fact that the charge is moving affects the value of the magnetic force, for without motion, there is no magnetic force. The electric force, in contrast, has the same value whether or not the charge is moving.

CALCULATIONS
An application of RHR-1 reveals that if the charge were positive, the magnetic force would point along the $+x$ axis. The charge is negative, however, so the magnetic force points along the $-x$ axis. According to Equation 21.1, the magnitude of the magnetic force is

$$F_{magnetic} = |q_0| vB \sin\theta$$

where $|q_0|$ is the magnitude of the charge. Since the velocity and the magnetic field are perpendicular, $\theta = 90°$, and we have $F_{magnetic} = |q_0| vB$.

The electric field points along the $-x$ axis and would apply a force that is in the same direction if the charge were positive. The charge is negative, however, so the electric force points in the opposite direction, or along the $+x$ axis. According to Equation 18.2, the magnitude of the electric force is $F_{electric} = |q_0| E$.

Using plus and minus signs to take into account the different directions of the magnetic and electric forces, we find that the net force is

$$\sum F = +F_{electric} - F_{magnetic} = +|q_0| E - |q_0| vB$$

$$= +\left|-2.80 \times 10^{-6}\ C\right|(123\ N/C) - \left|-2.80 \times 10^{-6}\ C\right|\left(4.80 \times 10^6\ m/s\right)\left(3.35 \times 10^{-5}\ T\right)$$

$$= \boxed{-1.06 \times 10^{-4}\ N}$$

where the negative sign indicates it points along the $-x$ axis.

CHAPTER 22 | *ELECTROMAGNETIC INDUCTION*

5. **REASONING AND SOLUTION** For the three rods in the drawing in the text, we have the following:

    Rod A: The motional emf is $\boxed{\text{zero}}$, because the velocity of the rod is parallel to the direction of the magnetic field, and the charges do not experience a magnetic force.

    Rod B: The motional emf ξ is, according to Equation 22.1,

    $$\xi = vBL = (2.7 \text{ m/s})(0.45 \text{ T})(1.3 \text{ m}) = \boxed{1.6 \text{ V}}$$

    The positive end of Rod B is $\boxed{\text{end 2}}$.

    Rod C: The motional emf is $\boxed{\text{zero}}$, because the magnetic force F on each charge is directed perpendicular to the length of the rod. For the ends of the rod to become charged, the magnetic force must be directed parallel to the length of the rod.

---

9. **REASONING** The minimum length d of the rails is the speed v of the rod times the time t, or $d = vt$. We can obtain the speed from the expression for the motional emf given in Equation 22.1. Solving this equation for the speed gives $v = \dfrac{\xi}{BL}$, where ξ is the motional emf, B is the magnitude of the magnetic field, and L is the length of the rod. Thus, the length of the rails is $d = vt = \left(\dfrac{\xi}{BL} \right) t$. While we have no value for the motional emf, we do know that the bulb dissipates a power of $P = 60.0$ W, and has a resistance of $R = 240 \; \Omega$. Power is related to the emf and the resistance according to $P = \dfrac{\xi^2}{R}$ (Equation 20.6c), which can be solved to show that $\xi = \sqrt{PR}$. Substituting this expression into the equation for d gives

    $$d = \left(\dfrac{\xi}{BL} \right) t = \left(\dfrac{\sqrt{PR}}{BL} \right) t$$

    SOLUTION Using the above expression for the minimum necessary length of the rails, we find that

    $$d = \left(\dfrac{\sqrt{PR}}{BL} \right) t = \left[\dfrac{\sqrt{(60.0 \text{ W})(240 \; \Omega)}}{(0.40 \text{ T})(0.60 \text{ m})} \right] (0.50 \text{ s}) = \boxed{250 \text{ m}}$$

---

11. **REASONING** The definition of magnetic flux Φ is $\Phi = BA\cos\phi$ (Equation 22.2), where B is the magnitude of the magnetic field, A is the area of the surface, and ϕ is the angle between the magnetic field vector and the normal to the surface. The values of B and A are the same for each of the surfaces, while the values for the angle ϕ are different. The z axis is the normal to the surface lying in the x, y plane, so that $\phi_{xy} = 35°$. The y axis is the normal to the surface lying in the x, z plane, so that $\phi_{xz} = 55°$. We can apply the definition of the flux to obtain the desired ratio directly.

SOLUTION Using Equation 22.2, we find that

$$\frac{\Phi_{xz}}{\Phi_{xy}} = \frac{BA\cos\phi_{xz}}{BA\cos\phi_{xy}} = \frac{\cos 55°}{\cos 35°} = \boxed{0.70}$$

---

13. **REASONING** The general expression for the magnetic flux through an area A is given by Equation 22.2: $\Phi = BA\cos\phi$, where B is the magnitude of the magnetic field and ϕ is the angle of inclination of the magnetic field $\mathbf{B}$ with respect to the normal to the area.

The magnetic flux through the door is a maximum when the magnetic field lines are perpendicular to the door and $\phi_1 = 0.0°$ so that $\Phi_1 = \Phi_{max} = BA(\cos 0.0°) = BA$.

SOLUTION When the door rotates through an angle ϕ_2, the magnetic flux that passes through the door decreases from its maximum value to one-third of its maximum value. Therefore, $\Phi_2 = \frac{1}{3}\Phi_{max}$, and we have

$$\Phi_2 = BA\cos\phi_2 = \frac{1}{3}BA \quad \text{or} \quad \cos\phi_2 = \frac{1}{3} \quad \text{or} \quad \phi_2 = \cos^{-1}\left(\frac{1}{3}\right) = \boxed{70.5°}$$

---

21. **REASONING** According to Equation 22.3, the average emf induced in a coil of N loops is $\xi = -N\Delta\Phi/\Delta t$.

SOLUTION For the circular coil in question, the flux through a single turn changes by

$$\Delta\Phi = BA\cos 45° - BA\cos 90° = BA\cos 45°$$

during the interval of $\Delta t = 0.010$ s. Therefore, for N turns, Faraday's law gives the magnitude of the emf as

$$|\xi| = \left| -N\frac{BA\cos 45°}{\Delta t} \right|$$

Since the loops are circular, the area A of each loop is equal to πr^2. Solving for B, we have

$$B = \frac{|\xi|\Delta t}{N\pi r^2 \cos 45°} = \frac{(0.065 \text{ V})(0.010 \text{ s})}{(950)\pi(0.060 \text{ m})^2\cos 45°} = \boxed{8.6\times10^{-5} \text{ T}}$$

27. **REASONING** According to Equation 22.3, the average emf ξ induced in a single loop ($N = 1$) is $\xi = -\Delta\Phi / \Delta t$. Since the magnitude of the magnetic field is changing, the area of the loop remains constant, and the direction of the field is parallel to the normal to the loop, the change in flux through the loop is given by $\Delta\Phi = (\Delta B)A$. Thus the magnitude $|\xi|$ of the induced emf in the loop is given by $|\xi| = |-(\Delta B) A / \Delta t|$.

Similarly, when the area of the loop is changed and the field B has a given value, we find the magnitude of the induced emf to be $|\xi| = |-B(\Delta A) / \Delta t|$.

SOLUTION
a. The magnitude of the induced emf when the field changes in magnitude is

$$|\xi| = \left|-A\left(\frac{\Delta B}{\Delta t}\right)\right| = (0.018 \text{ m}^2)(0.20 \text{ T/s}) = \boxed{3.6 \times 10^{-3} \text{ V}}$$

b. At a particular value of B (when B is changing), the rate at which the area must change can be obtained from

$$|\xi| = \left|-\frac{B\Delta A}{\Delta t}\right| \quad \text{or} \quad \frac{\Delta A}{\Delta t} = \frac{|\xi|}{B} = \frac{3.6 \times 10^{-3} \text{ V}}{1.8 \text{ T}} = \boxed{2.0 \times 10^{-3} \text{ m}^2 / \text{s}}$$

In order for the induced emf to be zero, the magnitude of the magnetic field and the area of the loop must change in such a way that the flux remains constant. Since the magnitude of the magnetic field is increasing, the area of the loop must decrease, if the flux is to remain constant. Therefore, $\boxed{\text{the area of the loop must be shrunk}}$.

---

31. **REASONING AND SOLUTION** Consider one revolution of either rod. The magnitude $|\xi|$ of the emf induced across the rod is

$$|\xi| = \left|-B\frac{\Delta A}{\Delta t}\right| = \frac{B(\pi L^2)}{\Delta t}$$

The angular speed of the rods is $\omega = 2\pi / \Delta t$, so $|\xi| = \frac{1}{2}BL^2\omega$. The rod tips have opposite polarity since they are rotating in opposite directions. Hence, the difference in potentials of the tips is

$$\Delta V = BL^2\omega$$

so

$$\omega = \frac{\Delta V}{BL^2} = \frac{4.5 \times 10^3 \text{ V}}{(4.7 \text{ T})(0.68 \text{ m})^2} = \boxed{2100 \text{ rad/s}}$$

33. **REASONING** The external magnetic field is perpendicular to the plane of the horizontal loop, so it must point either upward or downward. We will use Lenz's law to decide whether the external magnetic **B** field points up or down. This law predicts that the direction of the induced magnetic field **B$_{ind}$** opposes the change in the magnetic flux through the loop due to the external field.

SOLUTION The external magnetic field **B** is increasing in magnitude, so that the magnetic flux through the loop also *increases* with time. In order to oppose the increase in magnetic flux, the induced magnetic field **B$_{ind}$** must be directed *opposite* to the external magnetic field **B**. The drawing shows the loop as viewed from above, with an induced current I_{ind} flowing clockwise. According to Right-Hand Rule No. 2 (see Section 21.7), this induced current creates an induced magnetic field **B$_{ind}$** that is directed *into* the page at the center of the loop (and all other points of the loop's interior).

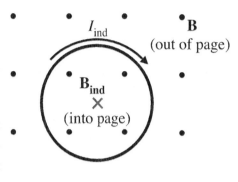

Therefore, the external magnetic field **B** must be directed *out* of the page. Because we are viewing the loop from above, "out of the page" corresponds to upward toward the viewer.

35. **REASONING** The current *I* produces a magnetic field, and hence a magnetic flux, that passes through the loops A and B. Since the current decreases to zero when the switch is opened, the magnetic flux also decreases to zero. According to Lenz's law, the current induced in each coil will have a direction such that the induced magnetic field will oppose the original flux change.

SOLUTION

a. The drawing in the text shows that the magnetic field at coil A is perpendicular to the plane of the coil and points down (when viewed from above the table top). When the switch is opened, the magnetic flux through coil A decreases to zero. According to Lenz's law, the induced magnetic field produced by coil A must oppose this change in flux. Since the magnetic field points down and is decreasing, the induced magnetic field must also point down. According to Right-Hand Rule No. 2 (RHR-2), the induced current must be clockwise around loop A.

b. The drawing in the text shows that the magnetic field at coil B is perpendicular to the plane of the coil and points up (when viewed from above the table top). When the switch is opened, the magnetic flux through coil B decreases to zero. According to Lenz's law, the induced magnetic field produced by coil B must oppose this change in flux. Since the magnetic field points up and is decreasing, the induced magnetic field must also point up. According to RHR-2, the induced current must be counterclockwise around loop B.

41. **REASONING** We can use the information given in the problem statement to determine the area of the coil A. Since it is square, the length of one side is $l = \sqrt{A}$.

SOLUTION According to Equation 22.4, the maximum emf ξ_0 induced in the coil is $\xi_0 = NAB\omega$. Therefore, the length of one side of the coil is

$$l = \sqrt{A} = \sqrt{\frac{\xi_0}{NB\omega}} = \sqrt{\frac{75.0 \text{ V}}{(248)(0.170 \text{ T})(79.1 \text{ rad/s})}} = \boxed{0.150 \text{ m}}$$

47. **REASONING** The peak emf ξ_0 produced by a generator is related to the number N of turns in the coil, the area A of the coil, the magnitude B of the magnetic field, and the angular speed ω_{coil} of the coil by $\xi_0 = NAB\omega_{coil}$ (see Equation 22.4). We are given that the angular speed ω_{coil} of the coil is 38 times as great as the angular speed ω_{tire} of the tire. Since the tires roll without slipping, the angular speed of a tire is related to the linear speed v of the bike by $\omega_{tire} = v/r$ (see Section 8.6), where r is the radius of a tire. The speed of the bike after 5.1 s can be found from its acceleration and the fact that the bike starts from rest.

SOLUTION The peak emf produced by the generator is

$$\xi_0 = NAB\omega_{coil} \tag{22.4}$$

Since the angular speed of the coil is 38 times as great as the angular speed of the tire, $\omega_{coil} = 38\omega_{tire}$. Substituting this expression into Equation 22.4 gives $\xi_0 = NAB\omega_{coil} = NAB(38\omega_{tire})$. Since the tire rolls without slipping, the angular speed of the tire is related to the linear speed v (the speed at which its axle is moving forward) by $\omega_{tire} = v/r$ (Equation 8.12), where r is the radius of the tire. Substituting this result into the expression for ξ_0 yields

$$\xi_0 = NAB(38\omega_{tire}) = NAB\left(38\frac{v}{r}\right) \tag{1}$$

The velocity of the car is given by $v = v_0 + at$ (Equation 2.4), where v_0 is the initial velocity, a is the acceleration and t is the time. Substituting this relation into Equation (1), and noting that $v_0 = 0$ m/s since the bike starts from rest, we find that

$$\xi_0 = NAB\left(38\frac{v}{r}\right) = NAB\left[38\frac{(v_0 + at)}{r}\right]$$

$$= (125)(3.86\times10^{-3} \text{ m}^2)(0.0900 \text{ T})(38)\left[\frac{0 \text{ m/s} + (0.550 \text{ m/s}^2)(5.10 \text{ s})}{0.300 \text{ m}}\right] = \boxed{15.4 \text{ V}}$$

49. **REASONING** The energy density is given by Equation 22.11 as

$$\text{Energy density} = \frac{\text{Energy}}{\text{Volume}} = \frac{1}{2\mu_0}B^2$$

The energy stored is the energy density times the volume.

SOLUTION The volume is the area A times the height h. Therefore, the energy stored is

$$\text{Energy} = \frac{B^2 Ah}{2\mu_0} = \frac{(7.0\times10^{-5}\ \text{T})^2(5.0\times10^8\ \text{m}^2)(1500\ \text{m})}{2(4\pi\times10^{-7}\ \text{T}\cdot\text{m/A})} = \boxed{1.5\times10^9\ \text{J}}$$

---

55. **REASONING AND SOLUTION** From the results of Example 13, the self-inductance L of a long solenoid is given by $L = \mu_0 n^2 Al$. Solving for the number of turns n per unit length gives

$$n = \sqrt{\frac{L}{\mu_0 Al}} = \sqrt{\frac{1.4\times10^{-3}\ \text{H}}{(4\pi\times10^{-7}\text{T}\cdot\text{m/A})(1.2\times10^{-3}\ \text{m}^2)(0.052\ \text{m})}} = 4.2\times10^3\ \text{turns/m}$$

Therefore, the total number of turns N is the product of n and the length l of the solenoid:

$$N = nl = (4.2\times10^3\ \text{turns/m})(0.052\ \text{m}) = \boxed{220\ \text{turns}}$$

---

65. **REASONING** The ratio $\left(I_s / I_p\right)$ of the current in the secondary coil to that in the primary coil is equal to the ratio $\left(N_p / N_s\right)$ of the number of turns in the primary coil to that in the secondary coil. This relation can be used directly to find the current in the primary coil.

SOLUTION Solving the relation $\left(I_s / I_p\right) = \left(N_p / N_s\right)$ (Equation 22.13) for I_p gives

$$I_p = I_s\left(\frac{N_s}{N_p}\right) = (1.6\ \text{A})\left(\frac{1}{8}\right) = \boxed{0.20\ \text{A}}$$

---

67. **REASONING** The power used to heat the wires is given by Equation 20.6b: $P = I^2 R$. Before we can use this equation, however, we must determine the total resistance R of the wire and the current that flows through the wire.

SOLUTION
a. The total resistance of one of the wires is equal to the resistance per unit length multiplied by the length of the wire. Thus, we have

$$(5.0 \times 10^{-2} \ \Omega/\text{km})(7.0 \ \text{km}) = 0.35 \ \Omega$$

and the total resistance of the transmission line is twice this value or 0.70 Ω. According to Equation 20.6a ($P = IV$), the current flowing into the town is

$$I = \frac{P}{V} = \frac{1.2 \times 10^6 \ \text{W}}{1200 \ \text{V}} = 1.0 \times 10^3 \ \text{A}$$

Thus, the power used to heat the wire is

$$P = I^2 R = \left(1.0 \times 10^3 \ \text{A}\right)^2 (0.70 \ \Omega) = \boxed{7.0 \times 10^5 \ \text{W}}$$

b. According to the transformer equation (Equation 22.12), the stepped-up voltage is

$$V_{\text{s}} = V_{\text{p}} \left(\frac{N_{\text{s}}}{N_{\text{p}}} \right) = (1200 \ \text{V}) \left(\frac{100}{1} \right) = 1.2 \times 10^5 \ \text{V}$$

According to Equation 20.6a ($P = IV$), the current in the wires is

$$I = \frac{P}{V} = \frac{1.2 \times 10^6 \ \text{W}}{1.2 \times 10^5 \ \text{V}} = 1.0 \times 10^1 \ \text{A}$$

The power used to heat the wires is now

$$P = I^2 R = \left(1.0 \times 10^1 \ \text{A}\right)^2 (0.70 \ \Omega) = \boxed{7.0 \times 10^1 \ \text{W}}$$

---

71. **REASONING** The peak emf ξ_0 of a generator is found from $\xi_0 = NAB\omega$ (Equation 22.4), where N is the number of turns in the generator coil, A is the coil's cross-sectional area, B is the magnitude of the uniform magnetic field in the generator, and ω is the angular frequency of rotation of the coil. In terms of the frequency f (in Hz), the angular frequency is given by $\omega = 2\pi f$ (Equation 10.6). Substituting Equation 10.6 into Equation 22.4, we obtain

$$\xi_0 = NAB(2\pi f) = 2\pi NABf \qquad (1)$$

When the rotational frequency f of the coil changes, the peak emf ξ_0 also changes. The quantities N, A, and B remain constant, however, because they depend on how the generator is constructed, not on how rapidly the coil rotates. We know the peak emf of the generator at one frequency, so we will use Equation (1) to determine the peak emf for a different frequency in part (a), and the frequency needed for a different peak emf in part (b).

SOLUTION

a. Solving Equation (1) for the quantities that do not change with frequency, we find that

$$\frac{\xi_0}{f} = \underbrace{2\pi NAB}_{\substack{\text{Same for all}\\\text{frequencies}}} \tag{2}$$

The peak emf is $\xi_{0,1} = 75$ V when the frequency is $f_1 = 280$ Hz. We wish to find the peak emf $\xi_{0,2}$ when the frequency is $f_2 = 45$ Hz. From Equation (2), we have that

$$\frac{\xi_{0,2}}{f_2} = \underbrace{2\pi NAB}_{\substack{\text{Same for all}\\\text{frequencies}}} = \frac{\xi_{0,1}}{f_1} \tag{3}$$

Solving Equation (3) for $\xi_{0,2}$, we obtain

$$\xi_{0,2} = \left(\frac{f_2}{f_1}\right)\xi_{0,1} = \left(\frac{45 \text{ Hz}}{280 \text{ Hz}}\right)(75 \text{ V}) = \boxed{12 \text{ V}}$$

b. Letting $\xi_{0,3} = 180$ V, Equation (2) yields

$$\frac{\xi_{0,3}}{f_3} = \frac{\xi_{0,1}}{f_1} \tag{4}$$

Solving Equation (4) for f_3, we find that

$$f_3 = \left(\frac{\xi_{0,3}}{\xi_{0,1}}\right)f_1 = \left(\frac{180 \text{ V}}{75 \text{ V}}\right)(280 \text{ Hz}) = \boxed{670 \text{ Hz}}$$

---

73. **REASONING** In solving this problem, we apply Lenz's law, which essentially says that the change in magnetic flux must be opposed by the induced magnetic field.

SOLUTION

a. The magnetic field due to the wire in the vicinity of position 1 is directed out of the paper. The coil is moving closer to the wire into a region of higher magnetic field, so the flux through the coil is increasing. Lenz's law demands that the induced field counteract this increase. The direction of the induced field, therefore, must be into the paper. The current in the coil must be $\boxed{\text{clockwise}}$.

b. At position 2 the magnetic field is directed into the paper and is decreasing as the coil moves away from the wire. The induced magnetic field, therefore, must be directed into the paper, so the current in the coil must be $\boxed{\text{clockwise}}$.

---

75. **REASONING AND SOLUTION** The energy stored in a capacitor is given by Equation 19.11b as $\frac{1}{2}CV^2$. The energy stored in an inductor is given by Equation 22.10 as $\frac{1}{2}LI^2$. Setting these two equations equal to each other and solving for the current I, we get

$$I = \sqrt{\frac{C}{L}}\, V = \sqrt{\frac{3.0 \times 10^{-6}\ \text{F}}{5.0 \times 10^{-3}\ \text{H}}}\,(35\ \text{V}) = \boxed{0.86\ \text{A}}$$

---

79. **REASONING** The energy dissipated in the resistance is given by Equation 6.10b as the power P dissipated multiplied by the time t, Energy $= Pt$. The power, according to Equation 20.6c, is the square of the induced emf ξ divided by the resistance R, $P = \xi^2/R$. The induced emf can be determined from Faraday's law of electromagnetic induction, Equation 22.3.

SOLUTION Expressing the energy consumed as Energy $= Pt$, and substituting in $P = \xi^2/R$, we find

$$\text{Energy} = Pt = \frac{\xi^2 t}{R}$$

The induced emf is given by Faraday's law as $\xi = -N\left(\Delta\Phi/\Delta t\right)$, and the resistance R is equal to the resistance per unit length ($3.3 \times 10^{-2}\ \Omega/\text{m}$) times the length of the circumference of the loop, $2\pi r$. Thus, the energy dissipated is

$$\text{Energy} = \frac{\left(-N\dfrac{\Delta\Phi}{\Delta t}\right)^2 t}{\left(3.3 \times 10^{-2}\ \Omega/\text{m}\right)2\pi r} = \frac{\left[-N\left(\dfrac{\Phi - \Phi_0}{t - t_0}\right)\right]^2 t}{\left(3.3 \times 10^{-2}\ \Omega/\text{m}\right)2\pi r}$$

$$= \frac{\left[-N\left(\dfrac{BA\cos\phi - B_0 A\cos\phi}{t - t_0}\right)\right]^2 t}{\left(3.3 \times 10^{-2}\ \Omega/\text{m}\right)2\pi r} = \frac{\left[-NA\cos\phi\left(\dfrac{B - B_0}{t - t_0}\right)\right]^2 t}{\left(3.3 \times 10^{-2}\ \Omega/\text{m}\right)2\pi r}$$

$$= \frac{\left[-(1)\pi(0.12\ \text{m})^2(\cos 0°)\left(\dfrac{0.60\ \text{T} - 0\ \text{T}}{0.45\ \text{s} - 0\ \text{s}}\right)\right]^2 (0.45\ \text{s})}{\left(3.3 \times 10^{-2}\ \Omega/\text{m}\right)2\pi(0.12\ \text{m})} = \boxed{6.6 \times 10^{-2}\ \text{J}}$$

83. ***REASONING*** According to the relation $\varepsilon = -L\left(\dfrac{\Delta I}{\Delta t}\right)$ (Equation 22.9), the emf ε induced in

the solenoid depends on the self-inductance L of the solenoid and the rate $\left(\dfrac{\Delta I}{\Delta t}\right)$ at which the

solenoidal current changes. The variables ΔI and Δt are known, and we can evaluate the self-inductance by using the definition, $L = N\Phi / I$ (Equation 22.8), where N is the number of turns in the coil, Φ is the magnetic flux that passes through one turn, and I is the current. The magnetic flux Φ is given by $\Phi = BA\cos\phi$ (Equation 22.2), and the magnetic field inside a solenoid has the value of $B = \mu_0 n I$ (Equation 21.8), where n is the number of turns per unit length. These relations will enable us to evaluate the emf induced in the solenoid.

The emf ε produced by self-induction is given by Faraday's law of induction in the form expressed in Equation 22.9: $\varepsilon = -L\dfrac{\Delta I}{\Delta t}$, where L is the self-inductance of the solenoid, ΔI is the change in current, and Δt is the change in time. Both ΔI and Δt are known, and L will be evaluated in the next step.

The self-inductance L is defined by $L = N\Phi / I$ (Equation 22.8), where N is the number of turns, Φ is the magnetic flux that passes through one turn, and I is the current in the coil. The number of turns is equal to the number n of turns per meter of length times the length ℓ of the solenoid, or $N = n\ell$. Substituting this expression for N into $L = N\Phi / I$ gives

$$L = \frac{n\ell\Phi}{I} \quad (1)$$

This result for the self-inductance can be substituted into Equation 22.9 above. Note that n and ℓ are known. The magnetic flux Φ is not known and, along with the current I, will be dealt with next.

According to the discussion in Section 22.3, the magnetic flux Φ is given by (Equation 22.2), $\Phi = BA\cos\phi$, where B is the magnitude of the magnetic field that penetrates the surface of area A, and ϕ is the angle between the magnetic field and the normal to the surface. In the case of a solenoid, the interior magnetic field is directed perpendicular to the plane of the loops (see Section 21.7), so $\phi = 0°$. Thus, the magnetic flux is $\Phi = BA\cos 0°$. According to the discussion in Section 21.7, the magnitude of the magnetic field inside a long solenoid is given by $B = \mu_0 n I$ (Equation 21.8). Substituting this expression for B into the equation for the flux yields:

$$\Phi = (\mu_0 n I)A\cos 0°.$$

All the variables on the right side of this equation, except the current, are known. We now substitute this expression into Equation 1 above. Note that the current I appears in both the numerator and denominator in the expression for the self-inductance L, so I will be algebraically eliminated from the final result. Algebraically combining the results of the three steps, we have:

$$\varepsilon = -L\frac{\Delta I}{\Delta t} = -\frac{n\ell\Phi}{I}\left(\frac{\Delta I}{\Delta t}\right) = -\frac{n\ell(\mu_0 nIA\cos 0°)}{I}\left(\frac{\Delta I}{\Delta t}\right) = -\mu_0 n^2 \ell A\left(\frac{\Delta I}{\Delta t}\right)$$

The self-inductance of the solenoid is $B = \mu_0 n^2 \ell A$, and the emf induced in the solenoid is:

$$\varepsilon = -\mu_0 n^2 \ell A\left(\frac{\Delta I}{\Delta t}\right)$$

$$= -\left(4\pi \times 10^{-7} \text{ T·m/A}\right)(6500 \text{ turns/m})^2 (8.0\times 10^{-2} \text{ m})(5.0\times 10^{-5} \text{ m}^2)\left(\frac{1.5 \text{ A}}{0.20 \text{ s}}\right)$$

$$= \boxed{-1.6\times 10^{-3} \text{ V}}$$

---

85. *CONCEPTS*

(i) Yes. The period is the time for the coil to rotate through one revolution, or cycle. During this time, the emf is positive for one half of the cycle and negative for the other half of the cycle.

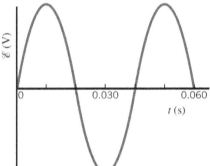

(ii) According to Equation 10.6, the angular frequency ω is related to the period T by $\omega = 2\pi/T$.

(iii) An examination of the graph shows that the emf reaches its maximum or peak value one-quarter of the way through a cycle. Since the time to complete one cycle is the period T, the time to reach the peak emf is $t = \frac{1}{4}T$.

(iv) The graph shows that in every cycle the emf has an interval of positive polarity followed by an interval of negative polarity. In other words, the polarity changes from positive to negative and then from negative to positive during each cycle. Thus the polarity changes twice during each cycle.

CALCULATIONS

(a) The period is the time for the generator to make one complete cycle. From the graph it can be seen that this time is

$$\boxed{T = 0.040 \text{ s}}$$

(b) The angular frequency ω is related to the period by $\omega = 2\pi/T$ so

$$\omega = \frac{2\pi}{T} = \frac{2\pi}{0.040 \text{ s}} = \boxed{160 \text{ rad/s}}$$

(c) When $t = \frac{1}{4}T$, the emf has reached its peak value. According to Equation 22.4, the peak emf is

$$\mathcal{E}_0 = NAB\omega = (10)(0.15 \text{ m}^2)(0.27 \text{ T})(160 \text{ rad/s}) = \boxed{65 \text{ V}}$$

(d) The emf produced by the generator as a function of time is given by Equation 22.4 as

$$\mathcal{E} = \mathcal{E}_0 \sin \omega t = (65 \text{ V})\sin\left[(160 \text{ rad/s})(0.025 \text{ s})\right] = \boxed{-49 \text{ V}}$$

CHAPTER 23 | ALTERNATING CURRENT CIRCUITS

1. **REASONING** As the frequency f of the generator increases, the capacitive reactance X_C of the capacitor decreases, according to $X_C = \dfrac{1}{2\pi f C}$ (Equation 23.2), where C is the capacitance of the capacitor. The decreasing capacitive reactance leads to an increasing rms current I_{rms}, as we see from $I_{rms} = \dfrac{V_{rms}}{X_C}$ (Equation 23.1), where V_{rms} is the constant rms voltage across the capacitor. We know that V_{rms} is constant, because it is equal to the constant rms generator voltage. The fuse is connected in series with the capacitor, so both have the same current I_{rms}. We will use Equations 23.1 and 23.2 to determine the frequency f at which the rms current is 15.0 A.

SOLUTION Solving Equation 23.2 for f, we obtain

$$f = \frac{1}{2\pi C X_C} \tag{1}$$

In terms of the rms current and voltage, Equation 23.1 gives the capacitive reactance as $X_C = \dfrac{V_{rms}}{I_{rms}}$. Substituting this relation into Equation (1) yields

$$f = \frac{1}{2\pi C X_C} = \frac{1}{2\pi C \left(\dfrac{V_{rms}}{I_{rms}} \right)} = \frac{I_{rms}}{2\pi C V_{rms}} = \frac{15.0 \text{ A}}{2\pi \left(63.0 \times 10^{-6} \text{ F} \right)\left(4.00 \text{ V} \right)} = \boxed{9470 \text{ Hz}}$$

5. **REASONING** The rms current in a capacitor is $I_{rms} = V_{rms}/X_C$, according to Equation 23.1. The capacitive reactance is $X_C = 1/(2\pi f C)$, according to Equation 23.2. For the first capacitor, we use $C = C_1$ in these expressions. For the two capacitors in parallel, we use $C = C_P$, where C_P is the equivalent capacitance from Equation 20.18 ($C_P = C_1 + C_2$). Taking the difference between the currents and using the given data, we can obtain the desired value for C_2. The capacitance C_1 is unknown, but it will be eliminated algebraically from the calculation.

SOLUTION Using Equations 23.1 and 23.2, we find that the current in a capacitor is

$$I_{rms} = \frac{V_{rms}}{X_C} = \frac{V_{rms}}{1/(2\pi f C)} = V_{rms} 2\pi f C$$

Applying this result to the first capacitor and the parallel combination of the two capacitors, we obtain

$$\underbrace{I_1 = V_{rms} 2\pi f C_1}_{\text{Single capacitor}} \quad \text{and} \quad \underbrace{I_{Combination} = V_{rms} 2\pi f (C_1 + C_2)}_{\text{Parallel combination}}$$

Subtracting I_1 from $I_{Combination}$, reveals that

$$I_{Combination} - I_1 = V_{rms} 2\pi f (C_1 + C_2) - V_{rms} 2\pi f C_1 = V_{rms} 2\pi f C_2$$

Solving for C_2 gives

$$C_2 = \frac{I_{Combination} - I_1}{V_{rms} 2\pi f} = \frac{0.18 \text{ A}}{(24 \text{ V}) 2\pi (440 \text{ Hz})} = \boxed{2.7 \times 10^{-6} \text{ F}}$$

---

9. **REASONING**   The rms current can be calculated from Equation 23.3, $I_{rms} = V_{rms}/X_L$, provided that the inductive reactance is obtained first. Then the peak value of the current I_0 supplied by the generator can be calculated from the rms current I_{rms} by using Equation 20.12, $I_0 = \sqrt{2} I_{rms}$.

SOLUTION At the frequency of $f = 620$ Hz, we find, using Equations 23.4 and 23.3, that

$$X_L = 2\pi f L = 2\pi (620 \text{ Hz})(8.2 \times 10^{-3} \text{ H}) = 32 \text{ } \Omega$$

$$I_{rms} = \frac{V_{rms}}{X_L} = \frac{10.0 \text{ V}}{32 \text{ } \Omega} = 0.31 \text{ A}$$

Therefore, from Equation 20.12, we find that the *peak value I_0* of the current supplied by the generator must be

$$I_0 = \sqrt{2} I_{rms} = \sqrt{2}(0.31 \text{ A}) = \boxed{0.44 \text{ A}}$$

---

15. **REASONING**
a. The inductive reactance X_L depends on the frequency f of the current and the inductance L through the relation $X_L = 2\pi f L$ (Equation 23.4). This equation can be used directly to find the frequency of the current.

b. The capacitive reactance X_C depends on the frequency f of the current and the capacitance C through the relation $X_C = 1/(2\pi f C)$ (Equation 23.2). By setting $X_C = X_L$ as specified in the problem statement, the capacitance can be found.

c. Since the inductive reactance is directly proportional to the frequency, tripling the frequency triples the inductive reactance.

d. The capacitive reactance is inversely proportional to the frequency, so tripling the frequency reduces the capacitive reactance by a factor of one-third.

SOLUTION
a. The frequency of the current is

$$f = \frac{X_L}{2\pi L} = \frac{2.10\times10^3\ \Omega}{2\pi\left(30.0\times10^{-3}\ \text{H}\right)} = \boxed{1.11\times10^4\ \text{Hz}} \tag{23.4}$$

b. The capacitance is

$$C = \frac{1}{2\pi f X_C} = \frac{1}{2\pi\left(1.11\times10^4\ \text{Hz}\right)\left(2.10\times10^3\ \Omega\right)} = \boxed{6.83\times10^{-9}\ \text{F}} \tag{23.2}$$

c. Since $X_L = 2\pi f L$, tripling the frequency f causes X_L to also triple:

$$X_L = 3\left(2.10\times10^3\ \Omega\right) = \boxed{6.30\times10^3\ \Omega}$$

d. Since $X_C = 1/(2\pi f C)$, tripling the frequency f causes X_C to decrease by a factor of 3:

$$X_C = \tfrac{1}{3}\left(2.10\times10^3\ \Omega\right) = \boxed{7.00\times10^2\ \Omega}$$

---

17. *REASONING* The voltage supplied by the generator can be found from Equation 23.6, $V_{rms} = I_{rms} Z$. The value of I_{rms} is given in the problem statement, so we must obtain the impedance of the circuit.

SOLUTION The impedance of the circuit is, according to Equation 23.7,

$$Z = \sqrt{R^2 + (X_L - X_C)^2} = \sqrt{(275\ \Omega)^2 + (648\ \Omega - 415\ \Omega)^2} = 3.60\times10^2\ \Omega$$

The rms voltage of the generator is

$$V_{rms} = I_{rms}Z = (0.233 \text{ A})(3.60 \times 10^2 \text{ }\Omega) = \boxed{83.9 \text{ V}}$$

19. ***REASONING***   We can use the equations for a series RCL circuit to solve this problem provided that we set $X_C = 0 \text{ }\Omega$ since there is no capacitor in the circuit. The current in the circuit can be found from Equation 23.6, $V_{rms} = I_{rms}Z$, once the impedance of the circuit has been obtained. Equation 23.8, $\tan\phi = (X_L - X_C)/R$, can then be used (with $X_C = 0 \text{ }\Omega$) to find the phase angle between the current and the voltage.

SOLUTION   The inductive reactance is (Equation 23.4)

$$X_L = 2\pi f L = 2\pi (106 \text{ Hz})(0.200 \text{ H}) = 133 \text{ }\Omega$$

The impedance of the circuit is

$$Z = \sqrt{R^2 + (X_L - X_C)^2} = \sqrt{R^2 + X_L^2} = \sqrt{(215 \text{ }\Omega)^2 + (133 \text{ }\Omega)^2} = 253 \text{ }\Omega$$

a.   The current through each circuit element is, using Equation 23.6,

$$I_{rms} = \frac{V_{rms}}{Z} = \frac{234 \text{ V}}{253 \text{ }\Omega} = \boxed{0.925 \text{ A}}$$

b.   The phase angle between the current and the voltage is, according to Equation 23.8 (with $X_C = 0 \text{ }\Omega$),

$$\tan\phi = \frac{X_L - X_C}{R} = \frac{X_L}{R} = \frac{133 \text{ }\Omega}{215 \text{ }\Omega} = 0.619 \qquad \text{or} \qquad \phi = \tan^{-1}(0.619) = \boxed{31.8°}$$

25. ***REASONING***   We can use the equations for a series RCL circuit to solve this problem, provided that we set $X_L = 0$ since there is no inductance in the circuit. Thus, according to Equations 23.6 and 23.7, the current in the circuit is $I_{rms} = V_{rms}/\sqrt{R^2 + X_C^2}$. When the frequency f is very large, the capacitive reactance is zero, or $X_C = 0$, in which case the current becomes $I_{rms}(\text{large } f) = V_{rms}/R$. When the current I_{rms} in the circuit is one-half the value of $I_{rms}(\text{large } f)$ that exists when the frequency is very large, we have

$$\frac{I_{rms}}{I_{rms}(\text{large } f)} = \frac{1}{2}$$

We can use these expressions to write the ratio above in terms of the resistance and the capacitive reactance. Once the capacitive reactance is known, the frequency can be determined.

SOLUTION The ratio of the currents is

$$\frac{I_{rms}}{I_{rms}(\text{large } f)} = \frac{V_{rms}/\sqrt{R^2 + X_C^2}}{V_{rms}/R} = \frac{R}{\sqrt{R^2 + X_C^2}} = \frac{1}{2} \quad \text{or} \quad \frac{R^2}{R^2 + X_C^2} = \frac{1}{4}$$

Taking the reciprocal of this result gives

$$\frac{R^2 + X_C^2}{R^2} = 4 \quad \text{or} \quad 1 + \frac{X_C^2}{R^2} = 4$$

Therefore,

$$\frac{X_C}{R} = \sqrt{3}$$

According to Equation 23.2, $X_C = 1/(2\pi f C)$, so it follows that

$$\frac{X_C}{R} = \frac{1/(2\pi f C)}{R} = \sqrt{3}$$

Thus, we find that

$$f = \frac{1}{2\pi RC\sqrt{3}} = \frac{1}{2\pi(85\ \Omega)(4.0 \times 10^{-6}\ \text{F})\sqrt{3}} = \boxed{270\ \text{Hz}}$$

---

31. **REASONING** The resonant frequency f_0 of a series RCL circuit depends on the inductance L and capacitance C through the relation $f_0 = \dfrac{1}{2\pi\sqrt{LC}}$ (Equation 23.10). Since all the variables are known except L, we can use this relation to find the inductance.

SOLUTION Solving Equation 23.10 for the inductance gives

$$L = \frac{1}{4\pi^2 f_0^2 C} = \frac{1}{4\pi^2(690 \times 10^3\ \text{Hz})^2(2.0 \times 10^{-9}\ \text{F})} = \boxed{2.7 \times 10^{-5}\ \text{H}}$$

---

33. **REASONING** The current in an RCL circuit is given by Equation 23.6, $I_{rms} = V_{rms} / Z$, where the impedance Z of the circuit is given by Equation 23.7 as $Z = \sqrt{R^2 + (X_L - X_C)^2}$. The current is a maximum when the impedance is a minimum for a given generator voltage. The minimum impedance occurs when the frequency is f_0, corresponding to the condition that $X_L = X_C$, or $2\pi f_0 L = 1/(2\pi f_0 C)$. Solving for the frequency f_0, called the resonant frequency, we find that

$$f_0 = \frac{1}{2\pi\sqrt{LC}}$$

Note that the resonant frequency depends on the inductance and the capacitance, but does not depend on the resistance.

SOLUTION
a. The frequency at which the current is a maximum is

$$f_0 = \frac{1}{2\pi\sqrt{LC}} = \frac{1}{2\pi\sqrt{(17.0\times10^{-3}\ H)(12.0\times10^{-6}\ F)}} = \boxed{352\ Hz}$$

b. The maximum value of the current occurs when $f = f_0$. This occurs when $X_L = X_C$, so that $Z = R$. Therefore, according to Equation 23.6, we have

$$I_{rms} = \frac{V_{rms}}{Z} = \frac{V_{rms}}{R} = \frac{155\ V}{10.0\ \Omega} = \boxed{15.5\ A}$$

---

39. **REASONING AND SOLUTION** At the resonant frequency f_0, we have $C = 1/(4\pi^2 f_0^2 L)$. We want to determine some series combination of capacitors whose equivalent capacitance C_s' is such that $f_0' = 3f_0$. Thus,

$$C_s' = \frac{1}{4\pi^2 f_0'^2 L} = \frac{1}{4\pi^2 (3f_0)^2 L} = \frac{1}{9}\left(\frac{1}{4\pi^2 f_0^2 L}\right) = \frac{1}{9}C$$

The equivalent capacitance of a series combination of capacitors is $1/C_s' = 1/C_1 + 1/C_2 + ...$ If we require that all the capacitors have the same capacitance C, the equivalent capacitance is

$$\frac{1}{C_s'} = \frac{1}{C} + \frac{1}{C} + \cdots = \frac{n}{C}$$

where n is the total number of identical capacitors. Using the result above, we find that

$$\frac{1}{C'_s} = \frac{1}{\frac{1}{9}C} = \frac{n}{C} \qquad \text{or} \qquad n = 9$$

Therefore, the number of *additional* capacitors that must be inserted in series in the circuit so that the frequency triples is $n' = n - 1 = \boxed{8}$.

---

43. **REASONING**   The voltage across the capacitor reaches its maximum instantaneous value when the generator voltage reaches its maximum instantaneous value. The maximum value of the capacitor voltage first occurs one-fourth of the way, or one-quarter of a period, through a complete cycle (see the voltage curve in Figure 23.4).

SOLUTION   The period of the generator is $T = 1/f = 1/(5.00 \text{ Hz}) = 0.200 \text{ s}$. Therefore, the least amount of time that passes before the instantaneous voltage across the capacitor reaches its maximum value is $\frac{1}{4}T = \frac{1}{4}(0.200 \text{ s}) = \boxed{5.00 \times 10^{-2} \text{ s}}$.

---

47. **REASONING**   The individual reactances are given by Equations 23.2 and 23.4, respectively,

Capacitive reactance $X_C = \dfrac{1}{2\pi f C}$

Inductive reactance $X_L = 2\pi f L$

When the reactances are equal, we have $X_C = X_L$, from which we find

$$\frac{1}{2\pi f C} = 2\pi f L \qquad \text{or} \qquad 4\pi^2 f^2 LC = 1$$

The last expression may be solved for the frequency f.

SOLUTION   Solving for f with $L = 52 \times 10^{-3}$ H and $C = 76 \times 10^{-6}$ F, we obtain

$$f = \frac{1}{2\pi\sqrt{LC}} = \frac{1}{2\pi\sqrt{(52\times10^{-3} \text{ H})(76\times10^{-6} \text{ F})}} = \boxed{8.0 \times 10^{1} \text{ Hz}}$$

---

51. **REASONING**   Since we know the values of the resonant frequency of the circuit, the capacitance, and the generator voltage, we can find the value of the inductance from Equation 23.10, the expression for the resonant frequency. The resistance can be found from energy considerations at resonance; the power factor is given by $\cos\phi$, where the phase angle ϕ is given by Equation 23.8, $\tan\phi = (X_L - X_C)/R$.

SOLUTION

a. Solving Equation 23.10 for the inductance L, we find that

$$L = \frac{1}{4\pi^2 f_0^2 C} = \frac{1}{4\pi^2 (1.30\times10^3 \text{ Hz})^2 (5.10\times10^{-6} \text{ F})} = \boxed{2.94\times10^{-3} \text{ H}}$$

b. At resonance, $f = f_0$, and the current is a maximum. This occurs when $X_L = X_C$, so that $Z = R$. Thus, the average power $\overline{P}$ provided by the generator is $\overline{P} = V_{rms}^2 / R$, and solving for R we find

$$R = \frac{V_{rms}^2}{\overline{P}} = \frac{(11.0 \text{ V})^2}{25.0 \text{ W}} = \boxed{4.84 \ \Omega}$$

c. When the generator frequency is 2.31 kHz, the individual reactances are

$$X_C = \frac{1}{2\pi f C} = \frac{1}{2\pi(2.31\times10^3 \text{ Hz})(5.10\times10^{-6} \text{ F})} = 13.5 \ \Omega$$

$$X_L = 2\pi f L = 2\pi(2.31\times10^3 \text{ Hz})(2.94\times10^{-3} \text{ H}) = 42.7 \ \Omega$$

The phase angle ϕ is, from Equation 23.8,

$$\phi = \tan^{-1}\left(\frac{X_L - X_C}{R}\right) = \tan^{-1}\left(\frac{42.7 \ \Omega - 13.5 \ \Omega}{4.84 \ \Omega}\right) = 80.6°$$

The power factor is then given by

$$\cos\phi = \cos 80.6° = \boxed{0.163}$$

---

53. **REASONING** The beat frequency is $|f_{oB} - f_{oA}|$, and to find it we need to determine the effect that the 1.000% increase of the inductance L_B has on the resonant frequency f_{oB}. We will do this by a direct application of $f_{oB} = \frac{1}{2\pi\sqrt{L_B C}}$ (Equation 23.10).

SOLUTION The beat frequency is the magnitude of the difference between the two resonant frequencies: Beat frequency $= |f_{oB} - f_{oA}|$ (1). The vertical bars, as usual, indicate that the magnitude or absolute value of the enclosed quantity is needed. The frequency f_{oA} is given, but the frequency f_{oB} is determined by the presence of the metal object and is unknown. We deal with it in the next step.

According to Equation 23.10, the resonant frequency of circuit B is:

$$f_{0B} = \frac{1}{2\pi\sqrt{L_B C}}$$

Initially, each inductor is adjusted so that $L_B = L_A$. Since the metal object causes the inductance of coil B to increase by 1.000%, the new value of L_B becomes $L_B = 1.01000 L_A$. Substituting this expression into Equation 23.10 gives:

$$f_{0B} = \frac{1}{2\pi\sqrt{L_B C}} = \frac{1}{2\pi\sqrt{1.01000 L_A C}}$$

Finally, we recognize that $f_{0A} = \dfrac{1}{2\pi\sqrt{L_A C}}$ and can then express the resonant frequency of circuit B in the presence of the metal object as follows:

$$f_{0B} = \left(\frac{1}{\sqrt{1.01000}}\right) f_{0A}$$

This result can now be substituted into Equation 1 above.

Combining the results of each step algebraically, we find that:

$$\text{Beat frequency} = |f_{0B} - f_{0A}| = \left|\left(\frac{1}{\sqrt{1.01000}}\right) f_{0A} - f_{0A}\right|$$

$$= \left|\left(\frac{1}{\sqrt{1.01000}}\right) - 1\right| f_{0A}$$

$$= \left|\left(\frac{1}{\sqrt{1.01000}}\right) - 1\right| (855.5 \text{ kHz}) = \boxed{4.2 \text{ kHz}}$$

---

54. *CONCEPTS*

(i) The capacitance of a capacitor is given by Equation 19.10 as $C = \kappa \varepsilon_0 A / d$, where κ is the dielectric constant of the material, ε_0 is the permittivity of free space, A is the area of each plate, and d is the separation between the plates. Since the dielectric constant and the plate are the same for each capacitor, the capacitance is inversely proportional to the plate separation. Therefore, capacitor 2, with the smaller plate separation, has the greater capacitance.

(ii) According to Equation 20.18, the equivalent capacitance of the two capacitors in parallel is $C_P = C_1 + C_2$, where C_1 and C_2 are the individual capacitances. Therefore, the value for C_P is greater than the value for C_1.

(iii) The capacitive reactance is given by $X_C = 1/2\pi f C$, according to equation 23.2, where f is the frequency. For a given frequency, the reactance is inversely proportional to the capacitance. Since the capacitance C_P is greater than C_1, the corresponding reactance is smaller.

(iv) According to Equation 23.1, the current is given by $I_{rms} = V_{rms}/X_C$, where V_{rms} is the rms voltage of the generator and X_C is the capacitive reactance. For a given voltage, the current is inversely proportional to the reactance. Since the reactance in the parallel case is smaller than for C_1 alone, the current in the parallel case is greater.

CALCULATIONS

Using Equation 23.1 to express the current as $I_{rms} = V_{rms}/X_C$ and Equation 23.2 to express the reactance as $X_C = 1/(2\pi f C)$, we find that the current is

$$I_{rms} = \frac{V_{rms}}{X_C} = \frac{V_{rms}}{1/(2\pi f C)} = V_{rms}(2\pi f C)$$

Applying this result to the case where C_1 is connected alone to the generator and to the case where two capacitors are connected in parallel, we obtain

$$\underbrace{I_{1,\,rms} = V_{rms}\,2\pi f C_1}_{C_1 \text{ alone}} \quad \text{and} \quad \underbrace{I_{P,\,rms} = V_{rms}\,2\pi f C_P}_{C_1 \text{ and } C_2 \text{ in parallel}}$$

Dividing these two equations gives

$$\frac{I_{P,\,rms}}{I_{1,\,rms}} = \frac{V_{rms}\,2\pi f C_P}{V_{rms}\,2\pi f C_1} = \frac{C_P}{C_1}$$

According to Equation 20.18, the effective capacitance of the two capacitors in parallel is $C_P = C_1 + C_2$, so the result for the current ratio becomes

$$\frac{I_{P,\,rms}}{I_{1,\,rms}} = \frac{C_1 + C_2}{C_1} = 1 + \frac{C_2}{C_1}$$

Since the capacitance of a capacitor is given by Equation 19.10 as $C = \kappa \varepsilon_0 A/d$, we find that

$$\frac{I_{P,\,\text{rms}}}{I_{1,\,\text{rms}}} = 1 + \frac{\kappa \varepsilon_0 A/d_2}{\kappa \varepsilon_0 A/d_1} = 1 + \frac{d_1}{d_2}$$

We know that $d_1 = 2d_2$, so that the current in the parallel case is

$$I_{P,\,\text{rms}} = I_{1,\,\text{rms}}\left(1 + \frac{d_1}{d_2}\right) = (0.60 \text{ A})\left(1 + \frac{2d_2}{d_2}\right) = \boxed{1.8 \text{ A}}$$

---

55. *CONCEPTS*

(i) When only the resistor is present, the current is given by Equation 20.14 as $I_{\text{rms}} = V_{\text{rms}}/R$, where V_{rms} is the rms voltage of the generator. When the resistor and the inductor are connected in series, the current is given by $I_{\text{rms}} = V_{\text{rms}}/Z$, according to Equation 23.6. In this expression, Z is the impedance and is given by Equation 23.7 as $Z = \sqrt{R^2 + X_L^2}$, where X_L is the inductive reactance and is given by Equation 23.4 as $X_L = 2\pi fL$. Since Z is greater than R, the current is greater when only the resistance is present.

(ii) Only the resistor in an ac circuit consumes power on average, the amount of the average power being $\overline{P} = I_{\text{rms}}^2 R$, according to Equation 20.15b. Since the resistance R is the same in both cases, a greater average power is consumed when the current I_{rms} is greater. Since the current is greater when only the resistor is present, more power is consumed in that case.

CALCULATIONS

When only the resistor is present, the average power is $\overline{P} = I_{\text{rms}}^2 R$, according to Equation 20.15b, and the current is given by $I_{\text{rms}} = V_{\text{rms}}/R$, according to Equation 20.14. Therefore, we find in this case that

Resistor only
$$\overline{P} = I_{\text{rms}}^2 R = \left(\frac{V_{\text{rms}}}{R}\right)^2 R = \frac{V_{\text{rms}}^2}{R}$$

When the inductor is also present, the average power is still $\overline{P} = I_{\text{rms}}^2 R$, but the current is now given by $I_{\text{rms}} = V_{\text{rms}}/Z$, according to Equation 23.6, where Z is the impedance. In this case, then, we have

Resistor and inductor
$$\overline{P} = I_{\text{rms}}^2 R = \left(\frac{V_{\text{rms}}}{Z}\right)^2 R = \frac{V_{\text{rms}}^2 R}{Z^2}$$

Dividing this equation by the analogous result for the resistor-only case, we obtain

$$\frac{\overline{P}_{\text{Resistor and inductor}}}{\overline{P}_{\text{Resistor only}}} = \frac{V_{\text{rms}}^2 R / Z^2}{V_{\text{rms}}^2 / R} = \frac{R^2}{Z^2}$$

In this expression, Z is the impedance and is given by Equation 23.7 as $Z = \sqrt{R^2 + X_L^2}$, where X_L is the inductive reactance and is given by Equation 23.4 as $X_L = 2\pi f L$. With these substitutions, the ratio of the powers becomes

$$\frac{\overline{P}_{\text{Resistor and inductor}}}{\overline{P}_{\text{Resistor only}}} = \frac{R^2}{R^2 + \left(2\pi f L\right)^2}$$

The power consumed in the series combination is

$$\overline{P}_{\text{Resistor and inductor}} = \overline{P}_{\text{Resistor only}} \left[\frac{R^2}{R^2 + \left(2\pi f L\right)^2} \right]$$

$$= \left(0.25 \text{ W}\right) \frac{\left(470 \text{ }\Omega\right)^2}{\left(470 \text{ }\Omega\right)^2 + \left[2\pi\left(1200 \text{ Hz}\right)\left(0.080 \text{ H}\right)\right]^2} = \boxed{0.094 \text{ W}}$$

As expected, more power is consumed in the resistor-only case.

CHAPTER 24 | ELECTROMAGNETIC WAVES

3. **REASONING** This is a standard exercise in the conversion of units. However, we will first need to determine the number of meters that light travels in one year, which is what we call a light-year. In the vacuum of space light travels at a speed of $c = 3.00 \times 10^8$ m/s. The constant speed c is given by $c = \dfrac{d}{t}$ (Equation 2.1), where d is the distance traveled in a time t. We can use this equation to determine d, since values are known for c and t. Then, armed with this value for one light-year, we will follow the usual procedure for converting units, as discussed in Section 1.3.

SOLUTION Solving Equation 2.1 for the distance d that light travels in one year, we find

$$d = ct = \left(3.00 \times 10^8 \text{ m/s}\right)\left(1.00 \text{ year}\right)\underbrace{\left(\frac{365.25 \text{ days}}{1 \text{ year}}\right)\left(\frac{24 \text{ hours}}{1 \text{ day}}\right)\left(\frac{3600 \text{ s}}{1 \text{ hour}}\right)}_{\text{Conversion of 1.00 year into seconds}} = 9.47 \times 10^{15} \text{ m}$$

Thus, 1 light-year $= 9.47 \times 10^{15}$ m. To find the distance to Alpha Centauri in meters, we follow the usual procedure and multiply the distance of 4.3 light-years by $1 = \dfrac{9.47 \times 10^{15} \text{ m}}{1 \text{ light-year}}$. In so doing, we obtain

$$\left(4.3 \text{ light-years}\right)\left(\frac{9.47 \times 10^{15} \text{ m}}{1 \text{ light-year}}\right) = \boxed{4.1 \times 10^{16} \text{ m}}$$

---

5. **REASONING** According to Equation 16.3, the displacement y of a wave that travels in the $+x$ direction and has amplitude A, frequency f, and wavelength λ is given by

$$y = A \sin\left(2\pi f t - \frac{2\pi x}{\lambda}\right)$$

This equation, with $y = E$, applies to the traveling electromagnetic wave in the problem, which is represented mathematically as

$$E = E_0 \sin\left[\left(1.5 \times 10^{10} \text{ s}^{-1}\right)t - \left(5.0 \times 10^1 \text{ m}^{-1}\right)x\right]$$

As E_0 is the maximum field strength, it represents the amplitude A of the wave. We can find the frequency and wavelength of this electromagnetic wave by comparing the mathematical form of the electric field with Equation 16.3.

SOLUTION

a. By inspection, we see that $2\pi f = 1.5 \times 10^{10}$ s^{-1}. Therefore, the frequency of the wave is

$$f = \frac{1.5 \times 10^{10} \text{ s}^{-1}}{2\pi} = \boxed{2.4 \times 10^9 \text{ Hz}}$$

b. As shown in Figure 17.15, the separation between adjacent nodes in any standing wave is one-half of a wavelength. By inspection of the mathematical form of the electric field and comparison with Equation 16.3, we infer that $2\pi / \lambda = 5.0 \times 10^1$ m^{-1}. Therefore,

$$\lambda = \frac{2\pi}{5.0 \times 10^1 \text{ m}^{-1}} = 0.126 \text{ m}$$

Therefore, the nodes in the standing waves formed by this electromagnetic wave are separated by $\lambda / 2 = \boxed{0.063 \text{ m}}$.

---

9. *REASONING* The frequency f of the UHF wave is related to its wavelength by $c = f\lambda$ (Equation 16.1), where c is the speed of light in a vacuum and λ is the wavelength. The electric and magnetic fields are both zero at the same positions, which are separated by a distance d equal to half a wavelength (see Figure 24.3). Therefore, we can express the wavelength in terms of the distance between adjacent positions of zero field as

$$\lambda = 2d \tag{1}$$

SOLUTION Solving $c = f\lambda$ (Equation 16.1) for f yields

$$f = \frac{c}{\lambda} \tag{2}$$

Substituting Equation (1) into Equation (2), we obtain

$$f = \frac{c}{\lambda} = \frac{c}{2d} = \frac{3.00 \times 10^8 \text{ m/s}}{2(0.34 \text{ m})} = \boxed{4.4 \times 10^8 \text{ Hz}}$$

---

17. *REASONING* We proceed by first finding the time t for sound waves to travel between the astronauts. Since this is the same time it takes for the electromagnetic waves to travel to earth, the distance between earth and the spaceship is $d_{\text{earth-ship}} = ct$.

SOLUTION The time it takes for sound waves to travel at 343 m/s through the air between the astronauts is

$$t = \frac{d_{\text{astronaut}}}{v_{\text{sound}}} = \frac{1.5 \text{ m}}{343 \text{ m/s}} = 4.4 \times 10^{-3} \text{ s}$$

Therefore, the distance between the earth and the spaceship is

$$d_{\text{earth-ship}} = ct = (3.0 \times 10^8 \text{ m/s})(4.4 \times 10^{-3} \text{ s}) = \boxed{1.3 \times 10^6 \text{ m}}$$

---

25. **REASONING** The rms value E_{rms} of the electric field is related to the average energy density $\bar{u}$ of the microwave radiation according to $\bar{u} = \varepsilon_0 E_{\text{rms}}^2$ (Equation 24.2b).

SOLUTION Solving for E_{rms} gives

$$E_{\text{rms}} = \sqrt{\frac{\bar{u}}{\varepsilon_0}} = \sqrt{\frac{4 \times 10^{-14} \text{ J/m}^3}{8.85 \times 10^{-12} \text{ C}^2/(\text{N} \cdot \text{m}^2)}} = \boxed{0.07 \text{ N/C}}$$

---

27. **REASONING** The average intensity of a wave is the average power per unit area that passes perpendicularly through a surface. Thus, the average power $\bar{P}$ of the wave is the product of the average intensity $\bar{S}$ and the area A. The average intensity is related to the rms-value E_{rms} of the electric field by $\bar{S} = c\varepsilon_0 E_{\text{rms}}^2$.

SOLUTION The average power is $\bar{P} = \bar{S}A$. Since $\bar{S} = c\varepsilon_0 E_{\text{rms}}^2$ (Equation 24.5b), we have

$$\bar{P} = \bar{S}A = \left(c\varepsilon_0 E_{\text{rms}}^2\right)A$$

$$= (3.00 \times 10^8 \text{ m/s})\left[8.85 \times 10^{-12} \text{ C}^2/(\text{N} \cdot \text{m}^2)\right](2.0 \times 10^9 \text{ N/C})^2 (1.6 \times 10^{-5} \text{ m}^2)$$

$$= \boxed{1.7 \times 10^{11} \text{ W}}$$

---

37. **REASONING** The Doppler effect for electromagnetic radiation is given by Equation 24.6, $f_o = f_s\left(1 \pm v_{\text{rel}}/c\right)$, where f_o is the observed frequency, f_s is the frequency emitted by the source, and v_{rel} is the speed of the source relative to the observer. As discussed in the text, the plus sign applies when the source and the observer are moving toward one another, while the minus sign applies when they are moving apart.

SOLUTION
a. At location A, the galaxy is moving away from the earth with a relative speed of

$$v_{\text{rel}} = (1.6 \times 10^6 \text{ m/s}) - (0.4 \times 10^6 \text{ m/s}) = 1.2 \times 10^6 \text{ m/s}$$

Therefore, the minus sign in Equation 24.6 applies and the observed frequency for the light from region A is

$$f_o = f_s \left(1 - \frac{v_{\text{rel}}}{c}\right) = (6.200 \times 10^{14} \text{ Hz})\left(1 - \frac{1.2 \times 10^6 \text{ m/s}}{3.0 \times 10^8 \text{ m/s}}\right) = \boxed{6.175 \times 10^{14} \text{ Hz}}$$

b. Similarly, at location B, the galaxy is moving away from the earth with a relative speed of

$$v_{\text{rel}} = (1.6 \times 10^6 \text{ m/s}) + (0.4 \times 10^6 \text{ m/s}) = 2.0 \times 10^6 \text{ m/s}$$

The observed frequency for the light from region B is

$$f_o = f_s \left(1 - \frac{v_{\text{rel}}}{c}\right) = (6.200 \times 10^{14} \text{ Hz})\left(1 - \frac{2.0 \times 10^6 \text{ m/s}}{3.0 \times 10^8 \text{ m/s}}\right) = \boxed{6.159 \times 10^{14} \text{ Hz}}$$

---

41. **REASONING** Malus' law, $\bar{S} = \bar{S}_0 \cos^2 \theta$ (Equation 24.7), relates the average intensity $\bar{S}_0$ of polarized light incident on the polarizing sheet to the average intensity $\bar{S}$ of light transmitted by the sheet, where θ is the angle between the polarization axis of the incident light and the transmission axis of the polarizing sheet. The incident light is horizontally polarized, so the angle θ is measured from the horizontal, and is, therefore, the angle we seek.

SOLUTION Solving $\bar{S} = \bar{S}_0 \cos^2 \theta$ (Equation 24.7) for θ, we obtain

$$\cos^2 \theta = \frac{\bar{S}}{\bar{S}_0} \quad \text{or} \quad \cos \theta = \sqrt{\frac{\bar{S}}{\bar{S}_0}} \quad \text{or} \quad \theta = \cos^{-1}\left(\sqrt{\frac{\bar{S}}{\bar{S}_0}}\right) \qquad (2)$$

Substituting the given values of the average incident and transmitted intensities yields

$$\theta = \cos^{-1}\left(\sqrt{\frac{0.764 \text{ W/m}^2}{0.883 \text{ W/m}^2}}\right) = \boxed{21.5^\circ}$$

---

47. **REASONING** The average intensity of light leaving each analyzer is given by Malus' Law (Equation 24.7). Thus, intensity of the light transmitted through the first analyzer is

$$\bar{S}_1 = \bar{S}_0 \cos^2 27^\circ$$

Similarly, the intensity of the light transmitted through the second analyzer is

$$\overline{S}_2 = \overline{S}_1 \cos^2 27° = \overline{S}_0 \cos^4 27°$$

And the intensity of the light transmitted through the third analyzer is

$$\overline{S}_3 = \overline{S}_2 \cos^2 27° = \overline{S}_0 \cos^6 27°$$

If we generalize for the Nth analyzer, we deduce that

$$\overline{S}_N = \overline{S}_{N-1} \cos^2 27° = \overline{S}_0 \cos^{2N} 27°$$

Since we want the light reaching the photocell to have an intensity that is reduced by a least a factor of one hundred relative to the first analyzer, we want $\overline{S}_N / \overline{S}_0 = 0.010$. Therefore, we need to find N such that $\cos^{2N} 27° = 0.010$. This expression can be solved for N.

SOLUTION Taking the common logarithm of both sides of the last expression gives

$$2N \log(\cos 27°) = \log 0.010 \quad \text{or} \quad N = \frac{\log 0.010}{2 \log (\cos 27°)} = \boxed{20}$$

---

51. **REASONING** The electromagnetic wave will be picked up by the radio when the resonant frequency f_0 of the circuit in Figure 24.4 is equal to the frequency of the broadcast wave, or f_0 = 1400 kHz. This frequency, in turn, is related to the capacitance C and inductance L of the circuit via $f_0 = 1/\left(2\pi\sqrt{LC}\right)$ (Equation 23.10). Since C is known, we can use this relation to find the inductance.

SOLUTION Solving the relation $f_0 = 1/\left(2\pi\sqrt{LC}\right)$ for the inductance L, we find that

$$L = \frac{1}{4\pi^2 f_0^2 C} = \frac{1}{4\pi^2 \left(1400\times10^3 \text{ Hz}\right)^2 \left(8.4\times10^{-11} \text{ F}\right)} = \boxed{1.5\times10^{-4} \text{ H}}$$

---

53. **REASONING AND SOLUTION**
a. According to Equation 24.5b, the average intensity is $\overline{S} = c\varepsilon_0 E_{rms}^2$. In addition, the average intensity is the average power $\overline{P}$ divided by the area A. Therefore,

$$E_{rms} = \sqrt{\frac{\overline{S}}{c\varepsilon_0}} = \sqrt{\frac{\overline{P}}{c\varepsilon_0 A}} = \sqrt{\frac{1.20\times10^4 \text{ W}}{(3.00\times10^8 \text{ m/s})[8.85\times10^{-12}\text{C}^2/(\text{N}\cdot\text{m}^2)](135 \text{ m}^2)}} = \boxed{183 \text{ N/C}}$$

b. Then, from $E_{rms} = cB_{rms}$ (Equation 24.3), we have

$$B_{rms} = \frac{E_{rms}}{c} = \frac{183 \text{ N/C}}{3.00 \times 10^8 \text{ m/s}} = \boxed{6.10 \times 10^{-7} \text{ T}}$$

---

55. **REASONING** In experiment 1, the light falling on the polarizer is unpolarized, so the average intensity $\overline{S}$ transmitted by it is $\overline{S} = \frac{1}{2}\overline{S}'$, where $\overline{S}'$ is the intensity of the incident light. The intensity of the light passing through the analyzer and reaching the photocell can be found by using Malus' law.

In the second experiment, all of the polarized light passes through the polarizer, so the average intensity reaching the analyzer is $\overline{S}'$. The intensity of the light passing through the analyzer and reaching the photocell can again be found by using Malus' law. By setting the intensities of the light reaching the photocells in experiments 1 and 2 equal, we can determine the number of additional degrees that the analyzer in experiment 2 must be rotated relative to that in experiment 1.

SOLUTION In experiment 1 the light intensity incident on the analyzer is $\overline{S}_0 = \frac{1}{2}\overline{S}'$. Malus' law (Equation 24.7) gives the average intensity $\overline{S}_1$ of the light reaching the photocell as

$$\overline{S}_1 = \overline{S}_0 \cos^2 60.0^\circ = \frac{1}{2}\overline{S}' \cos^2 60.0^\circ$$

In experiment 2 the incident light is polarized along the axis of the polarizer, so all the light is transmitted by the polarizer. Thus, the light intensity incident on the analyzer is $\overline{S}_0 = \overline{S}'$. Malus' law again gives the average intensity $\overline{S}_2$ of the light reaching the photocell as

$$\overline{S}_2 = \overline{S}_0 \cos^2 \theta = \overline{S}' \cos^2 \theta$$

Setting $\overline{S}_1 = \overline{S}_2$, we have

$$\tfrac{1}{2}\overline{S}' \cos^2 60.0^\circ = \overline{S}' \cos^2 \theta$$

Algebraically eliminating $\overline{S}'$ from this equation, we have that

$$\tfrac{1}{2}\cos^2 60.0^\circ = \cos^2 \theta \quad \text{or} \quad \theta = \cos^{-1}\left(\sqrt{\tfrac{1}{2}\cos^2 60.0^\circ}\right) = 69.3^\circ$$

The number of additional degrees that the analyzer must be rotated is $69.3^\circ - 60.0^\circ = \boxed{9.3^\circ}$. The angle θ is $\boxed{\text{increased}}$ by the additional rotation.

---

61. **REASONING**   The electromagnetic solar power that strikes an area $A_\perp$ oriented perpendicular to the direction in which the sunlight is radiated is $P = SA_\perp$, where S is the intensity of the sunlight. In the problem, the solar panels are not oriented perpendicular to the

direction of the sunlight, because it strikes the panels at an angle θ with respect to the normal. We wish to find the solar power that impinges on the solar panels when $\theta = 25°$, given that the incident power is 2600 W when $\theta = 65°$.

SOLUTION When the angle that the sunlight makes with the normal to the solar panel is θ, the power that strikes the solar panel is given by $P = SA \cos\theta$, where the area perpendicular to the sunlight is $A_\perp = A \cos\theta$ (see the drawing). Therefore we can write

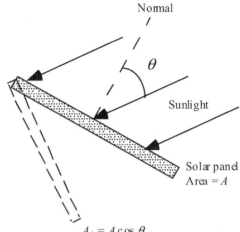

$$\frac{P_2}{P_1} = \frac{SA \cos\theta_2}{SA \cos\theta_1}$$

where the intensity S of the sunlight that reaches the panel, as well as the area A, are the same in both cases. Therefore, we have

$$\frac{P_2}{P_1} = \frac{\cos\theta_2}{\cos\theta_1}$$

Solving for P_2, we find that when $\theta_2 = 35°$, the solar power impinging on the panel is

$$P_2 = P_1 \left(\frac{\cos\theta_2}{\cos\theta_1} \right) = (2600 \text{ W}) \left(\frac{\cos 25°}{\cos 65°} \right) = \boxed{5600 \text{ W}}$$

---

63. **REASONING** As the car moves toward the wave emitted by the radar gun, the car intercepts a greater number of wave crests per second than it would if it were stationary. Thus, the car "observes" a wave frequency f_o that is greater than the frequency f_s emitted by the radar gun. The car then reflects the wave back toward the police car. In effect, the car becomes a moving source of radar waves that are emitted with a frequency f_o. Since the car is moving toward the police car, the reflected crests arrive at the police car with a frequency f_o' that is even greater than f_o. Thus, there are two frequency changes due to the Doppler effect, one associated with f_o and the other with f_o'. We will employ Equation 24.6 to determine these frequencies.

SOLUTION The desired frequency difference is $f_o' - f_s$, where f_s is the frequency emitted by the radar gun and is known; f_o' is the frequency of the wave returning to the police car after being reflected from the moving car and is not known. However, f_o' and the frequency f_o observed by the moving car are related by Equation 24.6:

$$f_\circ' = f_\circ\left(1+\frac{v_{rel}}{c}\right),$$

where v_{rel} is the relative speed between the moving car and the stationary police car. We have chosen the + sign in Equation 24.6, since the moving car and the police car are coming together. With this expression for $f_\circ'$, the desired frequency difference can be written as shown in Equation 1 at the right. The relative speed v_{rel} and the frequency f_s are known; the frequency $f_\circ$ will be determined in the next step.

The radar gun sends out a wave whose frequency is f_s. As the moving car approaches this wave, the car "observes" a frequency $f_\circ$ that is greater than f_s. The relation between these frequencies is given by Equation 24.6:

$$f_\circ = f_s\left(1+\frac{v_{rel}}{c}\right)$$

Again, the + sign has been chosen in Equation 24.6, since the moving car and the police car are coming together. All the variables on the right side of this equation are known, and we substitute this expression for $f_\circ$ into Equation 1, as indicated at the right. Algebraically combining the results of the two steps above, we obtain:

$$f_\circ' - f_s = f_\circ\left(1+\frac{v_{rel}}{c}\right) - f_s = f_s\left(1+\frac{v_{rel}}{c}\right)\left(1+\frac{v_{rel}}{c}\right) - f_s$$

Multiplying and combining terms in this equation yields:

$$f_\circ' - f_s = f_s\left[1+\frac{2v_{rel}}{c}+\left(\frac{v_{rel}}{c}\right)^2\right] - f_s = f_s\left(\frac{v_{rel}}{c}\right)\left(2+\frac{v_{rel}}{c}\right)$$

The value of v_{rel}/c is $(39\text{ m/s})/(3.0\times10^8\text{ m/s}) = 13\times10^{-8}$. This is an extremely small number when compared to the number 2, so the term $(2+v_{rel}/c)$ is very nearly equal to 2. Thus, we find that the difference between the frequency of the wave returning to the police car and that emitted by the radar gun is:

$$f_\circ' - f_s = f_s\left(\frac{v_{rel}}{c}\right)(2) = (8.0\times10^{-9}\text{ Hz})\left(\frac{39\text{ m/s}}{3.0\times10^8\text{ m/s}}\right)(2) = \boxed{2.1\times10^3\text{ Hz}}$$

65. *CONCEPTS*

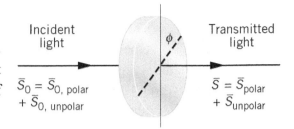

(i) When unpolarized light strikes a polarizer, one-half is absorbed and the other is transmitted, so that $\overline{S}_{\text{unpolar}} = \frac{1}{2}\overline{S}_{0,\text{unpolar}}$. This is true for any angle ϕ of the transmission axis.

(ii) When the polarized part of the incident beam passes through the polarizer, the intensity of its transmitted portion changes as the angle ϕ changes, ranging between some maximum value and 0 W/m², in accord with Malus' law. However, when the unpolarized part of the beam passes through the polarizer, its transmitted intensity does not change as the angle ϕ changes. Thus, the minimum intensity of the transmitted beam is 2.0 W/m², rather than 0 W/m², because of the unpolarized light.

CALCULATIONS

(a) The average intensity $\overline{S}$ of the transmitted light is (see the figure) $\overline{S} = \overline{S}_{\text{polar}} + \overline{S}_{\text{unpolar}}$. We are given that the minimum intensity is $\overline{S} = 2.0$ W/m² when $\phi = 20.0°$, and we know that $\overline{S}_{\text{polar}} = 0$ W/m² for this angle. Thus, $\overline{S}_{\text{unpolar}} = \overline{S} - \overline{S}_{\text{polar}} = 2.0$ W/m². The intensity $\overline{S}_{0,\text{unpolar}}$ of the unpolarized incident light is twice this amount because the polarizer absorbs one-half the incident light:

$$\overline{S}_{0,\text{unpolar}} = 2\overline{S}_{\text{unpolar}} = 2(2.0 \text{ W/m}^2) = \boxed{4.0 \text{ W/m}^2}$$

(b) When $\phi = \phi_{\text{max}}$, the intensity of the transmitted beam is at its maximum value of $\overline{S} = 8.0$ W/m². Writing the transmitted intensity as the sum of two parts, we have

$$\overline{S} = 8.0 \text{ W/m}^2 = \underbrace{\overline{S}_{0,\text{polar}} \cos^2 \theta}_{\overline{S}_{\text{polar}}} + \underbrace{2.0 \text{ W/m}^2}_{\overline{S}_{\text{unpolar}}}$$

The angle θ is the angle between the direction of polarization of the incident polarized light and the transmission axis of the polarizer. Since $\overline{S}_{\text{polar}}$ is a maximum, we know that $\theta = 0°$. Solving this equation for $\overline{S}_{0,\text{polar}}$ yields

$$\overline{S}_{0,\text{polar}} = 8.0 \text{ W/m}^2 - 2.0 \text{ W/m}^2 = \boxed{6.0 \text{ W/m}^2}$$

CHAPTER 25 | *THE REFLECTION OF LIGHT: MIRRORS*

3. ***REASONING*** The drawing shows the image of the bird in the plane mirror, as seen by the camera. Note that the image is as far behind the mirror as the bird is in front of it. We can find the distance d between the camera and the image by noting that this distance is the hypotenuse of a right triangle. The base of the triangle has a length of 3.7 m + 2.1 m and the height of the triangle is 4.3 m.

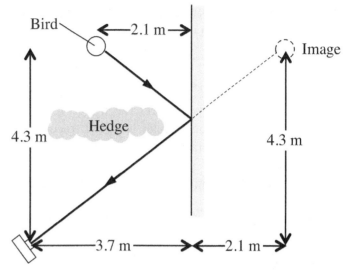

SOLUTION The distance d from the camera to the image of the bird can be obtained by using the Pythagorean theorem:

$$d = \sqrt{(3.7 \text{ m} + 2.1 \text{ m})^2 + (4.3 \text{ m})^2} = \boxed{7.2 \text{ m}}$$

5. ***REASONING*** The geometry is shown below. According to the law of reflection, the incident ray, the reflected ray, and the normal to the surface all lie in the same plane, and the angle of reflection θ_r equals the angle of incidence θ_i. We can use the law of reflection and the properties of triangles to determine the angle θ at which the ray leaves M_2.

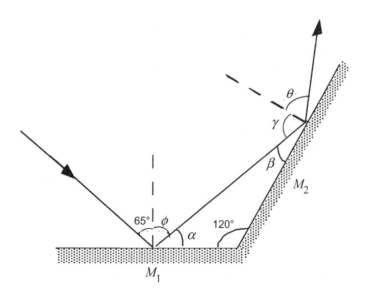

SOLUTION From the law of reflection, we know that $\phi = 65°$. We see from the figure that $\phi + \alpha = 90°$, or $\alpha = 90° – \phi = 90° – 65° = 25°$. From the figure and the fact that the sum of the interior angles in any triangle is $180°$, we have $\alpha + \beta + 120° = 180°$. Solving for β, we find that $\beta = 180° – (120° + 25°) = 35°$. Therefore, since $\beta + \gamma = 90°$, we find that the angle γ is given by $\gamma = 90° – \beta = 90° – 35° = 55°$. Since γ is the angle of incidence of the ray on mirror M_2, we know from the law of reflection that $\boxed{\theta = 55°}$.

15. **REASONING** The object distance ($d_0 = 11$ cm) is shorter than the focal length ($f = 18$ cm) of the mirror, so we expect the image to be virtual, appearing behind the mirror. Taking Figure 25.18a as our model, we will trace out: three rays from the tip of the object to the surface of the mirror, then three reflected rays, and finally three virtual rays extending behind the mirror and meeting at the tip of the image. The scale of the ray tracing will determine the location and height of the image. The three sets of rays are:

1. An incident ray from the object to the mirror, parallel to the principal axis and then reflected through the focal point F.

2. An incident ray from the object to the mirror, directly *away* from the focal point F and then reflected parallel to the principal axis. (The incident ray cannot pass from the object through the focal point, as this would take it away from the mirror, and it would not be reflected.)

3. An incident ray from the object to the mirror, directly away from the center of curvature C, then reflected back through C.

SOLUTION

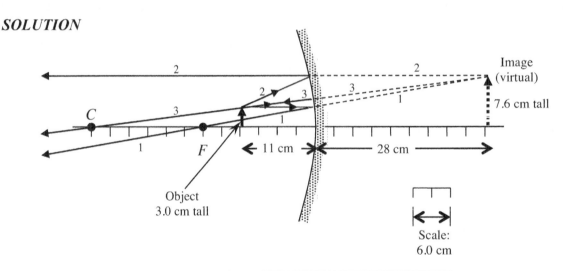

a. The ray diagram indicates that $\boxed{\text{the image is 28 cm behind the mirror}}$.

b. We see from the ray diagram that $\boxed{\text{the image is 7.6 cm tall}}$.

19. **REASONING** This problem can be solved using the mirror equation, Equation 25.3.

SOLUTION Using the mirror equation with $d_i = +26$ cm and $f = 12$ cm, we find

$$\frac{1}{d_o} = \frac{1}{f} - \frac{1}{d_i} = \frac{1}{12 \text{ cm}} - \frac{1}{26 \text{ cm}} \qquad \text{or} \qquad \boxed{d_o = +22 \text{ cm}}$$

23. **REASONING** Since the image is behind the mirror, the image is virtual, and the image distance is negative, so that $d_i = -34.0$ cm. The object distance is given as $d_o = 7.50$ cm. The mirror equation relates these distances to the focal length f of the mirror. If the focal length is positive, the mirror is concave. If the focal length is negative, the mirror is convex.

SOLUTION According to the mirror equation (Equation 25.3), we have

$$\frac{1}{f} = \frac{1}{d_o} + \frac{1}{d_i} \qquad \text{or} \qquad f = \frac{1}{\dfrac{1}{d_o} + \dfrac{1}{d_i}} = \frac{1}{\dfrac{1}{7.50 \text{ cm}} + \dfrac{1}{(-34.0 \text{ cm})}} = \boxed{9.62 \text{ cm}}$$

Since the focal length is positive, the mirror is $\boxed{\text{concave}}$.

27. **REASONING** When paraxial light rays that are parallel to the principal axis strike a convex mirror, the reflected rays diverge after being reflected, and appear to originate from the focal point F behind the mirror (see Figure 25.16). We can treat the sun as being infinitely far from the mirror, so it is reasonable to treat the incident rays as paraxial rays that are parallel to the principal axis.

SOLUTION
a. Since the sun is infinitely far from the mirror and its image is a virtual image that lies *behind* the mirror, we can conclude that the mirror is a $\boxed{\text{convex mirror}}$.

b. With $d_i = -12.0$ cm and $d_o = \infty$, the mirror equation (Equation 25.3) gives

$$\frac{1}{f} = \frac{1}{d_o} + \frac{1}{d_i} = \frac{1}{\infty} + \frac{1}{d_i} = \frac{1}{d_i}$$

Therefore, the focal length f lies 12.0 cm behind the mirror (this is consistent with the reasoning above that states that, after being reflected, the rays appear to originate from the focal point behind the mirror). In other words, $f = -12.0$ cm. Then, according to Equation 25.2, $f = -\frac{1}{2}R$, and the radius of curvature is

$$R = -2f = -2(-12.0 \text{ cm}) = \boxed{24.0 \text{ cm}}$$

35. *REASONING*

a. The image of the spacecraft appears a distance d_i beneath the surface of the moon, which we assume to be a convex spherical mirror of radius $R = 1.74\times10^6$ m. The image distance d_i is related to the focal length f of the moon and the object distance d_o by the mirror equation

$$\frac{1}{d_o}+\frac{1}{d_i}=\frac{1}{f} \tag{25.3}$$

The spacecraft is the object, so the object distance d_o is equal to the altitude of the spacecraft: $d_o = 1.22\times10^5$ m. The moon is assumed to be a convex mirror of radius R, so its focal length is given by

$$f=-\tfrac{1}{2}R \tag{25.2}$$

b. Both the spacecraft and its image would have the same angular speed ω about the center of the moon because both have the same orbital period. The linear speed v of an orbiting body is related to its angular speed ω by $v = r\omega$ (Equation 8.9), where r is the radius of the orbit. Therefore, the speeds of the spacecraft and its image are, respectively,

$$v_o = r_o\omega \qquad \text{and} \qquad v_i = r_i\omega \tag{1}$$

We have used v_o and r_o in Equation (1) to denote the orbital speed and orbital radius of the spacecraft, because it is the object. The symbols v_i and r_i denote orbital speed and orbital radius of the spacecraft's image. The orbital radius r_o of the spacecraft is the sum of the radius R of the moon and the spacecraft's altitude $d_o = 1.20\times10^5$ m above the moon's surface:

$$r_o = R+d_o \tag{2}$$

The "orbital radius" r_i of the image is its distance from the center of the moon. The image is below the surface, and the image distance d_i is negative. Therefore, the distance between the center of the moon and the image is found from

$$r_i = R+d_i \tag{3}$$

SOLUTION

a. Solving Equation 25.3 for $\dfrac{1}{d_i}$ and then taking the reciprocal of both sides, we obtain

$$\frac{1}{d_i}=\frac{1}{f}-\frac{1}{d_o}=\frac{d_o-f}{d_o f} \qquad \text{or} \qquad d_i=\frac{d_o f}{d_o-f} \tag{4}$$

Substituting Equation 25.2 into Equation (4), we find that

$$d_i = \frac{d_o f}{d_o - f} = \frac{d_o\left(-\frac{1}{2}R\right)}{d_o - \left(-\frac{1}{2}R\right)} = \frac{-\frac{1}{2}d_o R}{d_o + \frac{1}{2}R} = \frac{-\frac{1}{2}\left(1.22\times10^5 \text{ m}\right)\left(1.74\times10^6 \text{ m}\right)}{1.22\times10^5 \text{ m} + \frac{1}{2}\left(1.74\times10^6 \text{ m}\right)} = -1.07\times10^5 \text{ m}$$

Therefore, the image of the spacecraft would appear $\boxed{1.07\times10^5 \text{ m}}$ below the surface of the moon.

b. Solving the first of Equations (1) for ω yields $\omega = \dfrac{v_o}{r_o}$. Substituting this result into the second of Equations (1), we find that

$$v_i = r_i \omega = r_i \left(\frac{v_o}{r_o}\right) = \left(\frac{r_i}{r_o}\right) v_o \tag{5}$$

Substituting Equations (2) and (3) into Equation (5) yields

$$v_i = \left(\frac{r_i}{r_o}\right) v_o = \left(\frac{R + d_i}{R + d_o}\right) v_o = \left[\frac{1.74\times10^6 \text{ m} + \left(-1.07\times10^5 \text{ m}\right)}{1.74\times10^6 \text{ m} + 1.22\times10^5 \text{ m}}\right] (1620 \text{ m/s}) = \boxed{1420 \text{ m/s}}$$

---

37. **REASONING** The mirror equation relates the object and image distances to the focal length. Thus, we can apply the mirror equation once with the given object and image distances to determine the focal length. Then, we can use the mirror equation again with the focal length and the second object distance to determine the unknown image distance.

SOLUTION According to the mirror equation, we have

$$\underbrace{\frac{1}{d_{o1}} + \frac{1}{d_{i1}} = \frac{1}{f}}_{\text{First position of object}} \quad \text{and} \quad \underbrace{\frac{1}{d_{o2}} + \frac{1}{d_{i2}} = \frac{1}{f}}_{\text{Second position of object}}$$

Since the focal length is the same in both cases, it follows that

$$\frac{1}{d_{o1}} + \frac{1}{d_{i1}} = \frac{1}{d_{o2}} + \frac{1}{d_{i2}}$$

$$\frac{1}{d_{i2}} = \frac{1}{d_{o1}} + \frac{1}{d_{i1}} - \frac{1}{d_{o2}} = \frac{1}{25 \text{ cm}} + \frac{1}{\left(-17 \text{ cm}\right)} - \frac{1}{19 \text{ cm}} = -0.071 \text{ cm}^{-1}$$

$$d_{i2} = -14 \text{ cm}$$

The negative value for d_{i2} indicates that the image is located $\boxed{14 \text{ cm}}$ behind the mirror.

41. ***REASONING*** This problem can be solved by using the mirror equation, Equation 25.3, and the magnification equation, Equation 25.4.

    SOLUTION
    a. Using the mirror equation with $d_i = d_o$ and $f = R/2$, we have

    $$\frac{1}{d_o} = \frac{1}{f} - \frac{1}{d_i} = \frac{1}{R/2} - \frac{1}{d_o} \qquad \text{or} \qquad \frac{2}{d_o} = \frac{2}{R}$$

    Therefore, we find that $\boxed{d_o = R}$.

    b. According to the magnification equation, the magnification is

    $$m = -\frac{d_i}{d_o} = -\frac{d_o}{d_o} = \boxed{-1}$$

    c. Since the magnification m is negative, the image is $\boxed{\text{inverted}}$.

45. ***REASONING*** Since the size h_i of the image is one-fourth the size h_o of the object, we know from the magnification equation that

    $$\underset{\frac{1}{4}}{\underbrace{m}} = \frac{h_i}{h_o} = -\frac{d_i}{d_o} \qquad \text{so that} \qquad d_i = -\tfrac{1}{4}d_o \qquad\qquad (25.4)$$

    We can substitute this relation into the mirror equation to find the ratio d_o/f.

    SOLUTION Substituting $d_i = -\tfrac{1}{4}d_o$ into the mirror equation gives

    $$\frac{1}{d_o} + \frac{1}{d_i} = \frac{1}{f} \quad \text{or} \quad \frac{1}{d_o} + \frac{1}{-\tfrac{1}{4}d_0} = \frac{1}{f} \quad \text{or} \quad \frac{-3}{d_o} = \frac{1}{f} \qquad\qquad (25.3)$$

    Solving this equation for the ratio d_o/f yields $d_o/f = \boxed{-3}$.

47. **REASONING** The time of travel is proportional to the total distance for each path. Therefore, using the distances identified in the drawing, we have

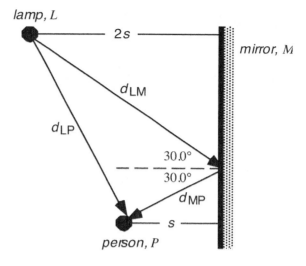

$$\frac{t_{\text{reflected}}}{t_{\text{direct}}} = \frac{d_{LM} + d_{MP}}{d_{LP}} \qquad (1)$$

We know that the angle of incidence is equal to the angle of reflection. We also know that the lamp L is a distance $2s$ from the mirror, while the person P is a distance s from the mirror.

Therefore, it follows that if $d_{MP} = d$, then $d_{LM} = 2d$. We may use the law of cosines (see Appendix E) to express the distance d_{LP} as

$$d_{LP} = \sqrt{(2d)^2 + d^2 - 2(2d)(d)\cos 2(30.0°)}$$

SOLUTION   Substituting the expressions for d_{MP}, d_{LM}, and d_{LP} into Equation (1), we find that the ratio of the travel times is

$$\frac{t_{\text{reflected}}}{t_{\text{direct}}} = \frac{2d + d}{\sqrt{(2d)^2 + d^2 - 2(2d)(d)\cos 2(30.0°)}} = \frac{3}{\sqrt{5 - 4\cos 60.0°}} = \boxed{1.73}$$

---

49. **CONCEPTS**

(i) A convex mirror always forms an image that is smaller than the object, as Figure 25.21a shows. Therefore, the mirror must be concave.

(ii) There are two places. Figure 25.18a illustrates that one of the places is between the center of curvature and the focal point. The enlarged image is real and inverted. Figure 25.19a shows that another possibility is between the focal point and the mirror, in which case the enlarged image is virtual and upright.

(iii) Since the image is inverted in Figure 25.18a, the magnification for this possibility is $m = -2$. In Figure 25.19a, however, the image is upright, so the magnification is $m = +2$ for this option. In either case, the image is twice the size of the ring.

CALCULATIONS

According to the mirror equation and the magnification equation, we have

$$\underbrace{\frac{1}{d_o} + \frac{1}{d_i} = \frac{1}{f}}_{\text{Mirror equation}} \qquad \text{and} \qquad \underbrace{m = -\frac{d_i}{d_o}}_{\text{Magnification equation}}$$

We can solve the magnification equation for the image distance and obtain $d_i = -md_o$. Substituting this expression into the mirror equation gives

$$\frac{1}{d_o} + \frac{1}{(-md_o)} = \frac{1}{f} \qquad \text{or} \qquad d_o = \frac{f(m-1)}{m}$$

Applying this result with the two magnifications (and noting from Equation 25.1 that $f = \frac{1}{2}R = 12 \text{ cm}$), we obtain:

$$\boldsymbol{m} = \quad d_o = \frac{f(m-1)}{m} = \frac{(12 \text{ cm})(-2-1)}{-2} = \boxed{18 \text{ cm}} \quad \boldsymbol{-2}$$

$$\boldsymbol{m} = \quad d_o = \frac{f(m-1)}{m} = \frac{(12 \text{ cm})(+2-1)}{+2} = \boxed{6.0 \text{ cm}} \quad \boldsymbol{+2}$$

CHAPTER 26 | THE REFRACTION OF LIGHT: LENSES AND OPTICAL INSTRUMENTS

1. **REASONING** Since the light will travel in glass at a constant speed v, the time it takes to pass perpendicularly through the glass is given by $t = d/v$, where d is the thickness of the glass. The speed v is related to the vacuum value c by Equation 26.1: $n = c/v$.

SOLUTION Substituting for v from Equation 26.1 and substituting values, we obtain

$$t = \frac{d}{v} = \frac{nd}{c} = \frac{(1.5)(4.0 \times 10^{-3} \text{ m})}{3.00 \times 10^8 \text{ m/s}} = \boxed{2.0 \times 10^{-11} \text{ s}}$$

5. **REASONING** The substance can be identified from Table 26.1 if its index of refraction is known. The index of refraction n is defined as the speed of light c in a vacuum divided by the speed of light v in the substance (Equation 26.1), both of which are known.

SOLUTION Using Equation 26.1, we find that

$$n = \frac{c}{v} = \frac{2.998 \times 10^8 \text{ m/s}}{2.201 \times 10^8 \text{ m/s}} = 1.362$$

Referring to Table 26.1, we see that the substance is $\boxed{\text{ethyl alcohol}}$.

11. **REASONING** The angle of refraction θ_2 is related to the angle of incidence θ_1 by Snell's law, $n_1 \sin \theta_1 = n_2 \sin \theta_2$ (Equation 26.2), where n_1 and n_2 are, respectively the indices of refraction of the incident and refracting media. For each case (ice and water), the variables θ_1, n_1 and n_2, are known, so the angles of refraction can be determined.

SOLUTION The ray of light impinges from air ($n_1 = 1.000$) onto either the ice or water at an angle of incidence of $\theta_1 = 60.0°$. Using $n_2 = 1.309$ for ice and $n_2 = 1.333$ for water, we find that the angles of refraction are

$$\sin \theta_2 = \frac{n_1 \sin \theta_1}{n_2} \quad \text{or} \quad \theta_2 = \sin^{-1}\left(\frac{n_1 \sin \theta_1}{n_2}\right)$$

Ice
$$\theta_{2,\ ice} = \sin^{-1}\left[\frac{(1.000)\sin 60.0°}{1.309}\right] = 41.4°$$

Water
$$\theta_{2,\ water} = \sin^{-1}\left[\frac{(1.000)\sin 60.0°}{1.333}\right] = 40.5°$$

The difference in the angles of refraction is $\theta_{2,\ ice} - \theta_{2,\ water} = 41.4° - 40.5° = \boxed{0.9°}$

---

13. **REASONING** We will use the geometry of the situation to determine the angle of incidence. Once the angle of incidence is known, we can use Snell's law to find the index of refraction of the unknown liquid. The speed of light v in the liquid can then be determined.

SOLUTION From the drawing in the text, we see that the angle of incidence at the liquid-air interface is

$$\theta_1 = \tan^{-1}\left(\frac{5.00\ cm}{6.00\ cm}\right) = 39.8°$$

The drawing also shows that the angle of refraction is 90.0°. Thus, according to Snell's law (Equation 26.2: $n_1 \sin\theta_1 = n_2 \sin\theta_2$), the index of refraction of the unknown liquid is

$$n_1 = \frac{n_2 \sin\theta_2}{\sin\theta_1} = \frac{(1.000)\ (\sin 90.0°)}{\sin 39.8°} = 1.56$$

From Equation 26.1 ($n = c/v$), we find that the speed of light in the unknown liquid is

$$v = \frac{c}{n_1} = \frac{3.00\times 10^8\ m/s}{1.56} = \boxed{1.92 \times 10^8\ m/s}$$

---

15. **REASONING** We begin by using Snell's law (Equation 26.2: $n_1 \sin\theta_1 = n_2 \sin\theta_2$) to find the index of refraction of the material. Then we will use Equation 26.1, the definition of the index of refraction ($n = c/v$) to find the speed of light in the material.

SOLUTION From Snell's law, the index of refraction of the material is

$$n_2 = \frac{n_1 \sin\theta_1}{\sin\theta_2} = \frac{(1.000)\sin 63.0°}{\sin 47.0°} = 1.22$$

Then, from Equation 26.1, we find that the speed of light v in the material is

$$v = \frac{c}{n_2} = \frac{3.00\times 10^8\ m/s}{1.22} = \boxed{2.46 \times 10^8\ m/s}$$

---

19. **REASONING** Following the discussion in Conceptual Example 4, we have the drawing at the right to use as a guide. In this drawing the symbol d refers to depths in the water, while the symbol h refers to heights in the air above the water. Moreover, symbols with a prime denote apparent distances, and unprimed symbols denote actual distances. We will use Equation 26.3 to relate apparent distances to actual distances. In so doing, we will use the fact that the refractive index of air is essentially $n_{air} = 1$ and denote the refractive index of water by $n_w = 1.333$ (see Table 26.1).

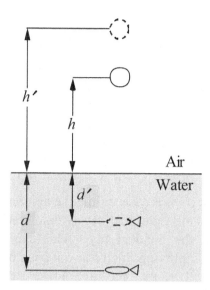

SOLUTION To the fish, the man appears to be a distance above the air-water interface that is given by Equation 26.3 as $h' = h(n_w / 1)$. Thus, measured above the eyes of the fish, the man appears to be located at a distance of

$$h' + d = h\left(\frac{n_w}{1}\right) + d \tag{1}$$

To the man, the fish appears to be a distance below the air-water interface that is given by Equation 26.3 as $d' = d(1/n_w)$. Thus, measured below the man's eyes, the fish appears to be located at a distance of

$$h + d' = h + d\left(\frac{1}{n_w}\right) \tag{2}$$

Dividing Equation (1) by Equation (2) and using the fact that $h = d$, we find

$$\frac{h' + d}{h + d'} = \frac{h\left(\dfrac{n_w}{1}\right) + d}{h + d\left(\dfrac{1}{n_w}\right)} = \frac{n_w + 1}{1 + \dfrac{1}{n_w}} = n_w \tag{3}$$

In Equation (3), $h' + d$ is the distance we seek, and $h + d'$ is given as 2.0 m. Thus, we find

$$h' + d = n_w\left(h + d'\right) = (1.333)(2.0 \text{ m}) = \boxed{2.7 \text{ m}}$$

21. ***REASONING*** The drawing at the right shows the geometry of the situation using the same notation as that in Figure 26.6. In addition to the text's notation, let t represent the thickness of the pane, let L represent the length of the ray in the pane, let x (shown twice in the figure) equal the displacement of the ray, and let the difference in angles $\theta_1 - \theta_2$ be given by ϕ.

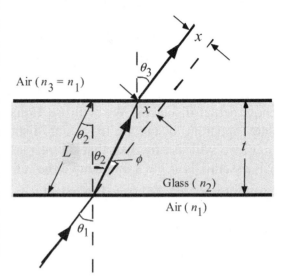

We wish to find the amount x by which the emergent ray is displaced relative to the incident ray. This can be done by applying Snell's law at each interface, and then making use of the geometric and trigonometric relations in the drawing.

SOLUTION If we apply Snell's law (see Equation 26.2) to the bottom interface we obtain $n_1 \sin \theta_1 = n_2 \sin \theta_2$. Similarly, if we apply Snell's law at the top interface where the ray emerges, we have $n_2 \sin \theta_2 = n_3 \sin \theta_3 = n_1 \sin \theta_3$. Comparing this with Snell's law at the bottom face, we see that $n_1 \sin \theta_1 = n_1 \sin \theta_3$, from which we can conclude that $\theta_3 = \theta_1$. Therefore, the emerging ray is parallel to the incident ray.

From the geometry of the ray and thickness of the pane, we see that $L \cos \theta_2 = t$, from which it follows that $L = t / \cos \theta_2$. Furthermore, we see that $x = L \sin \phi = L \sin \left(\theta_1 - \theta_2 \right)$. Substituting for L, we find

$$x = L \ \sin(\theta_1 - \theta_2) = \frac{t \ \sin(\theta_1 - \theta_2)}{\cos \theta_2}$$

Before we can use this expression to determine a numerical value for x, we must find the value of θ_2. Solving the expression for Snell's law at the bottom interface for θ_2, we have

$$\sin \theta_2 = \frac{n_1 \ \sin \theta_1}{n_2} = \frac{(1.000) \ (\sin 30.0°)}{1.52} = 0.329 \qquad \text{or} \qquad \theta_2 = \sin^{-1} 0.329 = 19.2°$$

Therefore, the amount by which the emergent ray is displaced relative to the incident ray is

$$x = \frac{t \ \sin \left(\theta_1 - \theta_2 \right)}{\cos \theta_2} = \frac{(6.00 \text{ mm}) \sin (30.0° - 19.2°)}{\cos 19.2°} = \boxed{1.19 \text{ mm}}$$

27. **REASONING** The light ray traveling in the oil can only penetrate into the water if it does not undergo total internal reflection at the boundary between the oil and the water. Total internal reflection will occur if the angle of incidence $\theta = 71.4°$ is greater than the critical angle θ_c for these two media. The critical angle is found from

$$\sin \theta_c = \frac{n_2}{n_1} \qquad (26.4)$$

where $n_2 = 1.333$ is the index of refraction of water (see Table 26.1), and $n_1 = 1.47$ is the index of refraction of the oil.

SOLUTION Solving Equation 26.4 for θ_c, we obtain

$$\theta_c = \sin^{-1}\left(\frac{n_2}{n_1}\right) = \sin^{-1}\left(\frac{1.333}{1.47}\right) = \boxed{65.1°}$$

Comparing this result to $\theta = 71.4°$, we see that the angle of incidence is greater than the critical angle $(\theta > \theta_c)$. Therefore, $\boxed{\text{the ray of light will not enter the water}}$; it will instead undergo total internal reflection within the oil.

---

33. **REASONING** Total internal reflection will occur at point P provided that the angle α in the drawing at the right exceeds the critical angle. This angle is determined by the angle θ_2 at which the light rays enter the quartz slab. We can determine θ_2 by using Snell's law of refraction and the incident angle, which is given as $\theta_1 = 34°$.

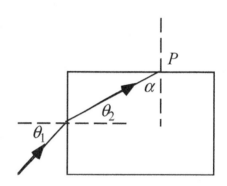

SOLUTION Using n for the refractive index of the fluid that surrounds the crystalline quartz slab and n_q for the refractive index of quartz and applying Snell's law give

$$n \sin \theta_1 = n_q \sin \theta_2 \qquad \text{or} \qquad \sin \theta_2 = \frac{n}{n_q} \sin \theta_1 \qquad (1)$$

But when α equals the critical angle, we have from Equation 26.4 that

$$\sin \alpha = \sin \theta_c = \frac{n}{n_q} \qquad (2)$$

According to the geometry in the drawing above, $\alpha = 90° - \theta_2$. As a result, Equation (2) becomes

$$\sin\left(90° - \theta_2\right) = \cos\theta_2 = \frac{n}{n_q} \qquad (3)$$

Squaring Equation (3), using the fact that $\sin^2\theta_2 + \cos^2\theta_2 = 1$, and substituting from Equation (1), we obtain

$$\cos^2\theta_2 = 1 - \sin^2\theta_2 = 1 - \frac{n^2}{n_q^2}\sin^2\theta_1 = \frac{n^2}{n_q^2} \qquad (4)$$

Solving Equation (4) for n and using the value given in Table 26.1 for the refractive index of crystalline quartz, we find

$$n = \frac{n_q}{\sqrt{1 + \sin^2\theta_1}} = \frac{1.544}{\sqrt{1 + \sin^2 34°}} = \boxed{1.35}$$

---

37. **REASONING** Since the light reflected from the coffee table is completely polarized parallel to the surface of the glass, the angle of incidence must be the Brewster angle ($\theta_B = 56.7°$) for the air-glass interface. We can use Brewster's law (Equation 26.5: $\tan\theta_B = n_2/n_1$) to find the index of refraction n_2 of the glass.

SOLUTION Solving Brewster's law for n_2, we find that the refractive index of the glass is

$$n_2 = n_1\tan\theta_B = (1.000)(\tan 56.7°) = \boxed{1.52}$$

---

39. **REASONING** Brewster's law (Equation 26.5: $\tan\theta_B = n_2/n_1$) relates the angle of incidence θ_B at which the reflected ray is completely polarized parallel to the surface to the indices of refraction n_1 and n_2 of the two media forming the interface. We can use Brewster's law for light incident from above to find the ratio of the refractive indices n_2/n_1. This ratio can then be used to find the Brewster angle for light incident from below on the same interface.

SOLUTION The index of refraction for the medium in which the incident ray occurs is designated by n_1. For the light striking from above $n_2/n_1 = \tan\theta_B = \tan 65.0° = 2.14$. The same equation can be used when the light strikes from below if the indices of refraction are interchanged

$$\theta_B = \tan^{-1}\left(\frac{n_1}{n_2}\right) = \tan^{-1}\left(\frac{1}{n_2/n_1}\right) = \tan^{-1}\left(\frac{1}{2.14}\right) = \boxed{25.0°}$$

45. **REASONING** Because the refractive index of the glass depends on the wavelength (i.e., the color) of the light, the rays corresponding to different colors are bent by different amounts in the glass. We can use Snell's law (Equation 26.2: $n_1 \sin \theta_1 = n_2 \sin \theta_2$) to find the angle of refraction for the violet ray and the red ray. The angle between these rays can be found by the subtraction of the two angles of refraction.

SOLUTION In Table 26.2 the index of refraction for violet light in crown glass is 1.538, while that for red light is 1.520. According to Snell's law, then, the sine of the angle of refraction for the violet ray in the glass is $\sin\theta_2 = (1.000/1.538) \sin 45.00° = 0.4598$, so that

$$\theta_2 = \sin^{-1}(0.4598) = 27.37°$$

Similarly, for the red ray, $\sin\theta_2 = (1.000/1.520) \sin 45.00° = 0.4652$, from which it follows that

$$\theta_2 = \sin^{-1}(0.4652) = 27.72°$$

Therefore, the angle between the violet ray and the red ray in the glass is

$$27.72° - 27.37° = \boxed{0.35°}$$

---

47. **REASONING** We can use Snell's law (Equation 26.2: $n_1 \sin \theta_1 = n_2 \sin \theta_2$) at each face of the prism. At the first interface where the ray enters the prism, $n_1 = 1.000$ for air and $n_2 = n_g$ for glass. Thus, Snell's law gives

$$(1)\sin 60.0° = n_g \sin \theta_2 \qquad \text{or} \qquad \sin \theta_2 = \frac{\sin 60.0°}{n_g} \qquad (1)$$

We will represent the angles of incidence and refraction at the second interface as θ_1' and θ_2', respectively. Since the triangle is an equilateral triangle, the angle of incidence at the second interface, where the ray emerges back into air, is $\theta_1' = 60.0° - \theta_2$. Therefore, at the second interface, where $n_1 = n_g$ and $n_2 = 1.000$, Snell's law becomes

$$n_g \sin (60.0° - \theta_2) = (1)\sin \theta_2' \qquad (2)$$

We can now use Equations (1) and (2) to determine the angles of refraction θ_2' at which the red and violet rays emerge into the air from the prism.

SOLUTION
 Red Ray The index of refraction of flint glass at the wavelength of red light is $n_g = 1.662$. Therefore, using Equation (1), we can find the angle of refraction for the red ray as it enters the prism:

$$\sin \theta_2 = \frac{\sin 60.0°}{1.662} = 0.521 \qquad \text{or} \qquad \theta_2 = \sin^{-1} 0.521 = 31.4°$$

Substituting this value for θ_2 into Equation (2), we can find the angle of refraction at which the red ray emerges from the prism:

$$\sin \theta_2' = 1.662 \sin (60.0° - 31.4°) = 0.796 \qquad \text{or} \qquad \theta_2' = \sin^{-1} 0.796 = \boxed{52.7°}$$

$\boxed{\text{Violet Ray}}$ For violet light, the index of refraction for glass is $n_g = 1.698$. Again using Equation (1), we find

$$\sin \theta_2 = \frac{\sin 60.0°}{1.698} = 0.510 \qquad \text{or} \qquad \theta_2 = \sin^{-1} 0.510 = 30.7°$$

Using Equation (2), we find

$$\sin \theta_2' = 1.698 \sin (60.0° - 30.7°) = 0.831 \qquad \text{or} \qquad \theta_2' = \sin^{-1} 0.831 = \boxed{56.2°}$$

---

49. **REASONING** The ray diagram is constructed by drawing the paths of two rays from a point on the object. For convenience, we choose the top of the object. The ray that is parallel to the principal axis will be refracted by the lens and pass through the focal point on the right side. The ray that passes through the center of the lens passes through undeflected. The image is formed at the intersection of these two rays on the right side of the lens.

SOLUTION The following ray diagram (to scale) shows that $\boxed{d_i = 18 \text{ cm}}$ and reveals a real, inverted, and enlarged image.

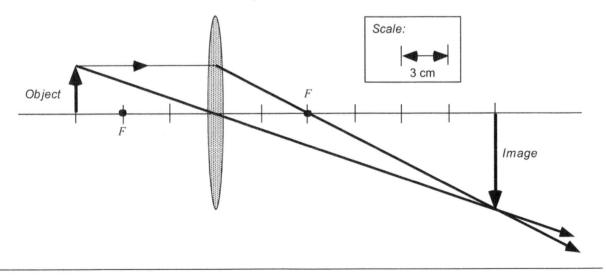

59. **REASONING** The optical arrangement is similar to that in Figure 26.26. We begin with the thin-lens equation, [Equation 26.6: $(1/d_o)+(1/d_i)=(1/f)$]. Since the distance between the moon and the camera is so large, the object distance d_o is essentially infinite, and $1/d_o = 1/\infty = 0$. Therefore the thin-lens equation becomes $1/d_i = 1/f$ or $d_i = f$. The diameter of the moon's imagine on the slide film is equal to the image height h_i, as given by the magnification equation (Equation 26.7: $h_i/h_o = -d_i/d_o$).

When the slide is projected onto a screen, the situation is similar to that in Figure 26.27. In this case, the thin-lens and magnification equations can be used in their usual forms.

SOLUTION
a. Solving the magnification equation for h_i gives

$$h_i = -h_o \frac{d_i}{d_o} = (-3.48\times10^6 \text{ m})\left(\frac{50.0\times10^{-3} \text{ m}}{3.85\times10^8 \text{ m}}\right) = -4.52\times10^{-4} \text{ m}$$

The diameter of the moon's image on the slide film is, therefore, $\boxed{4.52\times10^{-4} \text{ m}}$.

b. From the magnification equation, $h_i = -h_o(d_i/d_o)$. We need to find the ratio d_i/d_o. Beginning with the thin-lens equation, we have

$$\frac{1}{d_o}+\frac{1}{d_i}=\frac{1}{f} \quad \text{or} \quad \frac{1}{d_o}=\frac{1}{f}-\frac{1}{d_i} \quad \text{which leads to} \quad \frac{d_i}{d_o}=\frac{d_i}{f}-\frac{d_i}{d_i}=\frac{d_i}{f}-1$$

Therefore,

$$h_i = -h_o\left(\frac{d_i}{f}-1\right)=-\left(4.52\times10^{-4} \text{ m}\right)\left(\frac{15.0 \text{ m}}{110.0\times10^{-3} \text{ m}}-1\right)=-6.12\times10^{-2} \text{ m}$$

The diameter of the image on the screen is $\boxed{6.12\times10^{-2} \text{ m}}$.

---

61. **REASONING** The magnification equation (Equation 26.7) relates the object and image distances d_o and d_i, respectively, to the relative size of the of the image and object: $m = -(d_i/d_o)$. We consider two cases: in case 1, the object is placed 18 cm in front of a diverging lens. The magnification for this case is given by m_1. In case 2, the object is moved so that the magnification m_2 is reduced by a factor of 2 compared to that in case 1. In other words, we have $m_2 = \frac{1}{2}m_1$. Using Equation 26.7, we can write this as

$$-\frac{d_{i2}}{d_{o2}} = -\frac{1}{2}\left(\frac{d_{i1}}{d_{o1}}\right) \tag{1}$$

This expression can be solved for d_{o2}. First, however, we must find a numerical value for d_{i1}, and we must eliminate the variable d_{i2}.

SOLUTION
The image distance for case 1 can be found from the thin-lens equation [Equation 26.6: $(1/d_o)+(1/d_i)=(1/f)$]. The problem statement gives the focal length as $f=-12\,\text{cm}$. Since the object is 18 cm in front of the diverging lens, $d_{o1}=18\,\text{cm}$. Solving for d_{i1}, we find

$$\frac{1}{d_{i1}} = \frac{1}{f} - \frac{1}{d_{o1}} = \frac{1}{-12\,\text{cm}} - \frac{1}{18\,\text{cm}} \qquad \text{or} \qquad d_{i1} = -7.2\,\text{cm}$$

where the minus sign indicates that the image is virtual. Solving Equation (1) for d_{o2}, we have

$$d_{o2} = 2d_{i2}\left(\frac{d_{o1}}{d_{i1}}\right) \tag{2}$$

To eliminate d_{i2} from this result, we note that the thin-lens equation applied to case 2 gives

$$\frac{1}{d_{i2}} = \frac{1}{f} - \frac{1}{d_{o2}} = \frac{d_{o2}-f}{fd_{o2}} \qquad \text{or} \qquad d_{i2} = \frac{fd_{o2}}{d_{o2}-f}$$

Substituting this expression for d_{i2} into Equation (2), we have

$$d_{o2} = \left(\frac{2fd_{o2}}{d_{o2}-f}\right)\left(\frac{d_{o1}}{d_{i1}}\right) \qquad \text{or} \qquad d_{o2}-f = 2f\left(\frac{d_{o1}}{d_{i1}}\right)$$

Solving for d_{o2}, we find

$$d_{o2} = f\left[2\left(\frac{d_{o1}}{d_{i1}}\right)+1\right] = (-12\text{cm})\left[2\left(\frac{18\,\text{cm}}{-7.2\,\text{cm}}\right)+1\right] = \boxed{48\,\text{cm}}$$

69. **REASONING** The problem can be solved using the thin-lens equation [Equation 26.6: $(1/d_o)+(1/d_i)=(1/f)$] twice in succession. We begin by using the thin lens-equation to find the location of the image produced by the converging lens; this image becomes the object for the diverging lens.

SOLUTION

a. The image distance for the converging lens is determined as follows:

$$\frac{1}{d_{i1}} = \frac{1}{f} - \frac{1}{d_{o1}} = \frac{1}{12.0 \text{ cm}} - \frac{1}{36.0 \text{ cm}} \qquad \text{or} \qquad d_{i1} = 18.0 \text{ cm}$$

This image acts as the object for the diverging lens. Therefore,

$$\frac{1}{d_{i2}} = \frac{1}{f} - \frac{1}{d_{o2}} = \frac{1}{-6.00 \text{ cm}} - \frac{1}{(30.0 \text{ cm} - 18.0 \text{ cm})} \qquad \text{or} \qquad d_{i2} = -4.00 \text{ cm}$$

Thus, the final image is located $\boxed{4.00 \text{ cm to the left of the diverging lens}}$.

b. The magnification equation (Equation 26.7: $h_i / h_o = -d_i / d_o$) gives

$$\underbrace{m_c = -\frac{d_{i1}}{d_{o1}} = -\frac{18.0 \text{ cm}}{36.0 \text{ cm}} = -0.500}_{\text{Converging lens}} \qquad \underbrace{m_d = -\frac{d_{i2}}{d_{o2}} = -\frac{-4.00 \text{ cm}}{12.0 \text{ cm}} = 0.333}_{\text{Diverging lens}}$$

Therefore, the overall magnification is given by the product $m_c m_d = \boxed{-0.167}$.

c. Since the final image distance is negative, we can conclude that the image is $\boxed{\text{virtual}}$.

d. Since the overall magnification of the image is negative, the image is $\boxed{\text{inverted}}$.

e. The magnitude of the overall magnification is less than one; therefore, the final image is $\boxed{\text{smaller}}$.

---

71. **REASONING** We begin by using the thin-lens equation [Equation 26.6: $(1/d_o)+(1/d_i)=(1/f)$] to locate the image produced by the lens. This image is then treated as the object for the mirror.

SOLUTION

a. The image distance from the diverging lens can be determined as follows:

$$\frac{1}{d_i} = \frac{1}{f} - \frac{1}{d_o} = \frac{1}{-8.00 \text{ cm}} - \frac{1}{20.0 \text{ cm}} \qquad \text{or} \qquad d_i = -5.71 \text{ cm}$$

The image produced by the lens is 5.71 cm to the left of the lens. The distance between this image and the concave mirror is 5.71 cm + 30.0 cm = 35.7 cm. The mirror equation [Equation 25.3: $(1/d_o) + (1/d_i) = (1/f)$] gives the image distance from the mirror:

$$\frac{1}{d_i} = \frac{1}{f} - \frac{1}{d_o} = \frac{1}{12.0 \text{ cm}} - \frac{1}{35.7 \text{ cm}} \qquad \text{or} \qquad \boxed{d_i = 18.1 \text{ cm}}$$

b. The image is $\boxed{\text{real}}$, because d_i is a positive number, indicating that the final image lies to the left of the concave mirror.

c. The image is $\boxed{\text{inverted}}$, because a diverging lens always produces an upright image, and the concave mirror produces an inverted image when the object distance is greater than the focal length of the mirror.

---

77. **REASONING** The closest she can read the magazine is when the image formed by the contact lens is at the near point of her eye, or $d_i = -138$ cm. The image distance is negative because the image is a virtual image (see Section 26.10). Since the focal length is also known, the object distance can be found from the thin-lens equation.

SOLUTION The object distance d_o is related to the focal length f and the image distance d_i by the thin-lens equation:

$$\frac{1}{d_o} = \frac{1}{f} - \frac{1}{d_i} = \frac{1}{35.1 \text{ cm}} - \frac{1}{-138 \text{ cm}} \qquad \text{or} \qquad d_o = \boxed{28.0 \text{ cm}} \qquad (26.6)$$

---

79. **REASONING** The far point is 5.0 m from the right eye, and 6.5 m from the left eye. For an object infinitely far away ($d_o = \infty$), the image distances for the corrective lenses are then –5.0 m for the right eye and –6.5 m for the left eye, the negative sign indicating that the images are virtual images formed to the left of the lenses. The thin-lens equation [Equation 26.6: $(1/d_o) + (1/d_i) = (1/f)$] can be used to find the focal length. Then, Equation 26.8 can be used to find the refractive power for the lens for each eye.

SOLUTION Since the object distance d_o is essentially infinite, $1/d_o = 1/\infty = 0$, and the thin-lens equation becomes $1/d_i = 1/f$, or $d_i = f$. Therefore, for the right eye, $f = -5.0$ m, and the refractive power is (see Equation 26.8)

[Right eye] $\qquad \dfrac{\text{Refractive power}}{\text{(in diopters)}} = \dfrac{1}{f} = \dfrac{1}{(-5.0 \text{ m})} = \boxed{-0.20 \text{ diopters}}$

Similarly, for the left eye, $f = -6.5$ m, and the refractive power is

[Left eye] $\qquad \dfrac{\text{Refractive power}}{\text{(in diopters)}} = \dfrac{1}{f} = \dfrac{1}{(-6.5 \text{ m})} = \boxed{-0.15 \text{ diopters}}$

83. **REASONING** The angular magnification M of a magnifying glass is given by

$$M \approx \left(\frac{1}{f} - \frac{1}{d_i} \right) N \tag{26.10}$$

where f is the focal length of the lens, d_i is the image distance, and N is the near point of the eye. The focal length and the image distance are related to the object distance d_o by the thin-lens equation:

$$\frac{1}{f} - \frac{1}{d_i} = \frac{1}{d_o} \tag{26.6}$$

These two relations will allow us to determine the angular magnification.

SOLUTION Substituting Equation 26.6 into Equation 26.10 yields

$$M \approx \left(\frac{1}{f} - \frac{1}{d_i} \right) N = \frac{N}{d_o} = \frac{72 \text{ cm}}{4.0 \text{ cm}} = \boxed{18}$$

89. **REASONING AND SOLUTION** The information given allows us to determine the near point for this farsighted person. With $f = 45.4$ cm and $d_o = 25.0$ cm, we find from the thin-lens equation that

$$\frac{1}{d_i} = \frac{1}{f} - \frac{1}{d_o} = \frac{1}{45.4 \text{ cm}} - \frac{1}{25.0 \text{ cm}} \qquad \text{or} \qquad d_i = -55.6 \text{ cm}$$

Therefore, this person's near point, N, is 55.6 cm. We now need to find the focal length of the magnifying glass based on the near point for a normal eye, i.e., $M = N/f + 1$ (Equation 26.10, with $d_i = -N$), where $N = 25.0$ cm. Solving for the focal length gives

$$f = \frac{N}{M-1} = \frac{25.0 \text{ cm}}{7.50 - 1} = 3.85 \text{ cm}$$

We can now determine the maximum angular magnification for the farsighted person

$$M = \frac{N}{f} + 1 = \frac{55.6 \text{ cm}}{3.85 \text{ cm}} + 1 = \boxed{15.4}$$

91. **REASONING** The angular magnification of a compound microscope is given by Equation 26.11:

$$M \approx -\frac{(L - f_e)N}{f_o f_e}$$

where f_0 is the focal length of the objective, f_e is the focal length of the eyepiece, and L is the separation between the two lenses. This expression can be solved for f_0, the focal length of the objective.

SOLUTION Solving for f_0, we find that the focal length of the objective is

$$f_0 = -\frac{(L - f_e)N}{f_e M} = -\frac{(16.0 \text{ cm} - 1.4 \text{ cm})(25 \text{ cm})}{(1.4 \text{ cm})(-320)} = \boxed{0.81 \text{ cm}}$$

---

97. **REASONING** Knowing the angles subtended at the unaided eye and with the telescope will allow us to determine the angular magnification of the telescope. Then, since the angular magnification is related to the focal lengths of the eyepiece and the objective, we will use the known focal length of the eyepiece to determine the focal length of the objective.

SOLUTION From Equation 26.12, we have

$$M = \frac{\theta'}{\theta} = -\frac{f_0}{f_e}$$

where θ is the angle subtended by the unaided eye and θ' is the angle subtended when the telescope is used. We note that θ' is negative, since the telescope produces an inverted image. Thus, using Equation 26.12, we find

$$f_0 = -\frac{f_e \theta'}{\theta} = -\frac{(0.032 \text{ m})(-2.8 \times 10^{-3} \text{ rad})}{8.0 \times 10^{-5} \text{ rad}} = \boxed{1.1 \text{ m}}$$

---

99. **REASONING** The angular magnification M of an astronomical telescope is given by $M \approx -\frac{f_0}{f_e}$ (Equation 26.12), where f_0 is the focal length of the objective lens, and f_e is the focal length of the eyepiece. The length of the barrel must be adjusted so that the image formed by the objective appears very close to the focal point of the eyepiece. The result is that the length L of the barrel is approximately equal to the sum of the focal lengths of the telescope's two lenses: $L \approx f_0 + f_e$. Therefore, if the barrel can only be shortened by 0.50 cm, then the replacement eyepiece can have a focal length f_{e2} that is only 0.50 cm shorter than the focal length f_{e1} of the current eyepiece.

SOLUTION The focal length of the replacement eyepiece is

$$f_{e2} = f_{e1} - 0.50 \text{ cm} = 1.20 \text{ cm} - 0.50 \text{ cm} = 0.70 \text{ cm}$$

Therefore, from Equation 26.12, the angular magnification with the second eyepiece in place is

$$M \approx -\frac{f_o}{f_e} = -\frac{180 \text{ cm}}{0.70 \text{ cm}} = \boxed{-260}$$

105. **REASONING** The ray diagram is constructed by drawing the paths of two rays from a point on the object. For convenience, we will choose the top of the object. The ray that is parallel to the principal axis will be refracted by the lens so that it passes through the focal point on the right of the lens. The ray that passes through the center of the lens passes through undeflected. The image is formed at the intersection of these two rays. In this case, the rays do not intersect on the right of the lens. However, if they are extended backwards they intersect on the left of the lens, locating a virtual, upright, and enlarged image.

SOLUTION

a. The ray-diagram, drawn to scale, is shown below.

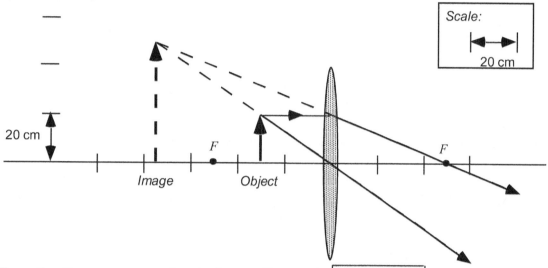

From the diagram, we see that the image distance is $\boxed{d_i = -75 \text{ cm}}$ and the magnification is $\boxed{+2.5}$. The negative image distance indicates that the image is virtual. The positive magnification indicates that the image is larger than the object.

b. From the thin-lens equation [Equation 26.6: $(1/d_o) + (1/d_i) = (1/f)$], we obtain

$$\frac{1}{d_i} = \frac{1}{f} - \frac{1}{d_o} = \frac{1}{50.0 \text{ m}} - \frac{1}{30.0 \text{ cm}} \quad \text{or} \quad \boxed{d_i = -75.0 \text{ cm}}$$

The magnification equation (Equation 26.7) gives the magnification to be

$$m = -\frac{d_i}{d_o} = -\frac{-75.0 \text{ cm}}{30.0 \text{ cm}} = \boxed{+2.50}$$

107. **REASONING** The refractive index n_{Liquid} of the liquid can be less than the refractive index of the glass n_{Glass}. However, we must consider the phenomenon of total internal reflection.

Some of the light will enter the liquid as long as the angle of incidence is less than or equal to the critical angle. At incident angles greater than the critical angle, total internal reflection occurs, and no light enters the liquid. Since the angle of incidence is 75.0°, the critical angle cannot be allowed to fall below 75.0°. The critical angle θ_c is determined according to Equation 26.4:

$$\sin\theta_c = \frac{n_{\text{Liquid}}}{n_{\text{Glass}}}$$

As n_{Liquid} decreases, the critical angle decreases. Therefore, n_{Liquid} cannot be less than the value calculated from this equation, in which $\theta_c = 75.0°$ and $n_{\text{Glass}} = 1.56$.

SOLUTION Using Equation 26.4, we find that

$$\sin\theta_c = \frac{n_{\text{Liquid}}}{n_{\text{Glass}}} \quad \text{or} \quad n_{\text{Liquid}} = n_{\text{Glass}}\sin\theta_c = (1.56)\sin 75.0° = \boxed{1.51}$$

109.**REASONING** The glasses form an image of a distant object at her far point, a distance $L = 0.690$ m from her eyes. This distance is equal to the magnitude $|d_i|$ of the image distance, which is measured from the glasses to the image, plus the distance s between her eyes and the glasses:

$$L = |d_i| + s \tag{1}$$

Because distant objects may be considered infinitely distant, the object distance in the thin-lens equation $\dfrac{1}{d_o} + \dfrac{1}{d_i} = \dfrac{1}{f}$ (Equation 26.6) is $d_o = \infty$, and we see that the image distance is equal to the focal length f of the eyeglasses:

$$\frac{1}{f} = \frac{1}{\infty} + \frac{1}{d_i} = 0 + \frac{1}{d_i} = \frac{1}{d_i} \quad \text{or} \quad d_i = f \tag{2}$$

The focal length f (in meters) is equal to the reciprocal of the refractive power (in diopters) of the glasses, according to $f = \dfrac{1}{\text{Refractive power}}$ (Equation 26.8).

SOLUTION Solving Equation (1) for s yields $s = L - |d_i|$. Substituting Equation (2) into this result, we obtain

$$s = L - |d_i| = L - |f| \tag{3}$$

Substituting $f = \dfrac{1}{\text{Refractive power}}$ (Equation 26.8) into Equation (3), we find that

$$s = L - |f| = L - \left| \dfrac{1}{\text{Refractive power}} \right| = 0.690 \text{ m} - \left| \dfrac{1}{-1.50 \text{ diopters}} \right| = \boxed{0.023 \text{ m}}$$

111. **REASONING AND SOLUTION** According to Equation 26.11, the angular magnification of the microscope is

$$M \approx -\dfrac{(L - f_e)N}{f_o f_e} = -\dfrac{(14.0 \text{ cm} - 2.5 \text{ cm})(25.0 \text{ cm})}{(0.50 \text{ cm})(2.5 \text{ cm})} = -2.3 \times 10^2$$

Now the new angle is

$$\theta' = M\theta = \left(-2.3 \times 10^2\right)\left(2.1 \times 10^{-5} \text{ rad}\right) = -4.8 \times 10^{-3} \text{ rad}$$

The magnitude of the angle is $\boxed{4.8 \times 10^{-3} \text{ rad}}$.

113. **REASONING** The drawing shows a ray of sunlight reaching the scuba diver (drawn as a black dot). The light reaching the scuba diver makes of angle of 28.0° with respect to the vertical. In addition, the drawing indicates that this angle is also the angle of refraction θ_2 of the light entering the water. The angle of incidence for this light is θ_1. These angles are related by Snell's law, $n_1 \sin\theta_1 = n_2 \sin\theta_2$ (Equation 26.2), where n_1 and n_2 are, respectively the indices of refraction of the air and water. Since θ_2, n_1, and n_2 are known, the angle of incidence can be determined.

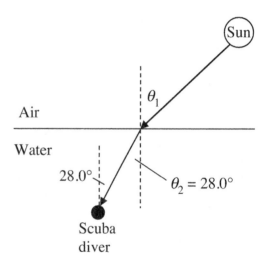

SOLUTION The angle of incidence of the light is given according to

$$\sin\theta_1 = \dfrac{n_2 \sin\theta_2}{n_1} \quad \text{or} \quad \theta_1 = \sin^{-1}\left(\dfrac{n_2 \sin\theta_2}{n_1} \right) = \sin^{-1}\left(\dfrac{1.333 \sin 28.0°}{1.000} \right) = \boxed{38.7°}$$

where the values $n_1 = 1.000$ and $n_2 = 1.333$ have been taken from Table 26.1.

119. *REASONING AND SOLUTION*
   a. A real image must be projected on the drum; therefore, the lens in the copier must be a
   converging lens .

   b. If the document and its copy have the same size, but are inverted with respect to one another, the magnification equation (Equation 26.7) indicates that $m = -d_i/d_o = -1$. Therefore, $d_i/d_o = 1$ or $d_i = d_o$. Then, the thin-lens equation (Equation 26.6) gives

$$\frac{1}{d_i} + \frac{1}{d_o} = \frac{1}{f} = \frac{2}{d_o} \qquad \text{or} \qquad d_o = d_i = 2f$$

   Therefore the document is located at a distance $\boxed{2f}$ from the lens.

   c. Furthermore, the image is located at a distance of $\boxed{2f}$ from the lens.

---

121. *REASONING* A contact lens is placed directly on the eye. Therefore, the object distance, which is the distance from the book to the lens, is 25.0 cm. The near point can be determined from the thin-lens equation [Equation 26.6: $(1/d_o) + (1/d_i) = (1/f)$].

   SOLUTION
   a. Using the thin-lens equation, we have

$$\frac{1}{d_i} = \frac{1}{f} - \frac{1}{d_o} = \frac{1}{65.0\text{ cm}} - \frac{1}{25.0\text{ cm}} \qquad \text{or} \qquad d_i = -40.6\text{ cm}$$

   In other words, at age 40, the man's near point is 40.6 cm. Similarly, when the man is 45, we have

$$\frac{1}{d_i} = \frac{1}{f} - \frac{1}{d_o} = \frac{1}{65.0\text{ cm}} - \frac{1}{29.0\text{ cm}} \qquad \text{or} \qquad d_i = -52.4\text{ cm}$$

   and his near point is 52.4 cm. Thus, the man's near point has changed by 52.4 cm − 40.6 cm = $\boxed{11.8\text{ cm}}$.

   b. With $d_o = 25.0$ cm and $d_i = -52.4$ cm, the focal length of the lens is found as follows:

$$\frac{1}{f} = \frac{1}{d_o} + \frac{1}{d_i} = \frac{1}{25.0\text{ cm}} + \frac{1}{(-52.4\text{ cm})} \qquad \text{or} \qquad \boxed{f = 47.8\text{ cm}}$$

123. **REASONING** The angular magnification of a refracting telescope is 32 800 times larger when you look through the correct end of the telescope than when you look through the wrong end. We wish to find the angular magnification, $M = -f_o / f_e$ (see Equation 26.12) of the telescope. Thus, we proceed by finding the ratio of the focal lengths of the objective and the eyepiece and using Equation 26.12 to find M.

SOLUTION When you look through the correct end of the telescope, the angular magnification of the telescope is $M_c = -f_o / f_e$. If you look through the wrong end, the roles of the objective and eyepiece lenses are interchanged, so that the angular magnification would be $M_w = -f_e / f_o$. Therefore,

$$\frac{M_c}{M_w} = \frac{-f_o / f_e}{-f_e / f_o} = \left(\frac{f_o}{f_e}\right)^2 = 32\,800 \qquad \text{or} \qquad \frac{f_o}{f_e} = \pm\sqrt{32\,800} = \pm181$$

The angular magnification of the telescope is negative, so we choose the positive root and obtain $M = -f_o / f_e = -(+181) = \boxed{-181}$.

___

127. **CONCEPTS**

(i) A converging lens can form either a real or a virtual image, depending on where the object is located relative to the focal point. If the object is to the left of the focal point as it is in this example, the image is real and falls to the right of the lens. Had the object been located between the focal point and the lens, the image would have been a virtual image located to the left of the converging lens.

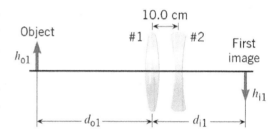

(ii) The image produced by the first lens acts as the object for the second lens.

(iii) According to the reasoning strategy in Section 26.8, the object located to the right of the lens is called a virtual object and is assigned a negative object distance.

(iv) Once the object distance for the second lens has been assigned a negative number, we can use the thin lens equation in the usual manner to find the image distance.

CALCULATIONS

(a) The distance d_{i1} of the image from the first lens can be found from the thin-lens equation, Equation 26.6. The focal length, $f_1 = +20.0$ cm, is positive because the lens is a converging lens, and the object distance, $d_{o1} = +45.0$ cm, is positive because it lies to the left of the lens:

$$\frac{1}{d_{i1}} = \frac{1}{f_1} - \frac{1}{d_{o1}} = \frac{1}{20.0 \text{ cm}} - \frac{1}{45.0 \text{ cm}} = 0.0278 \text{ cm}^{-1} \qquad \text{or} \qquad \boxed{d_{i1} = 36.0 \text{ cm}}$$

This image distance is positive, indicating that the image is real.

(b) The height h_{i1} of the image produced by the first lens can be obtained from the magnification equation, Equation 26.7:

$$h_{i1} = h_{o1}\left(-\frac{d_{i1}}{d_{o1}}\right) = (5.00 \text{ mm})\left(-\frac{36.0 \text{ cm}}{45.0 \text{ cm}}\right) = \boxed{-4.00 \text{ mm}}$$

The minus sign indicates that the first image is inverted with respect to the object (see the figure).

(c) The first image falls 36.0 cm to the right of the first lens. However, the second lens is located 10.0 cm to the right of the first lens, and therefore the first image is located $36.0 \text{ cm} - 10.0 \text{ cm} = 26.0 \text{ cm}$ to the right of the second lens. This image acts as the object for the second lens. Since the object for the second lens lies to the right of it, the object is a virtual object and is assigned a negative number: $d_{o2} = \boxed{-26.0 \text{ cm}}$

(d) The distance d_{i2} of the image from the second lens can be found from the thin-lens equation. The focal length, $f_2 = -15.0 \text{ cm}$, is negative because the lens is a diverging lens, and the object distance, $d_{o2} = -26.0 \text{ cm}$, is negative because it lies to the right of the lens:

$$\frac{1}{d_{i2}} = \frac{1}{f_2} - \frac{1}{d_{o2}} = \frac{1}{-15.0 \text{ cm}} - \frac{1}{-26.0 \text{ cm}} = -0.0282 \text{ cm}^{-1} \quad \text{or} \quad \boxed{d_{i2} = -35.5 \text{ cm}}$$

The negative sign for d_{i2} means that the image is formed to the left of the diverging lens and, hence, is a virtual image.

(e) The image height h_{i1} produced by the first lens becomes the object height h_{o2} for the second lens: $h_{o2} = h_{i1} = -4.00 \text{ mm}$. The height h_{i2} of the image produced by the second lens follows from the magnification equation:

$$h_{i2} = h_{o2}\left(-\frac{d_{i2}}{d_{o2}}\right) = (-4.00 \text{ mm})\left(-\frac{-35.5 \text{ cm}}{-26.0 \text{ cm}}\right) = \boxed{5.46 \text{ mm}}$$

CHAPTER 27 | *INTERFERENCE AND THE WAVE NATURE OF LIGHT*

1. **REASONING** The angles θ that determine the locations of the dark and bright fringes in a Young's double-slit experiment are related to the integers m that identify the fringes, the wavelength λ of the light, and the separation d between the slits. Since values are given for m, λ, and d, the angles can be calculated.

SOLUTION The expressions that specify θ in terms of m, λ, and d are as follows:

Bright fringes
$$\sin\theta = m\frac{\lambda}{d} \qquad m = 0, 1, 2, 3, \ldots \qquad (27.1)$$

Dark fringes
$$\sin\theta = \left(m+\tfrac{1}{2}\right)\frac{\lambda}{d} \qquad m = 0, 1, 2, 3, \ldots \qquad (27.2)$$

Applying these expressions gives the answers that we seek.

a.
$$\sin\theta = \left(m+\tfrac{1}{2}\right)\frac{\lambda}{d} \quad \text{or} \quad \theta = \sin^{-1}\left[\left(0+\tfrac{1}{2}\right)\frac{520\times10^{-9}\ \text{m}}{1.4\times10^{-6}\ \text{m}}\right] = \boxed{11^\circ}$$

b.
$$\sin\theta = m\frac{\lambda}{d} \quad \text{or} \quad \theta = \sin^{-1}\left[(1)\frac{520\times10^{-9}\ \text{m}}{1.4\times10^{-6}\ \text{m}}\right] = \boxed{22^\circ}$$

c.
$$\sin\theta = \left(m+\tfrac{1}{2}\right)\frac{\lambda}{d} \quad \text{or} \quad \theta = \sin^{-1}\left[\left(1+\tfrac{1}{2}\right)\frac{520\times10^{-9}\ \text{m}}{1.4\times10^{-6}\ \text{m}}\right] = \boxed{34^\circ}$$

d.
$$\sin\theta = m\frac{\lambda}{d} \quad \text{or} \quad \theta = \sin^{-1}\left[(2)\frac{520\times10^{-9}\ \text{m}}{1.4\times10^{-6}\ \text{m}}\right] = \boxed{48^\circ}$$

---

11. **REASONING** The light that travels through the plastic has a different path length than the light that passes through the unobstructed slit. Since the center of the screen now appears dark, rather than bright, destructive interference, rather than constructive interference occurs there. This means that the difference between the number of wavelengths in the plastic sheet and that in a comparable thickness of air is $\frac{1}{2}$.

SOLUTION The wavelength of the light in the plastic sheet is given by Equation 27.3 as

$$\lambda_{\text{plastic}} = \frac{\lambda_{\text{vacuum}}}{n} = \frac{586\times10^{-9}\ \text{m}}{1.60} = 366\times10^{-9}\ \text{m}$$

The number of wavelengths contained in a plastic sheet of thickness t is

$$N_{plastic} = \frac{t}{\lambda_{plastic}} = \frac{t}{366 \times 10^{-9} \text{ m}}$$

The number of wavelengths contained in an equal thickness of air is

$$N_{air} = \frac{t}{\lambda_{air}} = \frac{t}{586 \times 10^{-9} \text{ m}}$$

where we have used the fact that $\lambda_{air} \approx \lambda_{vacuum}$. Destructive interference occurs when the difference, $N_{plastic} - N_{air}$, in the number of wavelengths is $\frac{1}{2}$:

$$N_{plastic} - N_{air} = \frac{1}{2}$$

$$\frac{t}{366 \times 10^{-9} \text{ m}} - \frac{t}{586 \times 10^{-9} \text{ m}} = \frac{1}{2}$$

Solving this equation for t yields $t = 487 \times 10^{-9}$ m = $\boxed{487 \text{ nm}}$.

---

13. **REASONING**   To solve this problem, we must express the condition for destructive interference in terms of the film thickness t and the wavelength λ_{film} of the light as it passes through the magnesium fluoride coating. We must also take into account any phase changes that occur upon reflection.

SOLUTION   Since the coating is intended to be nonreflective, its thickness must be chosen so that destructive interference occurs between waves 1 and 2 in the drawing. For destructive interference, the combined phase difference between the two waves must be an odd integer number of half wavelengths. The phase change for wave 1 is equivalent to one-half of a wavelength, since this light travels from a smaller refractive index ($n_{air} = 1.00$) toward a larger refractive index ($n_{film} = 1.38$).

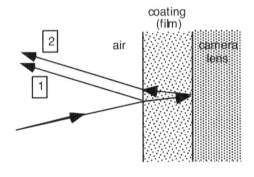

Similarly, there is a phase change when wave 2 reflects from the right surface of the film, since this light also travels from a smaller refractive index ($n_{film} = 1.38$) toward a larger one ($n_{lens} = 1.52$). Therefore, a phase change of one-half wavelength occurs at both boundaries, so the net phase change between waves 1 and 2 due to reflection is zero. Since wave 2 travels back and forth through the film and, and since the light is assumed to be at nearly normal incidence, the extra distance traveled by wave 2 compared to wave 1 is twice the film thickness, or $2t$. Thus, in this case, the minimum condition for destructive interference is

$$2t = \tfrac{1}{2}\lambda_{\text{film}}$$

The wavelength of light in the coating is

$$\lambda_{\text{film}} = \frac{\lambda_{\text{vacuum}}}{n} = \frac{565 \text{ nm}}{1.38} = 409 \text{ nm} \tag{27.3}$$

Solving the above expression for t, we find that the minimum thickness that the coating can have is

$$t = \tfrac{1}{4}\lambda_{\text{film}} = \tfrac{1}{4}(409 \text{ nm}) = \boxed{102 \text{ nm}}$$

---

19. **REASONING**   To solve this problem, we must express the condition for constructive interference in terms of the film thickness t and the wavelength λ_{film} of the light in the soap film. We must also take into account any phase changes that occur upon reflection.

SOLUTION   For the reflection at the top film surface, the light travels from air, where the refractive index is smaller ($n = 1.00$), toward the film, where the refractive index is larger ($n = 1.33$). Associated with this reflection there is a phase change that is equivalent to one-half of a wavelength. For the reflection at the bottom film surface, the light travels from the film, where the refractive index is larger ($n = 1.33$), toward air, where the refractive index is smaller ($n = 1.00$). Associated with this reflection, there is no phase change. As a result of these two reflections, there is a net phase change that is equivalent to one-half of a wavelength. To obtain the condition for constructive interference, this net phase change must be added to the phase change that arises because of the film thickness t, which is traversed twice by the light that penetrates it. For constructive interference we find that

$$2t + \tfrac{1}{2}\lambda_{\text{film}} = \lambda_{\text{film}}, 2\lambda_{\text{film}}, 3\lambda_{\text{film}}, \ldots$$

or

$$2t = \left(m + \tfrac{1}{2}\right)\lambda_{\text{film}}, \quad \text{where } m = 0, 1, 2, \ldots$$

Equation 27.3 indicates that $\lambda_{\text{film}} = \lambda_{\text{vacuum}}/n$. Using this expression and the fact that $m = 0$ for the minimum thickness t, we find that the condition for constructive interference becomes

$$2t = \left(m + \tfrac{1}{2}\right)\lambda_{\text{film}} = \left(0 + \tfrac{1}{2}\right)\left(\frac{\lambda_{\text{vacuum}}}{n}\right)$$

or

$$t = \frac{\lambda_{\text{vacuum}}}{4n} = \frac{611 \text{ nm}}{4(1.33)} = \boxed{115 \text{ nm}}$$

---

25. **REASONING**   This problem can be solved by using Equation 27.4 for the value of the angle θ when $m = 1$ (first dark fringe).

SOLUTION

a. When the slit width is $W = 1.8 \times 10^{-4}$ m and $\lambda = 675$ nm $= 675 \times 10^{-9}$ m, we find, according to Equation 27.4,

$$\theta = \sin^{-1}\left(m\frac{\lambda}{W}\right) = \sin^{-1}\left[(1)\frac{675 \times 10^{\pm 9}\,\text{m}}{1.8 \times 10^{\pm 4}\,\text{m}}\right] = \boxed{0.21°}$$

b. Similarly, when the slit width is $W = 1.8 \times 10^{-6}$ m and $\lambda = 675 \times 10^{-9}$ m, we find

$$\theta = \sin^{-1}\left[(1)\frac{675 \times 10^{\pm 9}\,\text{m}}{1.8 \times 10^{\pm 6}\,\text{m}}\right] = \boxed{22°}$$

---

27. **REASONING** The drawing shows a top view of the slit and screen, as well as the position of the central bright fringe and the third dark fringe. The distance y can be obtained from the tangent function as $y = L \tan\theta$. Since L is given, we need to find the angle θ before y can be determined. 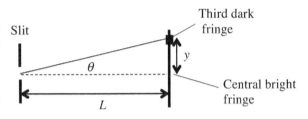 According to Equation 27.4, the angle θ is related to the wavelength λ of the light and the width W of the slit by $\sin\theta = m\lambda/W$, where $m = 3$ since we are interested in the angle for the third dark fringe.

SOLUTION We will first compute the angle between the central bright fringe and the third dark fringe using Equation 27.4 (with $m = 3$):

$$\theta = \sin^{-1}\left(\frac{m\lambda}{W}\right) = \sin^{-1}\left[\frac{3(668 \times 10^{-9}\,\text{m})}{6.73 \times 10^{-6}\,\text{m}}\right] = 17.3°$$

The vertical distance is

$$y = L \tan\theta = (1.85\text{ m})\tan 17.3° = \boxed{0.576\text{ m}}$$

---

31. **REASONING** The angle θ that specifies the location of the mth dark fringe is given by $\sin\theta = m\lambda/W$ (Equation 27.4), where λ is the wavelength of the light and W is the width of the slit. When θ has its maximum value of $90.0°$, the number of dark fringes that can be produced is a maximum. We will use this information to obtain a value for this number.

SOLUTION Solving Equation 27.4 for m, and setting $\theta = 90.0°$, we have

$$m = \frac{W \sin 90.0°}{\lambda} = \frac{(5.47 \times 10^{-6} \text{ m}) \sin 90.0°}{651 \times 10^{-9} \text{ m}} = 8.40$$

Therefore, the number of dark fringes is $\boxed{8}$.

---

35. ***REASONING*** According to Rayleigh's criterion, the two taillights must be separated by a distance s sufficient to subtend an angle $\theta_{min} \approx 1.22\lambda/D$ at the pupil of the observer's eye. Recalling that this angle must be expressed in radians, we relate θ_{min} to the distances s and L.

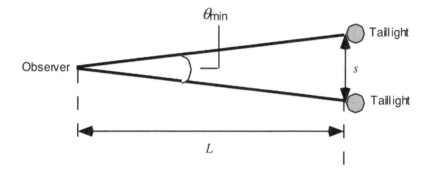

SOLUTION The wavelength λ is 660 nm. Therefore, we have from Equation 27.6

$$\theta_{min} \approx 1.22\frac{\lambda}{D} = 1.22\left(\frac{660 \times 10^{-9} \text{ m}}{7.0 \times 10^{-3} \text{ m}}\right) = 1.2 \times 10^{-4} \text{ rad}$$

According to Equation 8.1, the distance L between the observer and the taillights is

$$L = \frac{s}{\theta_{min}} = \frac{1.2 \text{ m}}{1.2 \times 10^{-4} \text{ rad}} = \boxed{1.0 \times 10^4 \text{ m}}$$

41. **REASONING** Assuming that the angle θ_{min} is small, the distance y between the blood cells is given by

$$y = f\theta_{min} \qquad (8.1)$$

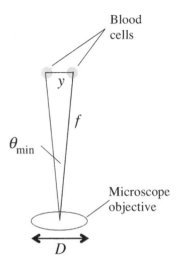

Blood cells

y

f

θ_{min}

Microscope objective

D

where f is the distance between the microscope objective and the cells (which is given as the focal length of the objective). However, the minimum angular separation θ_{min} of the cells is given by the Rayleigh criterion as $\theta_{min} = 1.22\ \lambda/D$ (Equation 27.6), where λ is the wavelength of the light and D is the diameter of the objective. These two relations can be used to find an expression for y in terms of λ.

SOLUTION

a. Substituting Equation 27.6 into Equation 8.1 yields

$$y = f\theta_{min} = f\left(\frac{1.22\lambda}{D}\right)$$

Since it is given that $f = D$, we see that $y = \boxed{1.22\ \lambda}$.

b. Because y is proportional to λ, the wavelength must be $\boxed{\text{shorter}}$ to resolve cells that are closer together.

43. **REASONING** The angle that specifies the third-order maximum of a diffraction grating is $\sin \theta = m\lambda/d$ (Equation 27.7), where $m = 3$, λ is the wavelength of the light, and d is the separation between the slits of the grating. The separation is equal to the width of the grating (1.50 cm) divided by the number of lines (2400).

SOLUTION Solving Equation 27.7 for the wavelength, we obtain

$$\lambda = \frac{d \sin \theta}{m} = \frac{\left(\dfrac{1.50\times10^{-2}\ \text{m}}{2400}\right)\sin 18.0°}{3} = 6.44\times10^{-7}\ \text{m} = \boxed{644\ \text{nm}}$$

47. ***REASONING AND SOLUTION*** The geometry of the situation is shown below.

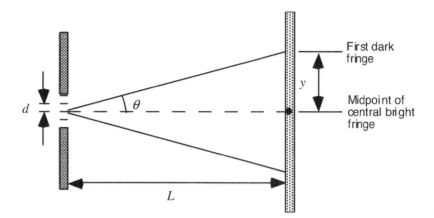

From the geometry, we have

$$\tan\theta = \frac{y}{L} = \frac{0.60 \text{ mm}}{3.0 \text{ mm}} = 0.20 \qquad \text{or} \qquad \theta = 11.3°$$

Then, solving Equation 27.7 with $m = 1$ for the separation d between the slits, we have

$$d = \frac{m\lambda}{\sin\theta} = \frac{(1)(780 \times 10^{-9} \text{ m})}{\sin 11.3°} = \boxed{4.0 \times 10^{-6} \text{ m}}$$

---

51. ***REASONING*** The angle θ that locates the first-order maximum produced by a grating with 3300 lines/cm is given by Equation 27.7, $\sin\theta = m\lambda/d$, with the order of the fringes given by $m = 0, 1, 2, 3, \ldots$ Any two of the diffraction patterns will overlap when their angular positions are the same.

SOLUTION Since the grating has 3300 lines/cm, we have

$$d = \frac{1}{3300 \text{ lines/cm}} = 3.0 \times 10^{-4} \text{ cm} = 3.0 \times 10^{-6} \text{ m}$$

a. In first order, $m = 1$; therefore, for violet light,

$$\theta = \sin^{-1}\left(m\frac{\lambda}{d}\right) = \sin^{-1}\left[(1)\left(\frac{410 \times 10^{-9} \text{ m}}{3.0 \times 10^{-6} \text{ m}}\right)\right] = \boxed{7.9°}$$

Similarly for red light,

$$\theta = \sin^{-1}\left(m\frac{\lambda}{d}\right) = \sin^{-1}\left[(1)\left(\frac{660 \times 10^{-9} \text{ m}}{3.0 \times 10^{-6} \text{ m}}\right)\right] = \boxed{13°}$$

b. Repeating the calculation for the second order maximum ($m = 2$), we find that

($m = 2$)	
for violet	$\theta = 16°$
for red	$\theta = 26°$

c. Repeating the calculation for the third order maximum ($m = 3$), we find that

($m = 3$)	
for violet	$\theta = 24°$
for red	$\theta = 41°$

d. Comparisons of the values for θ calculated in parts (a), (b) and (c) show that the second and third orders overlap .

---

55. **REASONING** In order for the two rays to interfere constructively and thereby form a bright interference fringe, the difference Δl between their path lengths must be an integral multiple m of the wavelength λ of the light:

$$\Delta l = m\lambda \tag{1}$$

In Equation (1), m can take on any integral value (m = 0, 1, 2, 3, ...). In this case, the rays meet at the eight-order bright fringe, so we have that $m = 8$.

SOLUTION Solving Equation (1) for λ, and substituting $m = 8$, we obtain

$$\lambda = \frac{\Delta l}{m} = \frac{4.57 \times 10^{-6} \text{ m}}{8} = 5.71 \times 10^{-7} \text{ m}$$

Using the equivalence 1 nm = 10^{-9} m, we convert this result to nanometers:

$$\lambda = \left(5.71 \times 10^{-7} \text{ m}\right)\left(\frac{1 \text{ nm}}{10^{-9} \text{ m}}\right) = \boxed{571 \text{ nm}}$$

---

57. **REASONING**   The slit separation d is given by Equation 27.1 with $m = 1$; namely $d = \lambda / \sin\theta$. As shown in Example 1 in the text, the angle θ is given by $\theta = \tan^{-1}(y/L)$.

SOLUTION   The angle θ is

$$\theta = \tan^{-1}\left(\frac{0.037\ \text{m}}{4.5\ \text{m}}\right) = 0.47°$$

Therefore, the slit separation d is

$$d = \frac{\lambda}{\sin\theta} = \frac{490 \times 10^{-9}\ \text{m}}{\sin 0.47°} = \boxed{6.0 \times 10^{-5}\ \text{m}}$$

---

61. **REASONING** For a diffraction grating, the angular position θ of a principal maximum on the screen is given by Equation 27.7 as $\sin\theta = m\lambda / d$ with $m = 0, 1, 2, 3, \ldots$

SOLUTION   When the fourth-order principal maximum of light A exactly overlaps the third-order principal maximum of light B, we have

$$\sin\theta_A = \sin\theta_B$$

$$\frac{4\lambda_A}{d} = \frac{3\lambda_B}{d} \qquad \text{or} \qquad \frac{\lambda_A}{\lambda_B} = \boxed{\frac{3}{4}}$$

---

67. **CONCEPTS**

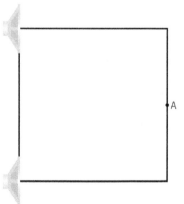

(i) The loudspeakers are in-phase sources of identical sound waves. The waves coming from one speaker travel a distance ℓ_1 in reaching point A, and the waves from the second speaker travel a distance ℓ_2. The condition that leads to constructive interference is $\ell_2 - \ell_1 = m\lambda$, where λ is the wavelength of the waves and $m = 0$, 1, 2, 3 .... Thus, the two distances are the same or differ by an integer number of wavelengths. Point A is the midpoint of a side of the square, so that ℓ_1 and ℓ_2 are the same, and constructive interference occurs.

(ii) The waves start out in phase, so the general condition that leads to destructive interference is $\ell_2 - \ell_1 = \left(m + \frac{1}{2}\right)\lambda$, where $m = 0.1, 2, 3, \ldots$

(iii) The general condition that leads to destructive interference is written as $\ell_2 - \ell_1 = \left(m + \frac{1}{2}\right)\lambda$, where $m = 0.1, 2, 3, \ldots$. There are many possible values of m, and we are being asked what the specific value is. As you walk from point A toward either empty corner, one of the distances decreases and the other increases; the loudness diminishes gradually and disappears at the empty corner, when the difference between the distances to the speakers has attained the smallest possible value of $\ell_2 - \ell_1 = \frac{1}{2}\lambda$. This means we are dealing with the specific case in which $m = 0$.

CALCULATIONS Consider the destructive interference that occurs at either empty corner. Using L to denote the length of a side of the square and taking advantage of the Pythagorean theorem, we have

$$\ell_1 = L \quad \text{and} \quad \ell_2 = \sqrt{L^2 + L^2} = \sqrt{2}L$$

The specific condition for destructive interference at the empty corner is

$$\ell_2 - \ell_1 = \sqrt{2}L - L = \frac{1}{2}\lambda$$

Solve for the wavelength of the waves gives

$$\lambda = 2L\left(\sqrt{2} - 1\right) = 2(3.5 \text{ m})\left(\sqrt{2} - 1\right) = \boxed{2.9 \text{ m}}$$

CHAPTER 28 | *SPECIAL RELATIVITY*

1. **REASONING**

   a. The two events in this problem are the creation of the pion and its subsequent decay (or breaking apart). Imagine a reference frame attached to the pion, so the pion is stationary relative to this reference frame. To a hypothetical person who is at rest with respect to this reference frame, these two events occur at the same place, namely, at the place where the pion is located. Thus, this hypothetical person measures the proper time interval Δt_0 for the decay of the pion. On the other hand, the person standing in the laboratory sees the two events occurring at different locations, since the pion is moving relative to that person. The laboratory person, therefore, measures a dilated time interval Δt. The relation between these two time intervals is given by $\Delta t = \Delta t_0 / \sqrt{1 - v^2/c^2}$ (Equation 28.1).

   b. According to the hypothetical person who is at rest in the reference frame attached to the moving pion, the distance x that the laboratory travels before the pion breaks apart is equal to the speed v of the laboratory relative to the pion times the proper time interval Δt_0, or $x = v \Delta t_0$. The speed of the laboratory relative to the pion is the same as the speed of the pion relative to the laboratory, namely, $0.990c$.

   SOLUTION

   a. The proper time interval is

   $$\Delta t_0 = \Delta t \sqrt{1 - \frac{v^2}{c^2}} = \left(3.5 \times 10^{-8} \text{ s}\right) \sqrt{1 - \frac{(0.990c)^2}{c^2}} = \boxed{4.9 \times 10^{-9} \text{ s}} \qquad (28.1)$$

   b. The distance x that the laboratory travels before the pion breaks apart, as measured by the hypothetical person, is

   $$x = v \Delta t_0 = \underbrace{(0.990)(3.00 \times 10^8 \text{ m/s})}_{0.990c}\left(4.9 \times 10^{-9} \text{ s}\right) = \boxed{1.5 \text{ m}}$$

3. **REASONING** The total time for the trip is one year. This time is the proper time interval Δt_0, because it is measured by an observer (the astronaut) who is at rest relative to the beginning and ending events (the times when the trip started and ended) and who sees them at the same location in spacecraft. On the other hand, the astronaut measures the clocks on earth to run at the dilated time interval Δt, which is the time interval of one hundred years. The relation between the two time intervals is given by Equation 28.1, which can be used to find the speed of the spacecraft.

SOLUTION The dilated time interval Δt is related to the proper time interval Δt_0 by $\Delta t = \Delta t_0 / \sqrt{1 - (v^2/c^2)}$. Solving this equation for the speed v of the spacecraft yields

$$v = c \sqrt{1 - \left(\frac{\Delta t_0}{\Delta t}\right)^2} = c \sqrt{1 - \left(\frac{1 \text{ yr}}{100 \text{ yr}}\right)^2} = \boxed{0.999\,95c} \qquad (28.1)$$

9. *REASONING*   All standard meter sticks at rest have a length of 1.00 m for observers who are at rest with respect to them. Thus, 1.00 m is the proper length L_0 of the meter stick. When the meter stick moves with speed v relative to an earth-observer, its length $L = 0.500$ m will be a contracted length. Since both L_0 and L are known, v can be found directly from Equation 28.2, $L = L_0 \sqrt{1 - (v^2/c^2)}$.

SOLUTION Solving Equation 28.2 for v, we find that

$$v = c \sqrt{1 - \left(\frac{L}{L_0}\right)^2} = (3.00 \times 10^8 \text{ m/s}) \sqrt{1 - \left(\frac{0.500 \text{ m}}{1.00 \text{ m}}\right)^2} = \boxed{2.60 \times 10^8 \text{ m/s}}$$

11. *REASONING* The tourist is moving at a speed of $v = 1.3$ m/s with respect to the path and, therefore, measures a contracted length L instead of the proper length of $L_0 = 9.0$ km. The contracted length is given by the length-contraction equation, Equation 28.2.

SOLUTION According to the length-contraction equation, the tourist measures a length that is

$$L = L_0 \sqrt{1 - \frac{v^2}{c^2}} = (9.0 \text{ km}) \sqrt{1 - \frac{(1.3 \text{ m/s})^2}{(3.0 \text{ m/s})^2}} = \boxed{8.1 \text{ km}}$$

21. *REASONING*   The height of the woman as measured by the observer is given by Equation 28.2 as $h = h_0 \sqrt{1 - (v/c)^2}$, where h_0 is her proper height. In order to use this equation, we must determine the speed v of the woman relative to the observer. We are given the magnitude of her relativistic momentum, so we can determine v from p.

SOLUTION According to Equation 28.3 $p = mv / \sqrt{1 - v^2/c^2}$, so $mv = p\sqrt{1 - v^2/c^2}$ Squaring both sides, we have

$$m^2 v^2 = p^2 (1 - v^2/c^2) \qquad \text{or} \qquad m^2 v^2 + p^2 \frac{v^2}{c^2} = p^2$$

$$v^2\left(m^2 + \frac{p^2}{c^2}\right) = p^2 \qquad \text{or} \qquad v^2 = \frac{p^2}{m^2 + \dfrac{p^2}{c^2}}$$

Solving for v and substituting values, we have

$$v = \frac{p}{\sqrt{m^2 + \dfrac{p^2}{c^2}}} = \frac{2.0\times10^{10}\ \text{kg}\cdot\text{m/s}}{\sqrt{(55\ \text{kg})^2 + \left(\dfrac{2.0\times10^{10}\ \text{kg}\cdot\text{m/s}}{3.00\times10^8\ \text{m/s}}\right)^2}} = 2.3\times10^8\ \text{m/s}$$

Then, the height that the observer measures for the woman is

$$h = h_0\sqrt{1-\left(\frac{v}{c}\right)^2} = (1.6\ \text{m})\sqrt{1-\left(\frac{2.3\times10^8\ \text{m/s}}{3.0\times10^8\ \text{m/s}}\right)^2} = \boxed{1.0\ \text{m}}$$

---

23. ***REASONING*** The magnitude p of the relativistic momentum of an object is given by $p = \dfrac{mv}{\sqrt{1-v^2/c^2}}$ (Equation 28.3), where m is the object's mass, v is the object's speed, and c is the speed of light in a vacuum. The principle of conservation of linear momentum (see Section 7.2) states that the total momentum of a system is conserved when no net external force acts on the system. This principle applies at speeds approaching the speed of light in a vacuum, provided that Equation 28.3 is used for the individual momenta of the objects that comprise the system.

SOLUTION The total momentum of the man/woman system is conserved, since friction is negligible, so that no net external force acts on the system. Therefore, the final total momentum $p_\text{m} + p_\text{w}$ must equal the initial total momentum, which is zero. As a result, $p_\text{m} = -p_\text{w}$ where Equation 28.3 must be used for the momenta p_m and p_w. Thus, we find

$$\frac{m_\text{m}v_\text{m}}{\sqrt{1-\left(v_\text{m}/c\right)^2}} = -\frac{m_\text{w}v_\text{w}}{\sqrt{1-\left(v_\text{w}/c\right)^2}} \tag{1}$$

We know that $m_\text{m} = 88$ kg, $m_\text{w} = 54$ kg, and $v_\text{w} = +2.5$ m/s. Remember that c has the hypothetical value of 3.0 m/s. Solving Equation (1) for v_m reveals that $v_\text{m} = \pm 2.0\ \text{m/s}$. We choose the negative value, since the man and woman recoil from one another and it is stated that the woman moves away in the positive direction. Therefore, we find that $\boxed{v_\text{m} = -2.0\ \text{m/s}}$.

25. **REASONING** According to the work-energy theorem, Equation 6.3, the work that must be done on the electron to accelerate it from rest to a speed of $0.990c$ is equal to the kinetic energy of the electron when it is moving at $0.990c$.

SOLUTION Using Equation 28.6, we find that

$$KE = mc^2 \left(\frac{1}{\sqrt{1-(v^2/c^2)}} - 1 \right)$$

$$= (9.11 \times 10^{-31} \text{ kg})(3.00 \times 10^8 \text{ m/s})^2 \left(\frac{1}{\sqrt{1-(0.990c)^2/c^2}} - 1 \right) = \boxed{5.0 \times 10^{-13} \text{ J}}$$

---

29. **REASONING AND SOLUTION**

a. In Section 28.6 it is shown that when the speed of a particle is $0.01c$ (or less), the relativistic kinetic energy becomes nearly equal to the nonrelativistic kinetic energy. Since the speed of the particle here is $0.001c$, the ratio of the relativistic kinetic energy to the nonrelativistic kinetic energy is $\boxed{1.0}$.

b. Taking the ratio of the relativistic kinetic energy, Equation 28.6, to the nonrelativistic kinetic energy, $\frac{1}{2}mv^2$, we find that

$$\frac{mc^2 \left(\frac{1}{\sqrt{1-(v^2/c^2)}} - 1 \right)}{\frac{1}{2}mv^2} = 2\left(\frac{c}{v}\right)^2 \left(\frac{1}{\sqrt{1-(v^2/c^2)}} - 1 \right)$$

$$= 2\left(\frac{c}{0.970c}\right)^2 \left(\frac{1}{\sqrt{1-(0.970c)^2/c^2}} - 1 \right) = \boxed{6.6}$$

---

35. ***REASONING*** Let's define the following relative velocities, assuming that the spaceship and exploration vehicle are moving in the positive direction.

v_{ES} = velocity of **E**xploration vehicle relative to the **S**paceship.

v_{EO} = velocity of **E**xploration vehicle relative to an **O**bserver on earth = +0.70c

v_{SO} = velocity of **S**paceship relative to an **O**bserver on earth = +0.50c

The velocity v_{ES} can be determined from the velocity-addition formula, Equation 28.8:

$$v_{ES} = \frac{v_{EO} + v_{OS}}{1 + \frac{v_{EO}v_{OS}}{c^2}}$$

The velocity v_{OS} of the observer on earth relative to the spaceship is not given. However, we know that v_{OS} is the negative of v_{SO}, so $v_{OS} = -v_{SO} = -(+0.50c) = -0.50c$.

SOLUTION The velocity of the exploration vehicle relative to the spaceship is

$$v_{ES} = \frac{v_{EO} + v_{OS}}{1 + \frac{v_{EO}v_{OS}}{c^2}} = \frac{+0.70c + (-0.50c)}{1 + \frac{(+0.70c)(-0.50c)}{c^2}} = +0.31c$$

The speed of the exploration vehicle relative to the spaceship is the magnitude of this result or $\boxed{0.31c}$.

---

39. ***REASONING*** *AND* ***SOLUTION***
In all parts of this problem, the direction of the intergalactic cruiser, the ions, and the laser light is taken to be the positive direction.

a. According to the second postulate of special relativity, all observers measure the speed of light to be c, regardless of their velocities relative to each other. Therefore, the aliens aboard the hostile spacecraft see the photons of the laser approach $\boxed{\text{at the speed of light, } c}$.

b. To find the velocity of the ions relative to the aliens, we define the relative velocities as follows:

v_{IS} = velocity of the **I**ons relative to the alien **S**pacecraft

v_{IC} = velocity of the **I**ons relative to the intergalactic **C**ruiser = +0.950c

v_{CS} = velocity of the intergalactic **C**ruiser relative to the alien **S**pacecraft = +0.800c

These velocities are related by the velocity-addition formula, Equation 28.8. The velocity of the ions relative to the alien spacecraft is:

$$v_{\text{IS}} = \frac{v_{\text{IC}} + v_{\text{CS}}}{1 + \dfrac{v_{\text{IC}} v_{\text{CS}}}{c^2}} = \frac{+0.950c + 0.800c}{1 + \dfrac{(+0.950c)(+0.800c)}{c^2}} = \boxed{+0.994c}$$

c. The aliens see the laser light (photons) moving with respect to the cruiser at a velocity

$$U = +1.000c - 0.800c = \boxed{+0.200c}$$

d. The aliens see the ions moving away from the cruiser at a velocity

$$U' = +0.994c - 0.800c = \boxed{+0.194c}$$

43. **REASONING** Assume that traveler A moves at a speed of $v_A = 0.70c$ and traveler B moves at a speed of $v_B = 0.90c$, both speeds being with respect to the earth. Each traveler is moving with respect to the earth and the distant star, so each measures a contracted length L_A or L_B for the distance traveled. However, an observer on earth is at rest with respect to the earth and the distant star (which is assumed to be stationary with respect to the earth), so he or she would measure the proper length L_0. For each traveler the contracted length is given by the length-contraction equation as stated in Equation 28.2:

$$L_A = L_0 \sqrt{1 - \frac{v_A^2}{c^2}} \quad \text{and} \quad L_B = L_0 \sqrt{1 - \frac{v_B^2}{c^2}}$$

It is important to note that the proper length L_0 is the same in each application of the length-contraction equation. Thus, we can combine the two equations and eliminate it. Then, since we are given values for L_A, v_A, and v_B, we will be able to determine L_B.

SOLUTION Dividing the expression for L_B by the expression for L_A and eliminating L_0, we obtain

$$\frac{L_B}{L_A} = \frac{L_0 \sqrt{1 - \dfrac{v_B^2}{c^2}}}{L_0 \sqrt{1 - \dfrac{v_A^2}{c^2}}} = \frac{\sqrt{1 - \dfrac{v_B^2}{c^2}}}{\sqrt{1 - \dfrac{v_A^2}{c^2}}}$$

$$L_B = L_A \frac{\sqrt{1 - \dfrac{v_B^2}{c^2}}}{\sqrt{1 - \dfrac{v_A^2}{c^2}}} = (6.5 \text{ light-years}) \frac{\sqrt{1 - \dfrac{(0.90c)^2}{c^2}}}{\sqrt{1 - \dfrac{(0.70c)^2}{c^2}}} = \boxed{4.0 \text{ light-years}}$$

47. **REASONING** Since the crew is initially at rest relative to the escape pod, the length of 45 m is the proper length L_0 of the pod. The length of the escape pod as determined by an observer on earth can be obtained from the relation for length contraction given by Equation 28.2, $L = L_0 \sqrt{1 - \left(v_{PE}^2 / c^2 \right)}$. The quantity v_{PE} is the speed of the escape pod relative to the earth, which can be found from the velocity-addition formula, Equation 28.8. The following are the relative velocities, assuming that the direction away from the earth is the positive direction:

v_{PE} = velocity of the escape **P**od relative to **E**arth.

v_{PR} = velocity of escape **P**od relative to the **R**ocket = –0.55c. This velocity is negative because the rocket is moving away from the earth (in the positive direction), and the escape pod is moving in an opposite direction (the negative direction) relative to the rocket.

v_{RE} = velocity of **R**ocket relative to **E**arth = +0.75c

These velocities are related by the velocity-addition formula, Equation 28.8.

SOLUTION The relative velocity of the escape pod relative to the earth is

$$v_{PE} = \frac{v_{PR} + v_{RE}}{1 + \dfrac{v_{PR} v_{RE}}{c^2}} = \frac{-0.55c + 0.75c}{1 + \dfrac{(-0.55c)(+0.75c)}{c^2}} = +0.34c$$

The speed of the pod relative to the earth is the magnitude of this result, or 0.34c. The length of the pod as determined by an observer on earth is

$$L = L_0 \sqrt{1 - \frac{v_{PE}^2}{c^2}} = (45 \text{ m}) \sqrt{1 - \frac{(0.34c)^2}{c^2}} = \boxed{42 \text{ m}}$$

---

51. **CONCEPTS**

(i) The rest energy is the energy that an object has when its speed is zero. According to special relativity, the relation between the rest energy E_0 and the mass m is given by equation 28.5 as $E_0 = mc^2$, where c is the speed of light in vacuum. Thus, the rest energy is directly proportional to the mass. From the table it can be seen that the particles a and b have identical rest energies, so they have identical

Particle	Rest Energy	Total Energy
a	E'	$2E'$
b	E'	$4E'$
c	$5E'$	$6E'$

masses. Particle c has the greatest rest energy, so it has the greatest mass. The ranking of the masses, largest first, is c, then a and b (a tie).

(ii) No, because the expression $KE = \frac{1}{2}mv^2$ applies only when the speed of the object is much, much less than the speed of light. According to special relativity, the kinetic energy is the difference between the total energy E and the rest energy E_0, so $KE = E - E_0$. Therefore, we can examine the table and determine the kinetic energy of each particle in terms of E'. The kinetic energies of particles a, b, and c are, respectively, $2E' - E' = E'$, $4E' - E' = 3E'$, and $6E' - 5E' = E'$. The ranking of the kinetic energies, largest first, is b, then a and c (a tie).

CALCULATIONS

(a) The mass of particle a can be determined from its rest energy $E_0 = mc^2$. Since $E_0 = E'$ (see the table), its mass is

$$m_a = \frac{E'}{c^2} = \frac{5.98 \times 10^{-10} \text{ J}}{\left(3.00 \times 10^8 \text{ m/s}\right)^2} = \boxed{6.64 \times 10^{-27} \text{ kg}}$$

In a similar manner, we find that the masses of particles b and c are

$$\boxed{m_b = 6.64 \times 10^{-27} \text{ kg}} \quad \text{and} \quad \boxed{m_c = 33.2 \times 10^{-27} \text{ kg}}$$

As expected, the ranking is $m_c > m_a = m_c$.

(b) According to Equation 2.6, the kinetic energy KE is equal to the total energy E minus the rest energy E_0: $KE = E - E_0$. For particle a, its total energy is $E = 2E'$ and its rest energy is $E_0 = E'$, so its kinetic energy is

$$KE_a = 2E' - E' = E' = \boxed{5.98 \times 10^{-10} \text{ J}}$$

The kinetic energies of particles b and c can be determined in a similar fashion:

$$\boxed{KE_b = 17.9 \times 10^{-10} \text{ J}} \quad \text{and} \quad \boxed{KE_c = 5.98 \times 10^{-10} \text{ J}}$$

As anticipated, the ranking is $KE_b > KE_a = KE_c$.

CHAPTER 29 | *PARTICLES AND WAVES*

3. **REASONING** According to Equation 29.3, the work function W_0 is related to the photon energy hf and the maximum kinetic energy KE_{max} by $W_0 = hf - KE_{max}$. This expression can be used to find the work function of the metal.

    SOLUTION KE_{max} is 6.1 eV. The photon energy (in eV) is, according to Equation 29.2,

    $$hf = \left(6.63 \times 10^{-34} \text{ J} \cdot \text{s}\right)\left(3.00 \times 10^{15} \text{ Hz}\right)\left(\frac{1 \text{ eV}}{1.60 \times 10^{\pm 19} \text{ J}}\right) = 12.4 \text{ eV}$$

    The work function is, therefore,

    $$W_0 = hf - KE_{max} = 12.4 \text{ eV} - 6.1 \text{ ev} = \boxed{6.3 \text{ eV}}$$

---

5. **REASONING** The energy of the photon is related to its frequency by Equation 29.2, $E = hf$. Equation 16.1, $v = f\lambda$, relates the frequency and the wavelength for any wave.

    SOLUTION   Combining Equations 29.2 and 16.1, and noting that the speed of a photon is c, the speed of light in a vacuum, we have

    $$\lambda = \frac{c}{f} = \frac{c}{(E/h)} = \frac{hc}{E} = \frac{(6.63 \times 10^{-34} \text{ J} \cdot \text{s})(3.0 \times 10^8 \text{ m/s})}{6.4 \times 10^{-19} \text{ J}} = 3.1 \times 10^{-7} \text{ m} = \boxed{310 \text{ nm}}$$

---

9. **REASONING AND SOLUTION** The number of photons per second, N, entering the owl's eye is $N = SA/E$, where S is the intensity of the beam, A is the area of the owl's pupil, and E is the energy of a single photon. Assuming that the owl's pupil is circular, $A = \pi r^2 = \pi\left(\frac{1}{2}d\right)^2$, where d is the diameter of the owl's pupil. Combining Equations 29.2 and 16.1, we have $E = hf = hc/\lambda$. Therefore,

    $$N = \frac{SA\lambda}{hc} = \frac{(5.0 \times 10^{-13} \text{ W/m}^2)\pi\left[\frac{1}{2}\left(8.5 \times 10^{\pm 3} \text{m}\right)\right]^2 (510 \times 10^{\pm 9} \text{ m})}{(6.63 \times 10^{-34} \text{ J} \cdot \text{s})(3.0 \times 10^8 \text{ m/s})} = \boxed{73 \text{ photons/s}}$$

13. *REASONING AND SOLUTION*

a. According to Equation 24.5b, the electric field can be found from $E = \sqrt{S/(\varepsilon_0 c)}$. The intensity S of the beam is

$$S = \frac{\text{Energy per unit time}}{A} = \frac{Nhf}{A} = \frac{Nh}{A}\left(\frac{c}{\lambda}\right)$$

$$= \frac{\left(1.30 \times 10^{18}\ \text{photons/s}\right)\left(6.63 \times 10^{-34}\ \text{J·s}\right)}{\pi\left(1.00 \times 10^{-3}\ \text{m}\right)^2}\left(\frac{3.00 \times 10^8\ \text{m/s}}{514.5 \times 10^{-9}\ \text{m}}\right)$$

$$= 1.60 \times 10^5\ \text{W/m}^2$$

where N is the number of photons per second emitted. Then,

$$E = \sqrt{S/(\varepsilon_0 c)} = \boxed{7760\ \text{N/C}}$$

b. According to Equation 24.3, the average magnetic field is

$$B = E/c = \boxed{2.59 \times 10^{-5}\ \text{T}}$$

---

17. *REASONING* The angle θ through which the X-rays are scattered is related to the difference between the wavelength λ' of the scattered X-rays and the wavelength λ of the incident X-rays by Equation 29.7 as

$$\lambda' - \lambda = \frac{h}{mc}(1 - \cos\theta)$$

where h is Planck's constant, m is the mass of the electron, and c is the speed of light in a vacuum. We can use this relation directly to find the angle, since all the other variables are known.

SOLUTION Solving Equation 29.7 for the angle θ, we obtain

$$\cos\theta = 1 - \frac{mc}{h}(\lambda' - \lambda)$$

$$= 1 - \frac{\left(9.11 \times 10^{-31}\ \text{kg}\right)\left(3.00 \times 10^8\ \text{m/s}\right)}{6.63 \times 10^{-34}\ \text{J·s}}\left(0.2703 \times 10^{-9}\ \text{m} - 0.2685 \times 10^{-9}\ \text{m}\right) = 0.26$$

$$\theta = \cos^{-1}(0.26) = \boxed{75°}$$

21. **REASONING** The change in wavelength that occurs during Compton scattering is given by Equation 29.7:

$$\lambda' - \lambda = \frac{h}{mc}(1 - \cos\theta) \quad \text{or} \quad (\lambda' - \lambda)_{\text{max}} = \frac{h}{mc}(1 - \cos 180°) = \frac{2h}{mc}$$

$(\lambda' - \lambda)_{\text{max}}$ is the maximum change in the wavelength, and to calculate it we need a value for the mass m of a nitrogen molecule. This value can be obtained from the mass per mole M of nitrogen (N_2) and Avogadro's number N_A, according to $m = M / N_A$ (see Section 14.1).

SOLUTION Using a value of $M = 0.0280$ kg/mol, we obtain the following result for the maximum change in the wavelength:

$$(\lambda' - \lambda)_{\text{max}} = \frac{2h}{mc} = \frac{2h}{\left(\dfrac{M}{N_A}\right)c} = \frac{2\left(6.63 \times 10^{-34} \text{ J·s}\right)}{\left(\dfrac{0.0280 \text{ kg/mol}}{6.02 \times 10^{23} \text{ mol}^{-1}}\right)\left(3.00 \times 10^8 \text{ m/s}\right)}$$

$$= \boxed{9.50 \times 10^{-17} \text{ m}}$$

27. **REASONING** In order for the person to diffract to the same extent as the sound wave, the de Broglie wavelength of the person must be equal to the wavelength of the sound wave.

SOLUTION
a. Since the wavelengths are equal, we have that

$$\lambda_{\text{sound}} = \lambda_{\text{person}}$$

$$\lambda_{\text{sound}} = \frac{h}{m_{\text{person}} v_{\text{person}}}$$

Solving for v_{person}, and using the relation $\lambda_{\text{sound}} = v_{\text{sound}} / f_{\text{sound}}$ (Equation 16.1), we have

$$v_{\text{person}} = \frac{h}{m_{\text{person}}\left(v_{\text{sound}} / f_{\text{sound}}\right)} = \frac{h f_{\text{sound}}}{m_{\text{person}} v_{\text{sound}}}$$

$$= \frac{(6.63 \times 10^{-34} \text{ J·s})(128 \text{ Hz})}{(55.0 \text{ kg})(343 \text{ m/s})} = \boxed{4.50 \times 10^{-36} \text{ m/s}}$$

b.   At the speed calculated in part (a), the time required for the person to move a distance of one meter is

$$t = \frac{x}{v} = \frac{1.0 \text{ m}}{4.50 \times 10^{\pm 36} \text{ m/s}} \underbrace{\left(\frac{1.0 \text{ h}}{3600 \text{ s}} \right) \left(\frac{1 \text{ day}}{24.0 \text{ h}} \right) \left(\frac{1 \text{ year}}{365.25 \text{ days}} \right)}_{\substack{\text{Factors to convert} \\ \text{seconds to years}}} = \boxed{7.05 \times 10^{27} \text{ years}}$$

---

31. ***REASONING***   The de Broglie wavelength λ is related to Planck's constant h and the magnitude p of the particle's momentum. The magnitude of the momentum can be related to the particle's kinetic energy. Thus, using the given wavelength and the fact that the kinetic energy doubles, we will be able to obtain the new wavelength.

SOLUTION   The de Broglie wavelength is

$$\lambda = \frac{h}{p} \qquad\qquad (29.8)$$

The kinetic energy and the magnitude of the momentum are

$$\text{KE} = \frac{1}{2} mv^2 \qquad (6.2) \qquad p = mv \qquad (7.2)$$

where m and v are the mass and speed of the particle. Substituting Equation 7.2 into Equation 6.2, we can relate the kinetic energy and momentum as follows:

$$\text{KE} = \frac{1}{2} mv^2 = \frac{m^2 v^2}{2m} = \frac{p^2}{2m} \qquad \text{or} \qquad p = \sqrt{2m(\text{KE})}$$

Substituting this result for p into Equation 29.8 gives

$$\lambda = \frac{h}{p} = \frac{h}{\sqrt{2m(\text{KE})}}$$

Applying this expression for the final and initial wavelengths λ_f and λ_i, we obtain

$$\lambda_f = \frac{h}{\sqrt{2m(\text{KE})_f}} \qquad \text{and} \qquad \lambda_i = \frac{h}{\sqrt{2m(\text{KE})_i}}$$

Dividing the two equations and rearranging reveals that

$$\frac{\lambda_f}{\lambda_i} = \frac{\dfrac{h}{\sqrt{2m(KE)_f}}}{\dfrac{h}{\sqrt{2m(KE)_i}}} = \sqrt{\frac{(KE)_i}{(KE)_f}} \qquad \text{or} \qquad \lambda_f = \lambda_i\sqrt{\frac{(KE)_i}{(KE)_f}}$$

Using the given value for λ_i and the fact that $KE_f = 2(KE_i)$, we find

$$\lambda_f = \lambda_i\sqrt{\frac{(KE)_i}{(KE)_f}} = \left(2.7\times10^{-10}\ \text{m}\right)\sqrt{\frac{\cancel{KE_i}}{2\left(\cancel{KE_i}\right)}} = \boxed{1.9\times10^{-10}\ \text{m}}$$

---

35. ***REASONING*** The de Broglie wavelength λ of the electron is related to the magnitude p of its momentum by $\lambda = h/p$ (Equation 29.8), where h is Planck's constant. If the speed of the electron is much less than the speed of light, the magnitude of the electron's momentum is given by $p = mv$ (Equation 7.2). Thus, the de Broglie wavelength can be written as $\lambda = h/(mv)$.

When the electron is at rest, it has electric potential energy, but no kinetic energy. The electric potential energy EPE is given by EPE $= eV$ (Equation 19.3), where e is the magnitude of the charge on the electron and V is the potential difference. When the electron reaches its maximum speed, it has no potential energy, but its kinetic energy is $\frac{1}{2}mv^2$. The conservation of energy states that the final total energy of the electron equals the initial total energy:

$$\underbrace{\tfrac{1}{2}mv^2}_{\substack{\text{Final total}\\ \text{energy}}} = \underbrace{eV}_{\substack{\text{Initial total}\\ \text{energy}}}$$

Solving this equation for the final speed gives $v = \sqrt{2eV/m}$. Substituting this expression for v into $\lambda = h/(mv)$ gives $\lambda = h/\sqrt{2meV}$.

SOLUTION After accelerating through the potential difference, the electron has a de Broglie wavelength of

$$\lambda = \frac{h}{\sqrt{2meV}} = \frac{6.63\times10^{-34}\ \text{J}\cdot\text{s}}{\sqrt{2\left(9.11\times10^{-31}\ \text{kg}\right)\left(1.60\times10^{-19}\ \text{C}\right)\left(418\ \text{V}\right)}} = \boxed{6.01\times10^{-11}\ \text{m}}$$

37. **REASONING** Suppose the object is moving along the $+y$ axis. The uncertainty in the object's position is $\Delta y = 2.5$ m. The minimum uncertainty Δp_y in the object's momentum is specified by the Heisenberg uncertainty principle (Equation 29.10) in the form $(\Delta p_y)(\Delta y) = h/(4\pi)$. Since momentum is mass m times velocity v, the uncertainty in the velocity Δv is related to the uncertainty in the momentum by $\Delta v = (\Delta p_y)/m$.

SOLUTION
a. Using the uncertainty principle, we find the minimum uncertainty in the momentum as follows:

$$\left(\Delta p_y\right)\left(\Delta y\right) = \frac{h}{4\pi}$$

$$\Delta p_y = \frac{h}{4\pi\,\Delta y} = \frac{6.63\times10^{-34}\ \text{J}\cdot\text{s}}{4\pi\left(2.5\ \text{m}\right)} = \boxed{2.1\times10^{-35}\ \text{kg}\cdot\text{m/s}}$$

b. For a golf ball this uncertainty in momentum corresponds to an uncertainty in velocity that is given by

$$\Delta v_y = \frac{\Delta p_y}{m} = \frac{2.1\times10^{-35}\ \text{kg}\cdot\text{m/s}}{0.045\ \text{kg}} = \boxed{4.7\times10^{-34}\ \text{m/s}}$$

c. For an electron this uncertainty in momentum corresponds to an uncertainty in velocity that is given by

$$\Delta v_y = \frac{\Delta p_y}{m} = \frac{2.1\times10^{-35}\ \text{kg}\cdot\text{m/s}}{9.11\times10^{-31}\ \text{kg}} = \boxed{2.3\times10^{-5}\ \text{m/s}}$$

---

39. **REASONING** The Heisenberg uncertainty principle specifies the relationship between the uncertainty Δy in a particle's position and the uncertainty Δp_y in the particle's linear momentum. This principle is stated as follows: $(\Delta p_y)(\Delta y) \geq h/(4\pi)$ (Equation 29.10), where $h = 6.63 \times 10^{-34}$ J·s is Planck's constant. Since we seek the minimum uncertainty in the speed (which is proportional to the momentum) of the oxygen molecule, we will use the equals sign in Equation 29.10 and omit the greater-than symbol ($>$). For our purposes in this problem, then, the uncertainty principle becomes

$$(\Delta p_y)(\Delta y) = \frac{h}{4\pi} \tag{1}$$

We can use Equation (1) to calculate the minimum uncertainty in the speed, because the magnitude p_y of the momentum is related to the speed v_y according to $p_y = mv_y$ (Equation 7.2), where m is the mass of the oxygen molecule.

SOLUTION Substituting $p_y = mv_y$ (Equation 7.2) into Equation (1), we obtain

$$(\Delta p_y)(\Delta y) = \frac{h}{4\pi} \qquad \text{or} \qquad (m\Delta v_y)(\Delta y) = \frac{h}{4\pi} \qquad (2)$$

Solving Equation (2) for the uncertainty Δv_y in the speed, we find that

$$\Delta v_y = \frac{h}{4\pi m \Delta y} = \frac{6.63 \times 10^{-34} \text{ J} \cdot \text{s}}{4\pi \left(5.3 \times 10^{-26} \text{ kg}\right)\left(0.12 \times 10^{-3} \text{ m}\right)} = \boxed{8.3 \times 10^{-6} \text{ m/s}}$$

---

45. ***REASONING AND SOLUTION*** The de Broglie wavelength λ is given by Equation 29.8 as $\lambda = h/p$, where p is the magnitude of the momentum of the particle. The magnitude of the momentum is $p = mv$, where m is the mass and v is the speed of the particle. Using this expression in Equation 29.8, we find that $\lambda = h/(mv)$, or

$$v = \frac{h}{m\lambda} = \frac{6.63 \times 10^{-34} \text{ J} \cdot \text{s}}{\left(1.67 \times 10^{-27} \text{ kg}\right)\left(1.30 \times 10^{-14} \text{ m}\right)} = 3.05 \times 10^7 \text{ m/s}$$

The kinetic energy of the proton is

$$KE = \tfrac{1}{2}mv^2 = \tfrac{1}{2}(1.67 \times 10^{-27} \text{ kg})(3.05 \times 10^7 \text{ m/s})^2 = \boxed{7.77 \times 10^{-13} \text{ J}}$$

---

53. ***CONCEPTS***

(i) According to Equation 29.8, the de Broglie wavelength is given by $\lambda = h/p$, where h is Planck's constant. The greater the momentum, the smaller is the de Broglie wavelength, and vice versa.

(ii) According to Equation 7.2, the magnitude of the momentum is $p = mv$, where v is the speed. Equation 6.2 gives the kinetic energy as $KE = \tfrac{1}{2}mv^2$, which can be solved to show that $v = \sqrt{2(KE)/m}$. Substituting this result into Equation 7.2 shows that

$$p = m\sqrt{\frac{2(KE)}{m}} = \sqrt{2m(KE)}$$

(iii) According to $\lambda = h/p$, the particle with the greater wavelength is the one with the smaller momentum. However $p = \sqrt{2m(KE)}$ indicates that, for a given kinetic energy, the particle with the smaller momentum is the one with the smaller mass. The masses of the electron and proton are $m_{\text{electron}} = 9.11 \times 10^{-31}$ kg and $m_{\text{proton}} = 1.67 \times 10^{-27}$ kg. Thus, the electron, with its smaller mass, has the greater de Broglie wavelength.

CALCULATIONS

Using Equation 29.8 for the de Broglie wavelength and the fact that the magnitude of the momentum is related to the kinetic energy by $p = \sqrt{2m(\text{KE})}$, we have

$$\lambda = \frac{h}{p} = \frac{h}{\sqrt{2m(\text{KE})}}$$

Applying this result to the electron and the proton gives

$$\frac{\lambda_{\text{electron}}}{\lambda_{\text{proton}}} = \frac{h/\sqrt{2m_{\text{electron}}(\text{KE})}}{h/\sqrt{2m_{\text{proton}}(\text{KE})}} = \sqrt{\frac{m_{\text{proton}}}{m_{\text{electron}}}} = \sqrt{\frac{1.67 \times 10^{-27} \text{ kg}}{9.11 \times 10^{-31} \text{ kg}}} = \boxed{42.8}$$

As expected, the wavelength for the electron is greater than that for the proton.

CHAPTER 30 | *THE NATURE OF THE ATOM*

1. ***REASONING*** Assuming that the hydrogen atom is a sphere of radius r_{atom}, its volume V_{atom} is given by $\frac{4}{3}\pi r_{atom}^3$. Similarly, if the radius of the nucleus is $r_{nucleus}$, the volume $V_{nucleus}$ is given by $\frac{4}{3}\pi r_{nucleus}^3$.

SOLUTION
a. According to the given data, the nuclear dimensions are much smaller than the orbital radius of the electron; therefore, we can treat the nucleus as a point about which the electron orbits. The electron is normally at a distance of about 5.3×10^{-11} m from the nucleus, so we can treat the atom as a sphere of radius $r_{atom} = 5.3 \times 10^{-11}$ m. The volume of the atom is

$$V_{atom} = \frac{4}{3}\pi r_{atom}^3 = \frac{4}{3}\pi \left(5.3 \times 10^{-11}\ \text{m}\right)^3 = \boxed{6.2 \times 10^{-31}\ \text{m}^3}$$

b. Similarly, since the nucleus has a radius of approximately $r_{nucleus} = 1 \times 10^{-15}$ m, its volume is

$$V_{nucleus} = \frac{4}{3}\pi r_{nucleus}^3 = \frac{4}{3}\pi \left(1 \times 10^{-15}\ \text{m}\right)^3 = \boxed{4 \times 10^{-45}\ \text{m}^3}$$

c. The percentage of the atomic volume occupied by the nucleus is

$$\frac{V_{nucleus}}{V_{atom}} \times 100\% = \frac{\frac{4}{3}\pi r_{nucleus}^3}{\frac{4}{3}\pi r_{atom}^3} \times 100\% = \left(\frac{1 \times 10^{-15}\ \text{m}}{5.3 \times 10^{-11}\ \text{m}}\right)^3 \times 100\% = \boxed{7 \times 10^{-13}\ \%}$$

---

5. ***REASONING*** The distance of closest approach can be obtained by setting the kinetic energy KE of the α particle equal to the electric potential energy EPE of the α particle. According to Equation 19.3, we have that EPE = (2*e*)*V*, where 2*e* is the charge on the α particle and *V* is the electric potential created by a gold nucleus. According to Equation 19.6, the electric potential of the gold nucleus (charge = *Ze* = 79*e*) is *V* = *k*(79*e*)/*r*, where *r* is the distance between the α particle and the gold nucleus. Therefore, we have that

$$\text{EPE} = (2e)V = (2e)\frac{k(79e)}{r} \tag{1}$$

In this expression, we note that $k = 8.99 \times 10^9\ \text{N} \cdot \text{m}^2/\text{C}^2$ and $e = 1.602 \times 10^{-19}$ C.

SOLUTION Solving Equation (1) for the distance *r* we obtain

$$r = \frac{(2e)k(79e)}{\text{EPE}} = \frac{(8.99\times10^9 \ \text{N}\cdot\text{m}^2/\text{C}^2)2(79)(1.602\times10^{-19} \ \text{C})^2}{5.0\times10^{-13} \ \text{J}} = \boxed{7.3\times10^{-14} \ \text{m}}$$

7. **REASONING** According to the Bohr model, the energy E_n (in eV) of the electron in an orbit is given by Equation 30.13: $E_n = -13.6(Z^2/n^2)$. In order to find the principal quantum number of the state in which the electron in a doubly ionized lithium atom Li^{2+} has the same total energy as a ground state electron in a hydrogen atom, we equate the right hand sides of Equation 30.13 for the hydrogen atom and the lithium ion. This gives

$$-(13.6)\left(\frac{Z^2}{n^2}\right)_H = -(13.6)\left(\frac{Z^2}{n^2}\right)_{Li} \qquad \text{or} \qquad n_{Li}^2 = \left(\frac{n^2}{Z^2}\right)_H Z_{Li}^2$$

This expression can be evaluated to find the desired principal quantum number.

SOLUTION For hydrogen, $Z=1$, and $n=1$ for the ground state. For lithium Li^{2+}, $Z=3$. Therefore,

$$n_{Li}^2 = \left(\frac{n^2}{Z^2}\right)_H Z_{Li}^2 = \left(\frac{1^2}{1^2}\right)(3^2) \qquad \text{or} \qquad \boxed{n_{Li} = 3}$$

11. **REASONING** According to Equation 30.14, the wavelength λ emitted by the hydrogen atom when it makes a transition from the level with n_i to the level with n_f is given by

$$\frac{1}{\lambda} = \frac{2\pi^2 mk^2 e^4}{h^3 c} (Z^2)\left(\frac{1}{n_f^2} - \frac{1}{n_i^2}\right) \qquad \text{with} \quad n_i, n_f = 1, \ 2, \ 3,\ldots \quad \text{and} \quad n_i > n_f$$

where $2\pi^2 mk^2 e^4/(h^3 c) = 1.097\times10^7 \ \text{m}^{-1}$ and $Z=1$ for hydrogen. Once the wavelength for the particular transition in question is determined, Equation 29.2 ($E=hf=hc/\lambda$) can be used to find the energy of the emitted photon.

SOLUTION In the Paschen series, $n_f = 3$. Using the above expression with $Z=1$, $n_i=7$ and $n_f = 3$, we find that

$$\frac{1}{\lambda} = (1.097\times10^7 \ \text{m}^{-1})(1^2)\left(\frac{1}{3^2} - \frac{1}{7^2}\right) \qquad \text{or} \qquad \lambda = 1.005\times10^{-6} \ \text{m}$$

The photon energy is

$$E = \frac{hc}{\lambda} = \frac{(6.63\times10^{-34} \ \text{J}\cdot\text{s})(3.00\times10^8 \ \text{m/s})}{1.005\times10^{-6} \ \text{m}} = \boxed{1.98\times10^{-19} \ \text{J}}$$

17. **REASONING** A wavelength of 410.2 nm is emitted by the hydrogen atoms in a high-voltage discharge tube. This transition lies in the visible region (380–750 nm) of the hydrogen spectrum. Thus, we can conclude that the transition is in the Balmer series and, therefore, that $n_f = 2$. The value of n_i can be found using Equation 30.14, according to which the Balmer series transitions are given by

$$\frac{1}{\lambda} = R\left(1^2\right)\left(\frac{1}{2^2} - \frac{1}{n_i^2}\right) \qquad n = 3, 4, 5, \ldots$$

This expression may be solved for n_i for the energy transition that produces the given wavelength.

SOLUTION Solving for n_i, we find that

$$n_i = \frac{1}{\sqrt{\frac{1}{2^2} - \frac{1}{R\lambda}}} = \frac{1}{\sqrt{\frac{1}{2^2} - \frac{1}{(1.097\times10^7 \text{ m}^{-1})(410.2\times10^{-9} \text{ m})}}} = 6$$

Therefore, the initial and final states are identified by $\boxed{n_i = 6 \text{ and } n_f = 2}$.

19. **REASONING** For either series of lines, the wavelengths λ can be obtained from $\frac{1}{\lambda} = \left(1.097\times10^7 \text{ m}^{-1}\right)(Z)^2\left(\frac{1}{n_f^2} - \frac{1}{n_i^2}\right)$ (Equation 30.14), where $Z = 1$ is the number of protons in the hydrogen nucleus, $n_i, n_f = 1, 2, 3, \ldots$, and $n_i > n_f$.

SOLUTION For the Paschen series, $n_f = 3$. The range of wavelengths occurs for values of $n_i = 4$ to $n_i = \infty$. Using Equation 30.14, we find that the shortest wavelength occurs for $n_i = \infty$ and is

$$\frac{1}{\lambda} = \left(1.097\times10^7 \text{ m}^{-1}\right)(1)^2\left(\frac{1}{n_f^2} - \frac{1}{n_i^2}\right) = \left(1.097\times10^7 \text{ m}^{-1}\right)\left(\frac{1}{3^2}\right) \quad \text{or} \quad \underline{\lambda = 8.204\times10^{-7} \text{ m}}$$

Shortest wavelength in Paschen series

The longest wavelength in the Paschen series occurs for $n_i = 4$ and is

$$\frac{1}{\lambda} = \left(1.097\times10^7 \text{ m}^{-1}\right)\left(\frac{1}{3^2} - \frac{1}{4^2}\right) \quad \text{or} \quad \underline{\lambda = 1.875\times10^{-6} \text{ m}}$$

Longest wavelength in Paschen series

For the Brackett series, $n_f = 4$. The range of wavelengths occurs for values of $n_i = 5$ to $n_i = \infty$. Using Equation 30.14, we find that the shortest wavelength occurs for $n_i = \infty$ and is

$$\frac{1}{\lambda} = \left(1.097 \times 10^7 \text{ m}^{-1}\right)(1)^2 \left(\frac{1}{n_f^2} - \frac{1}{n_i^2}\right) = \left(1.097 \times 10^7 \text{ m}^{-1}\right)\left(\frac{1}{4^2}\right) \quad \text{or} \quad \underbrace{\lambda = 1.459 \times 10^{-6} \text{ m}}_{\substack{\text{Shortest wavelength in} \\ \text{Brackett series}}}$$

The longest wavelength in the Brackett series occurs for $n_i = 5$ and is

$$\frac{1}{\lambda} = \left(1.097 \times 10^7 \text{ m}^{-1}\right)\left(\frac{1}{4^2} - \frac{1}{5^2}\right) \quad \text{or} \quad \underbrace{\lambda = 4.051 \times 10^{-6} \text{ m}}_{\substack{\text{Longest wavelength in} \\ \text{Brackett series}}}$$

Since the longest wavelength in the Paschen series falls within the Brackett series, the wavelengths of the two series overlap.

---

27. **REASONING** The maximum value for the magnetic quantum number is $m_l = 1$; thus, in state A, $l = 2$, while in state B, $l = 1$. According to the quantum mechanical theory of angular momentum, the magnitude of the orbital angular momentum for a state of given l is $L = \sqrt{l(l+1)}\,(h/2\pi)$ (Equation 30.15). This expression can be used to form the ratio L_A / L_B of the magnitudes of the orbital angular momenta for the two states.

SOLUTION Using Equation 30.15, we find that

$$\frac{L_A}{L_B} = \frac{\sqrt{2(2+1)}\,\dfrac{h}{2\pi}}{\sqrt{1(1+1)}\,\dfrac{h}{2\pi}} = \sqrt{\frac{6}{2}} = \sqrt{3} = \boxed{1.732}$$

---

29. **REASONING** The total energy E_n for a hydrogen atom in the quantum mechanical picture is the same as in the Bohr model and is given by Equation 30.13:

$$E_n = -(13.6 \text{ eV})\frac{1}{n^2} \tag{30.13}$$

Thus, we need to determine values for the principal quantum number n if we are to calculate the three smallest possible values for E. Since the maximum value of the orbital quantum number ℓ is $n - 1$, we can obtain a minimum value for n as $n_{min} = \ell + 1$. But how to obtain ℓ? It can be obtained, because the problem statement gives the maximum value of L_z, the z component of the angular momentum. According to Equation 30.16, L_z is

$$L_z = m_1 \frac{h}{2\pi} \qquad (30.16)$$

where m_ℓ is the magnetic quantum number and h is Planck's constant. For a given value of ℓ the allowed values for m_ℓ are as follows: $-\ell, \ldots, -2, -1, 0, +1, +2, \ldots, +\ell$. Thus, the maximum value of m_ℓ is ℓ, and we can use Equation 30.16 to calculate the maximum value of m_ℓ from the maximum value given for L_z.

SOLUTION Solving Equation 30.16 for m_ℓ gives

$$m_1 = \frac{2\pi L_z}{h} = \frac{2\pi\left(4.22\times10^{-34} \text{ J}\cdot\text{s}\right)}{6.63\times10^{-34} \text{ J}\cdot\text{s}} = 4$$

As explained in the **REASONING**, this maximum value for m_ℓ indicates that $\ell = 4$. Therefore, a minimum value for n is

$$n_{min} = 1 + 1 = 4 + 1 = 5 \quad \text{or} \quad n \geq 5$$

This means that the three energies we seek correspond to $n = 5$, $n = 6$, and $n = 7$. Using Equation 30.13, we find them to be

[n = 5] $\qquad E_5 = -\left(13.6 \text{ eV}\right)\frac{1}{5^2} = \boxed{-0.544 \text{ eV}}$

[n = 6] $\qquad E_6 = -\left(13.6 \text{ eV}\right)\frac{1}{6^2} = \boxed{-0.378 \text{ eV}}$

[n = 7] $\qquad E_7 = -\left(13.6 \text{ eV}\right)\frac{1}{7^2} = \boxed{-0.278 \text{ eV}}$

---

35. **REASONING** In the theory of quantum mechanics, there is a selection rule that restricts the initial and final values of the orbital quantum number l . The selection rule states that when an electron makes a transition between energy levels, the value of l may not remain the same or increase or decrease by more than one. In other words, the rule requires that $\Delta l = \pm 1$.

    SOLUTION
    a. For the transition $2s \rightarrow 1s$, the electron makes a transition from the 2s state $\left(n = 2, l = 0\right)$ to the 1s state $\left(n = 1, l = 0\right)$. Since the value of l is the same in both states, $\Delta l = 0$, and we can conclude that this energy level transition is $\boxed{\text{not allowed}}$.

b. For the transition $2p \rightarrow 1s$, the electron makes a transition from the 2p state $(n=2, l=1)$ to the 1s state $(n=1, l=0)$. The value of l changes so that $\Delta l = 0 - 1 = -1$, and we can conclude that this energy level transition is allowed .

c. For the transition $4p \rightarrow 2p$, the electron makes a transition from the 4p state $(n=4, l=1)$ to the 2p state $(n=2, l=1)$. Since the value of l is the same in both states, $\Delta l = 0$, and we can conclude that this energy level transition is not allowed .

d. For the transition $4s \rightarrow 2p$, the electron makes a transition from the 4s state $(n=4, l=0)$ to the 2p state $(n=2, l=1)$. The value of l changes so that $\Delta l = 1 - 0 = +1$, and we can conclude that this energy level transition is allowed .

e. For the transition $3d \rightarrow 3s$, the electron makes a transition from the 3d state $(n=3, l=2)$ to the 3s state $(n=3, l=0)$. The value of l changes so that $\Delta l = 0 - 2 = -2$, and we can conclude that this energy level transition is not allowed .

41. ***REASONING*** As discussed in text Example 11, the wavelengths of the K_α electrons depend upon the quantity $Z - 1$, where Z is the atomic number of the target. The Bohr model predicts that the energies of the electrons are given by $E_n = -(13.6 \text{ eV})\dfrac{Z^2}{n^2}$ (Equation 30.13), where n can take on any integer value greater than zero. However, the K_α photons correspond to a transition to the $n = 1$ shell from the $n = 2$ shell, and during this transition the nuclear charge is partly screened by the electron that remains in the $n = 1$ shell, as discussed in text Example 10. We will replace Z with $Z - 1$ in Equation 30.13 to account for this screening. Therefore, the energy of the electron's initial and final states are calculated from

$$E_n = -(13.6 \text{ eV})\frac{(Z-1)^2}{n^2} \qquad (1)$$

The energy E of the emitted X-ray is equal to the higher initial energy E_2 of the electron minus the lower final energy E_1:

$$E = E_2 - E_1 \qquad (2)$$

SOLUTION Substituting Equation (1) into Equation (2), we obtain

$$E = -(13.6 \text{ eV})\frac{(Z-1)^2}{2^2} - \left[-(13.6 \text{ eV})\frac{(Z-1)^2}{1^2} \right] = (13.6 \text{ eV})(Z-1)^2 \left(\frac{1}{1^2} - \frac{1}{2^2} \right) \qquad (3)$$

$$= \frac{3}{4}(13.6 \text{ eV})(Z-1)^2$$

Solving Equation (3) for $(Z-1)^2$ and taking the square root of both sides yields

$$(Z-1)^2 = \frac{4E}{3(13.6 \text{ eV})} \qquad \text{or} \qquad Z = \sqrt{\frac{4E}{3(13.6 \text{ eV})}} + 1$$

The prediction of the Bohr model is, then,

$$Z = \sqrt{\frac{4(9890 \text{ eV})}{3(13.6 \text{ eV})}} + 1 = 32.1$$

The closest integer to this result is $Z = \boxed{32}$; the atomic number of $\boxed{\text{germanium (Ge)}}$.

---

43. ***REASONING*** In the spectrum of X-rays produced by the tube, the cutoff wavelength λ_0 and the voltage V of the tube are related according to Equation 30.17, $V = hc/(e\lambda_0)$. Since the voltage is increased from zero until the K_α X-ray just appears in the spectrum, it follows that $\lambda_0 = \lambda_\alpha$ and $V = hc/(e\lambda_\alpha)$. Using Equation 30.14 for $1/\lambda_\alpha$, we find that

$$V = \frac{hc}{e\lambda_\alpha} = \frac{hcR(Z-1)^2}{e}\left(\frac{1}{1^2} - \frac{1}{2^2}\right)$$

In this expression we have replaced Z with $Z-1$, in order to account for shielding, as explained in Example 11 in the text.

SOLUTION The desired voltage is, then,

$$V = \frac{\left(6.63\times10^{-34} \text{ J·s}\right)\left(3.00\times10^8 \text{ m/s}\right)\left(1.097\times10^7 \text{ m}^{-1}\right)(47-1)^2}{\left(1.60\times10^{-19} \text{ C}\right)}\left(\frac{1}{1^2} - \frac{1}{2^2}\right) = \boxed{21\ 600 \text{ V}}$$

---

45. ***REASONING*** The number of photons emitted by the laser will be equal to the total energy carried in the beam divided by the energy per photon.

SOLUTION The total energy carried in the beam is, from the definition of power,

$$E_{\text{total}} = Pt = (1.5 \text{ W})(0.050 \text{ s}) = 0.075 \text{ J}$$

The energy of a single photon is given by Equations 29.2 and 16.1 as

$$E_{\text{photon}} = hf = \frac{hc}{\lambda} = \frac{\left(6.63 \times 10^{-34} \text{ J} \cdot \text{s}\right)\left(3.00 \times 10^8 \text{ m/s}\right)}{514 \times 10^{-9} \text{ m}} = 3.87 \times 10^{-19} \text{ J}$$

where we have used the fact that 514 nm $= 514 \times 10^{-9}$ m. Therefore, the number of photons emitted by the laser is

$$\frac{E_{\text{total}}}{E_{\text{photon}}} = \frac{0.075 \text{ J}}{3.87 \times 10^{-19} \text{ J/photon}} = \boxed{1.9 \times 10^{17} \text{ photons}}$$

---

55. **REASONING AND SOLUTION**
a. The longest wavelength in the Pfund series occurs for the transition $n = 6$ to $n = 5$, so that according to Equation 30.14 with $Z = 1$, we have

$$\frac{1}{\lambda} = R\left(\frac{1}{5^2} - \frac{1}{n^2}\right) = \left(1.097 \times 10^7 \text{ m}^{-1}\right)\left(\frac{1}{5^2} - \frac{1}{6^2}\right) \quad \text{or} \quad \boxed{\lambda = 7458 \text{ nm}}$$

b. The shortest wavelength occurs when $1/n^2 = 0$, so that

$$\frac{1}{\lambda} = R\left(\frac{1}{5^2} - \frac{1}{n^2}\right) = \left(1.097 \times 10^7 \text{ m}^{-1}\right)\left(\frac{1}{5^2}\right) \quad \text{or} \quad \boxed{\lambda = 2279 \text{ nm}}$$

c. The lines in the Pfund series occur in the $\boxed{\text{infrared region}}$.

---

59. **REASONING** Singly ionized helium, He$^+$, is a hydrogen-like species with $Z = 2$. The wavelengths of the series of lines produced when the electron makes a transition from higher energy levels into the $n_f = 4$ level are given by Equation 30.14 with $Z = 2$ and $n_f = 4$:

$$\frac{1}{\lambda} = \left(1.097 \times 10^7 \text{ m}^{-1}\right)\left(2^2\right)\left(\frac{1}{4^2} - \frac{1}{n_i^2}\right)$$

SOLUTION Solving this expression for n_i gives

$$n_i = \left[\frac{1}{4^2} - \frac{1}{4\lambda\left(1.097 \times 10^7 \text{ m}^{-1}\right)}\right]^{-1/2}$$

Evaluating this expression at the limits of the range for λ, we find that $n_i = 19.88$ for $\lambda = 380$ nm, and $n_i = 5.58$ for $\lambda = 750$ nm. Therefore, the values of n_i for energy levels from which the electron makes the transitions that yield wavelengths in the range between 380 nm and 750 nm are $\boxed{6 \leq n_i \leq 19}$.

61. **REASONING AND SOLUTION** The shortest wavelength, λ_s, occurs when $n_i = \infty$, so that $1/n_i = 0$. In that case Equation 30.14 becomes

$$\frac{1}{\lambda_s} = \frac{RZ^2}{n_f^2} \quad \text{or} \quad RZ^2 = \frac{n_f^2}{\lambda_s}$$

The longest wavelength in the series occurs when $n_i = n_f + 1$.

$$\frac{1}{\lambda_1} = RZ^2\left(\frac{1}{n_f^2} - \frac{1}{n_i^2}\right) = \left(\frac{n_f^2}{\lambda_s}\right)\left(\frac{1}{n_f^2} - \frac{1}{(n_f+1)^2}\right)$$

A little algebra gives n_f as follows:

$$\frac{\lambda_1}{\lambda_s} = \frac{41.02\times10^{-9}\text{ m}}{22.79\times10^{-9}\text{ m}} = \frac{(n_f+1)^2}{2n_f+1}$$

Rearranging this equation gives

$$n_f^2 - 1.600\,n_f - 0.800 = 0$$

Solving the quadratic equation yields only one positive root, which is $n_f = 2$. Therefore, $1/\lambda_s = 1/(22.79 \times 10^{-9}\text{ m}) = RZ^2/n_f^2 = RZ^2/4$, which gives $Z = 4$. As a result, the next-to-the-longest wavelength is:

$$\frac{1}{\lambda} = R(4)^2\left(\frac{1}{2^2} - \frac{1}{4^2}\right) \quad \text{or} \quad \lambda = 30.39\times10^{-9}\text{ m} = \boxed{30.39\text{ nm}}$$

63. **CONCEPTS**

(i) A K_α photon is produced when an electron in a metal atom jumps from the higher-energy $n = 2$ state to the lower energy $n = 1$ state (see transition A in the figure). The energy E of the photon is the difference between the energies of these two states: $E = E_2 - E_1$.

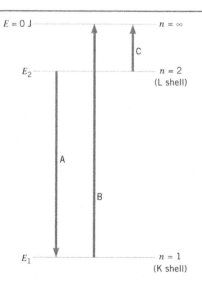

(ii) It takes energy to produce a K_α photon, and this energy comes from electrons striking the target of the X-ray tube. According to Equation 19.3, the energy possessed by each electron when it arrives at the target is eV, where e is the magnitude of the electron's charge and V is the potential difference or voltage across the tube. To cause the target to emit a K_α X-ray photon, the impinging electron must have enough energy to create a vacancy in the K shell. The vacancy is created because the impinging electron provides energy to move a K-shell electron into a higher energy level or to remove it entirely from the atom. In the present case, with a vacancy in the L shell, the impinging electron can create the K-shell vacancy if it has the energy needed to elevate a K-shell electron to the L shell, the energy needed being $E_2 - E_1$ (see the figure). This, then, is the minimum energy that the electron must have. Correspondingly, the minimum voltage across the tube must be such that $eV_{min} = E_2 - E_1$, or $V_{min} = (E_2 - E_1)/e$.

(iii) The K-shell ionization energy is the energy needed to remove an electron in the K shell ($n_i = 1$) completely from the atom (see transition B in the figure). The removed electron is assumed to have no kinetic energy and is infinitely far away ($n_f = \infty$), so that it has no electrical potential energy. Thus the total energy of the removed electron is zero. Similarly, the L-shell ionization energy is the energy needed to remove an L-shell electron ($n_i = 2$) completely from the atom (see transition C in the figure).

(iv) As transitions B and C in the figure suggest, the difference between the two ionization energies is equal to the difference between the total energy E_2 of the L shell and the total energy E_1 of the K shell.

CALCULATIONS

(a) The minimum voltage across the X-ray tube is $V_{min} = (E_2 - E_1)/e$. But the difference $E_2 - E_1$ in energies between the $n = 2$ and $n = 1$ states is equal to the difference in energies between the K-shell and L-shell ionization energies, or $8979 \text{ eV} - 951 \text{ eV} = 8028 \text{ eV}$. Thus, the minimum voltage is

$$V_{min} = \frac{E_2 - E_1}{e} = \frac{(8028 \text{ eV})\left(\dfrac{1.602 \times 10^{-19} \text{ J}}{1 \text{ eV}}\right)}{1.602 \times 10^{-19} \text{ C}} = \boxed{8028 \text{ V}}$$

(b) The wavelength of the K_α photon is $\lambda = c/f$ (Equation 16.1), where f is its frequency. The frequency is related to the energy $E_2 - E_1$ of the photon by Equation 29.2, $f = (E_2 - E_1)/h$, where h is Planck's constant. Combining these relations gives:

$$\lambda = \frac{c}{f} = \frac{c}{(E_2 - E_1)/h} = \frac{hc}{E_2 - E_1}$$

$$= \frac{\left(6.626 \times 10^{-34} \text{ J} \cdot \text{s}\right)\left(2.998 \times 10^{8} \text{ m/s}\right)}{\left(8028 \text{ eV}\right)\left(\dfrac{1.602 \times 10^{-19} \text{ J}}{1 \text{ eV}}\right)} = \boxed{1.545 \times 10^{-10} \text{ m}}$$

CHAPTER 31 | NUCLEAR PHYSICS AND RADIOACTIVITY

1. **REASONING** For an element whose chemical symbol is X, the symbol for the nucleus is $_Z^A X$, where A represents the total number of protons and neutrons (the nucleon number) and Z represents the number of protons in the nucleus (the atomic number). The number of neutrons N is related to A and Z by Equation 31.1: $A = Z + N$.

SOLUTION For the nucleus $_{82}^{208} Pb$, we have $Z = 82$ and $A = 208$.

a. The net electrical charge of the nucleus is equal to the total number of protons multiplied by the charge on a single proton. Since the $_{82}^{208} Pb$ nucleus contains 82 protons, the net electrical charge of the $_{82}^{208} Pb$ nucleus is

$$q_{net} = (82)(+1.60 \times 10^{-19} C) = \boxed{+1.31 \times 10^{-17} C}$$

b. The number of neutrons is $N = A - Z = 208 - 82 = \boxed{126}$.

c. By inspection, the number of nucleons is $A = \boxed{208}$.

d. The approximate radius of the nucleus can be found from Equation 31.2, namely

$$r = (1.2 \times 10^{-15} \text{ m}) A^{1/3} = (1.2 \times 10^{-15} \text{ m})(208)^{1/3} = \boxed{7.1 \times 10^{-15} \text{ m}}$$

e. The nuclear density is the mass per unit volume of the nucleus. The total mass of the nucleus can be found by multiplying the mass $m_{nucleon}$ of a single nucleon by the total number A of nucleons in the nucleus. Treating the nucleus as a sphere of radius r, the nuclear density is

$$\rho = \frac{m_{total}}{V} = \frac{m_{nucleon} A}{\frac{4}{3} \pi r^3} = \frac{m_{nucleon} A}{\frac{4}{3} \pi \left[(1.2 \times 10^{-15} \text{ m}) A^{1/3} \right]^3} = \frac{m_{nucleon}}{\frac{4}{3} \pi (1.2 \times 10^{-15} \text{ m})^3}$$

Therefore,

$$\rho = \frac{1.67 \times 10^{-27} \text{ kg}}{\frac{4}{3} \pi (1.2 \times 10^{-15} \text{ m})^3} = \boxed{2.3 \times 10^{17} \text{ kg} / \text{m}^3}$$

5. ***REASONING***

a. The number of protons in a given nucleus $^A_Z X$ is specified by its atomic number Z.

b. The number N of neutrons in a given nucleus $^A_Z X$ is equal to the nucleon number A (the number of protons and neutrons) minus the atomic number Z (the number of protons): $N = A - Z$ (Equation 31.1).

c. In an electrically neutral niobium atom, the number of electrons in orbit about the nucleus is equal to the number of protons in the nucleus.

SOLUTION

a. The number of protons in the uranium $^{238}_{92} U$ nucleus is $Z = \boxed{92}$.

b. The number N of neutrons in the $^{202}_{80} Hg$ nucleus is $N = A - Z = 202 - 80 = \boxed{122}$.

c. The number of electrons that orbit the $^{93}_{41} Nb$ nucleus in the neutral niobium atom is equal to the number of protons in the nucleus, or $\boxed{41}$.

---

9. ***REASONING*** According to Equation 31.2, the radius of a nucleus in meters is $r = \left(1.2 \times 10^{-15} \text{ m}\right) A^{1/3}$, where A is the nucleon number. If we treat the neutron star as a uniform sphere, its density (Equation 11.1) can be written as

$$\rho = \frac{M}{V} = \frac{M}{\frac{4}{3}\pi r^3}$$

Solving for the radius r, we obtain,

$$r = \sqrt[3]{\frac{M}{\frac{4}{3}\pi\rho}}$$

This expression can be used to find the radius of a neutron star of mass M and density ρ.

SOLUTION As discussed in Conceptual Example 1, nuclear densities have the same approximate value in all atoms. If we consider a uniform spherical nucleus, then the density of nuclear matter is approximately given by

$$\rho = \frac{M}{V} \approx \frac{A \times (\text{mass of a nucleon})}{\frac{4}{3}\pi r^3} = \frac{A \times (\text{mass of a nucleon})}{\frac{4}{3}\pi \left[(1.2 \times 10^{-15} \text{ m}) A^{1/3} \right]^3}$$

$$= \frac{1.67 \times 10^{-27} \text{ kg}}{\frac{4}{3}\pi (1.2 \times 10^{-15} \text{ m})^3} = 2.3 \times 10^{17} \text{ kg/m}^3$$

The mass of the sun is 1.99×10^{30} kg (see inside of the front cover of the text). Substituting values into the expression for r determined above, we find

$$r = \sqrt[3]{\frac{(0.40)(1.99 \times 10^{30} \text{ kg})}{\frac{4}{3}\pi(2.3 \times 10^{17} \text{ kg/m}^3)}} = \boxed{9.4 \times 10^3 \text{ m}}$$

11. **REASONING** To obtain the binding energy, we will calculate the mass defect and then use the fact that 1 u is equivalent to 931.5 MeV. The atomic mass given for $^7_3$Li includes the 3 electrons in the neutral atom. Therefore, when computing the mass defect, we must account for these electrons. We do so by using the atomic mass of 1.007 825 u for the hydrogen atom $^1_1$H, which also includes the single electron, instead of the atomic mass of a proton.

SOLUTION Noting that the number of neutrons is $7 - 3 = 4$, we obtain the mass defect Δm as follows:

$$\Delta m = \underbrace{3(1.007\ 825\ \text{u})}_{\substack{\text{3 hydrogen atoms} \\ \text{(protons plus electrons)}}} + \underbrace{4(1.008\ 665\ \text{u})}_{\text{4 neutrons}} - \underbrace{7.016\ 003\ \text{u}}_{\substack{\text{Intact lithium atom} \\ \text{(including 3 electrons)}}} = 4.2132 \times 10^{-2}\ \text{u}$$

Since 1 u is equivalent to 931.5 MeV, the binding energy is

$$\text{Binding energy} = \left(4.2132 \times 10^{-2}\ \text{u}\right)\left(\frac{931.5\ \text{MeV}}{1\ \text{u}}\right) = \boxed{39.25\ \text{MeV}}$$

17. **REASONING** Since we know the difference in binding energies for the two isotopes, we can determine the corresponding mass defect. Also knowing that the isotope with the larger binding energy contains one more neutron than the other isotope gives us enough information to calculate the atomic mass difference between the two isotopes.

SOLUTION The mass defect corresponding to a binding energy difference of 5.03 MeV is

$$(5.03\ \text{MeV})\left(\frac{1\ \text{u}}{931.5\ \text{MeV}}\right) = 0.005\ 40\ \text{u}$$

Since the isotope with the larger binding energy has one more neutron ($m = 1.008\ 665\ \text{u}$) than the other isotope, the difference in atomic mass between the two isotopes is

$$1.008\ 665\ \text{u} - 0.005\ 40\ \text{u} = \boxed{1.003\ 27\ \text{u}}$$

19. **REASONING** As discussed in Section 31.4, β^+ decay occurs when the nucleus emits a positron, which has the same mass as an electron but carries a charge of $+e$ instead of $-e$. The general form for β^+ decay is

$$\underbrace{^A_Z P}_{\substack{\text{Parent}\\\text{nucleus}}} \rightarrow \underbrace{^A_{Z-1} D}_{\substack{\text{Daughter}\\\text{nucleus}}} + \underbrace{^0_{+1} e}_{\substack{\beta^+ \text{ particle}\\\text{(positron)}}}$$

SOLUTION

a. Therefore, the β^+ decay process for $^{18}_9 F$ is $\boxed{^{18}_9 F \rightarrow {}^{18}_8 O + {}^0_{+1} e}$.

b. Similarly, the β^+ decay process for $^{15}_8 O$ is $\boxed{^{15}_8 O \rightarrow {}^{15}_7 N + {}^0_{+1} e}$.

21. **REASONING** The reaction and the atomic masses are:

$$\underbrace{^{191}_{76} Os}_{190.960\ 920\ u} \rightarrow \underbrace{^{191}_{77} Ir + {}^0_{-1} e}_{190.960\ 584\ u}$$

When the $^{191}_{76} Os$ nucleus is converted into an iridium $^{191}_{77} Ir$ nucleus, the number of orbital electrons remains the same, so the resulting iridium atom is missing one orbital electron. However, the given mass includes all 77 electrons of a neutral iridium atom. In effect, then, the value of 190.960 584 u for $^{191}_{77} Ir$ already includes the mass of the β^- particle. Since energy is released during the decay, the combined mass of the iridium $^{191}_{77} Ir$ daughter nucleus and the β^- particle is less than the mass of the osmium $^{191}_{76} Os$ parent nucleus. The difference in mass is equivalent to the energy released. To obtain the energy released in MeV, we will use the fact that 1 u is equivalent to 931.5 MeV.

SOLUTION The mass decrease that accompanies the β^- decay of osmium $^{191}_{76} Os$ is 190.960 920 u − 190. 960 584 u = 3.36×10^{-4} u. Since 1 u is equivalent to 931.5 MeV, the energy released is

$$\text{Energy released} = \left(3.36 \times 10^{-4}\ u\right)\left(\frac{931.5\ \text{MeV}}{1\ u}\right) = \boxed{0.313\ \text{MeV}}$$

31. **REASONING** Energy is released during the β decay. To find the energy released, we determine how much the mass has decreased because of the decay and then calculate the equivalent energy. The reaction and masses are shown below:

$$\underbrace{^{22}_{11}\text{Na}}_{21.994\ 434\ u} \rightarrow \underbrace{^{22}_{10}\text{Ne}}_{21.991\ 383\ u} + \underbrace{^{0}_{+1}e}_{5.485\ 799 \times 10^{-4}\ u}$$

SOLUTION   The decrease in mass is

$$21.994\ 434\ u - \left(21.991\ 383\ u + 5.485\ 799 \times 10^{\pm 4}\ u + 5.485\ 799 \times 10^{\pm 4}\ u\right) = 0.001\ 954\ u$$

where the extra electron mass takes into account the fact that the atomic mass for sodium includes the mass of 11 electrons, whereas the atomic mass for neon includes the mass of only 10 electrons.

Since 1 u is equivalent to 931.5 MeV, the released energy is

$$\left(0.001\ 954\ u\right)\left(\frac{931.5\ \text{MeV}}{1\ u}\right) = \boxed{1.82\ \text{MeV}}$$

---

33. **REASONING**   The basis of our solution is the fact that only one-half of the radioactive nuclei present initially remain after a time equal to one half-life. After a time period that equals two half-lives, the number of nuclei remaining is $\frac{1}{2} \times \frac{1}{2}$ or $\frac{1}{4}$ of the initial number. After three half-lives, the number remaining is $\frac{1}{2} \times \frac{1}{2} \times \frac{1}{2}$ or $\frac{1}{8}$ of the initial number, and so on. Thus, to determine the fraction of nuclei remaining after a given time period, we need only to know how many half-lives that period contains.

SOLUTION   For sample A, the number of nuclei remaining is $\frac{1}{4} = \frac{1}{2} \times \frac{1}{2}$ of the initial number. We can see, then, that the time period involved is equal to two half lives of radioactive isotope A. But we know that $T_{1/2,\ B} = \frac{1}{2} T_{1/2,\ A}$, so that this time period must be equal to four half-lives of radioactive isotope B. The fraction f of the B nuclei that remain, therefore, is

$$f = \frac{1}{2} \times \frac{1}{2} \times \frac{1}{2} \times \frac{1}{2} = \boxed{\frac{1}{16}}$$

---

37. **REASONING**   We can find the decay constant from Equation 31.5, $N = N_0 e^{-\lambda t}$. If we multiply both sides by the decay constant λ, we have

$$\lambda N = \lambda N_0 e^{-\lambda t} \qquad \text{or} \qquad A = A_0 e^{-\lambda t}$$

where A_0 is the initial activity and A is the activity after a time t. Once the decay constant is known, we can use the same expression to determine the activity after a total of six days.

SOLUTION Solving the expression above for the decay constant λ, we have

$$\lambda = -\frac{1}{t}\ln\left(\frac{A}{A_0}\right) = -\frac{1}{2\text{ days}}\ln\left(\frac{285\text{ disintegrations/min}}{398\text{ disintegrations/min}}\right) = 0.167\text{ days}^{-1}$$

Then the activity four days after the second day is

$$A = (285\text{ disintegrations/min})\,e^{-(0.167\text{ days}^{-1})(4.00\text{ days})} = \boxed{146\text{ disintegrations/min}}$$

---

39. **REASONING AND SOLUTION** According to Equation 31.5, $N = N_0 e^{-\lambda t}$, the decay constant is

$$\lambda = -\frac{1}{t}\ln\left(\frac{N}{N_0}\right) = -\frac{1}{20\text{ days}}\ln\left(\frac{8.14\times10^{14}}{4.60\times10^{15}}\right) = 0.0866\text{ days}^{-1}$$

The half-life is, from Equation 31.6,

$$T_{1/2} = \frac{0.693}{\lambda} = \frac{0.693}{0.0866\text{ days}^{-1}} = \boxed{8.00\text{ days}}$$

---

45. **REASONING** According to Equation 31.5, the number of nuclei remaining after a time t is $N = N_0 e^{-\lambda t}$. Using this expression, we find the ratio N_A/N_B as follows:

$$\frac{N_A}{N_B} = \frac{N_{0A}e^{-\lambda_A t}}{N_{0B}e^{-\lambda_B t}} = e^{-(\lambda_A - \lambda_B)t}$$

where we have used the fact that initially the numbers of the two types of nuclei are equal $(N_{0A} = N_{0B})$. Taking the natural logarithm of both sides of the equation above shows that

$$\ln(N_A/N_B) = -(\lambda_A - \lambda_B)t \qquad \text{or} \qquad \lambda_A - \lambda_B = \frac{-\ln(N_A/N_B)}{t}$$

SOLUTION Since $N_A/N_B = 3.00$ when $t = 3.00$ days, it follows that

$$\lambda_A - \lambda_B = \frac{-\ln(3.00)}{3.00\text{ days}} = -0.366\text{ days}^{-1}$$

But we need to find the half-life of species B, so we use Equation 31.6, which indicates that $\lambda = 0.693/T_{1/2}$. With this expression for λ, the result for $\lambda_A - \lambda_B$ becomes

$$0.693 \left(\frac{1}{T_{1/2}^A} - \frac{1}{T_{1/2}^B} \right) = -0.366 \text{ days}^{-1}$$

Since $T_{1/2}^B = 1.50$ days, the result above can be solved to show that $\boxed{T_{1/2}^A = 7.23 \text{ days}}$.

---

47. ***REASONING*** According to Equation 31.5, the number of nuclei remaining after a time t is $N = N_0 e^{-\lambda t}$. If we multiply both sides of this equation by the decay constant λ, we have $\lambda N = \lambda N_0 e^{-\lambda t}$. Recognizing that λN is the activity A, we have $A = A_0 e^{-\lambda t}$, where A_0 is the activity at time $t = 0$. A_0 can be determined from the fact that we know the mass of the specimen, and that the activity of one gram of carbon in a living organism is 0.23 Bq. The decay constant λ can be determined from the value of 5730 yr for the half-life of $^{14}_{6}$C using Equation 31.6. With known values for A_0 and λ, the given activity of 1.6 Bq can be used to determine the age t of the specimen.

SOLUTION For $^{14}_{6}$C, the decay constant is

$$\lambda = \frac{0.693}{T_{1/2}} = \frac{0.693}{5730 \text{ yr}} = 1.21 \times 10^{-4} \text{ yr}^{-1}$$

The activity at time $t = 0$ is $A_0 = (9.2 \text{ g})(0.23 \text{ Bq/g}) = 2.1 \text{ Bq}$. Since $A = 1.6 \text{ Bq}$ and $A_0 = 2.1 \text{ Bq}$, the age of the specimen can be determined from

$$A = 1.6 \text{ Bq} = (2.1 \text{ Bq}) \, e^{-(1.21 \times 10^{-4} \text{ yr}^{-1}) t}$$

Taking the natural logarithm of both sides leads to

$$\ln \left(\frac{1.6 \text{ Bq}}{2.1 \text{ Bq}} \right) = -(1.21 \times 10^{-4} \text{ yr}^{-1}) t$$

Therefore, the age of the specimen is

$$t = \frac{\ln \left(\dfrac{1.6 \text{ Bq}}{2.1 \text{ Bq}} \right)}{-1.21 \times 10^{-4} \text{ yr}^{-1}} = \boxed{2.2 \times 10^3 \text{ yr}}$$

51. **REASONING** According to Equation 31.5, $N = N_0 e^{-\lambda t}$. If we multiply both sides by the decay constant λ, we have

$$\lambda N = \lambda N_0 e^{-\lambda t} \qquad \text{or} \qquad A = A_0 e^{-\lambda t}$$

where A_0 is the initial activity and A is the activity after a time t. The decay constant λ is related to the half-life through Equation 31.6: $\lambda = 0.693 / T_{1/2}$. We can find the age of the fossils by solving for the time t. The maximum error can be found by evaluating the limits of the accuracy as given in the problem statement.

SOLUTION The age of the fossils is

$$t = -\frac{T_{1/2}}{0.693} \ln\left(\frac{A}{A_0}\right) = -\frac{5730 \text{ yr}}{0.693} \ln\left(\frac{0.100 \text{ Bq}}{0.23 \text{ Bq}}\right) = \boxed{6900 \text{ yr}}$$

The maximum error can be found as follows. When there is an error of $+10.0$ %, $A = 0.100 \text{ Bq} + 0.0100 \text{ Bq} = 0.110 \text{ Bq}$, and we have

$$t = -\frac{5730 \text{ yr}}{0.693} \ln\left(\frac{0.110 \text{ Bq}}{0.23 \text{ Bq}}\right) = 6100 \text{ yr}$$

Similarly, when there is an error of -10.0 %, $A = 0.100 \text{ Bq} - 0.0100 \text{ Bq} = 0.090 \text{ Bq}$, and we have

$$t = -\frac{5730 \text{ yr}}{0.693} \ln\left(\frac{0.090 \text{ Bq}}{0.23 \text{ Bq}}\right) = 7800 \text{ yr}$$

The maximum error in the age of the fossils is $7800 \text{ yr} - 6900 \text{ yr} = \boxed{900 \text{ yr}}$.

---

59. **REASONING AND SOLUTION** As shown in Figure 31.17, if the first dynode produces 3 electrons, the second produces 9 electrons (3^2), the third produces 27 electrons (3^3), so the Nth produces 3^N electrons. The number of electrons that leaves the 14th dynode and strikes the 15th dynode is

$$3^{14} = \boxed{4\ 782\ 969 \text{ electrons}}$$

---

63. **REASONING** The wavelength of the photon is related to the speed of light and the frequency of the photon. The frequency is not given, but it can be obtained from the 0.186-MeV energy of the photon. The photon is emitted with this energy when the nucleus changes from one energy state to a lower energy state. The energy is the difference ΔE between the two nuclear

energy levels, in a way very similar to that discussed in Section 30.3 for the energy levels of the electron in the hydrogen atom. In that section, we saw that the energy difference ΔE is related to the frequency f and Planck's constant h, so that we will be able to obtain the frequency from the given energy value.

SOLUTION The photon wavelength λ is related to the photon frequency f and the speed c of light in a vacuum according to Equation 16.1: $\lambda = \dfrac{c}{f}$ (1), as shown at the right. We have no value for the frequency, so we evaluate it in the next step.

Section 30.3 discusses the fact that the photon emitted when the electron in a hydrogen atom changes from a higher to a lower energy level has an energy ΔE, which is the difference between the energy levels. A similar situation exists here when the nucleus changes from a higher to a lower energy level. The γ-ray photon that is emitted has an energy ΔE given by $\Delta E = hf$ (Equation 30.4). Solving for the frequency, we obtain:

$$f = \frac{\Delta E}{h}.$$

Combining the results of each step algebraically, we find that:

$$\lambda = \frac{c}{f} = \frac{c}{\Delta E / h}$$

The wavelength of the gamma photon is:

$$\lambda = \frac{hc}{\Delta E} = \frac{(6.63 \times 10^{-34} \ \text{J} \cdot \text{s})(3.00 \times 10^{8} \ \text{m/s})}{(0.186 \times 10^{6} \ \text{eV})\left(\dfrac{1.60 \times 10^{-19} \ \text{J}}{1 \ \text{eV}}\right)} = \boxed{6.68 \times 10^{-12} \ \text{m}}$$

Note that we have converted the value of $\Delta E = 0.186 \times 10^{6} \ \text{eV}$ into joules by using the fact that $1 \ \text{eV} = 1.60 \times 10^{-19} \ \text{J}$.

---

65. **CONCEPTS**

(i) According to Equation 12.4, the heat needed is $Q = cm\Delta T$, where the specific heat capacity of water is $c = 4186 \ \text{J/(kg} \cdot \text{C}^{\circ})$.

(ii) The total energy released is just the number of disintegrations times the energy for each one, or $E_{\text{Total}} = nE$.

(iii) The number of radioactive nuclei that are remaining after a time t is given by Equations 31.5 and 31.6 as

$$N = N_0 e^{-0.693t/T_{1/2}}$$

where N_0 is the number of radioactive nuclei present at $t = 0$ s and $T_{1/2}$ is the half-life for the decay. Therefore, the number n of disintegrations that occur during the time t is $N - N_0$ or

$$n = N_0 - N = N_0\left(1 - e^{-0.693t/T_{1/2}}\right) \tag{1}$$

CALCULATIONS

According to Equation 12.4, the heat needed to change the temperature of the water is $Q = cm\Delta T$. This heat is provided by the total energy released in one hour of radioactive decay, or $E_{\text{Total}} = nE$. Setting $Q = E_{\text{Total}}$, we obtain $nE = cm\Delta T$, which can be solved for the change in temperature ΔT :

$$\Delta T = \frac{nE}{cm}$$

Substituting Equation (1) for n into the above expression, we obtain

$$\Delta T = \frac{N_0\left(1 - e^{-0.693t/T_{1/2}}\right)E}{cm}$$

$$= \frac{\left(2.64 \times 10^{21}\right)\left[1 - e^{-0.693(1.000\ \text{h})/\left(1.677 \times 10^4\ \text{h}\right)}\right]\left(8.83 \times 10^{-13}\ \text{J}\right)}{\left[4186\ \text{J}/\left(\text{kg} \cdot \text{C}°\right)\right]\left(3.8\ \text{kg}\right)} = \boxed{6.1\ \text{C}°}$$

CHAPTER 32 | *IONIZING RADIATION, NUCLEAR ENERGY, AND ELEMENTARY PARTICLES*

1. **REASONING** The biologically equivalent dose (in rems) is the product of the absorbed dose (in rads) and the relative biological effectiveness (RBE), according to Equation 32.4. We can apply this equation to each type of radiation. Since the biologically equivalent doses of the neutrons and α particles are equal, we can solve for the unknown RBE.

   SOLUTION Applying Equation 32.4 to each type of particle and using the fact that the biological equivalent doses are equal, we find that

   $$\underbrace{\left(\text{Absorbed dose}\right)_\alpha \text{RBE}_\alpha}_{\substack{\text{Biologically equivalent dose} \\ \text{of } \alpha \text{ particles}}} = \underbrace{\left(\text{Absorbed dose}\right)_{\text{neutrons}} \text{RBE}_{\text{neutrons}}}_{\substack{\text{Biologically equivalent dose} \\ \text{of neutrons}}}$$

   Solving for RBE_α and noting that $(\text{Absorbed dose})_{\text{neutrons}} = 6(\text{Absorbed dose})_\alpha$, we have

   $$\text{RBE}_\alpha = \frac{\left(\text{Absorbed dose}\right)_{\text{neutrons}}}{\left(\text{Absorbed dose}\right)_\alpha}\left(\text{RBE}_{\text{neutrons}}\right) = \frac{6\left(\text{Absorbed dose}\right)_\alpha}{\left(\text{Absorbed dose}\right)_\alpha}(2.0) = \boxed{12}$$

5. **REASONING AND SOLUTION** According to Equation 32.2, the absorbed dose (AD) is equal to the energy absorbed by the tumor divided by its mass:

   $$\text{AD} = \frac{\text{Energy absorbed}}{\text{Mass}} = \frac{(25\ \text{s})\left(1.6\times10^{10}\ \text{s}^{-1}\right)\left(4.0\times10^6\ \text{eV}\right)\left(\dfrac{1.60\times10^{-19}\ \text{J}}{1\ \text{eV}}\right)}{0.015\ \text{kg}}$$

   $$= 1.7\times10^1\ \text{Gy} = 1.7\times10^3\ \text{rad}$$

   The biologically equivalent dose (BED) is equal to the product of the absorbed dose (AD) and the RBE (see Equation 32.4):

   $$\text{BED} = \text{AD} \times \text{RBE} = (1.7 \times 10^3\ \text{rad})(14) = \boxed{2.4 \times 10^4\ \text{rem}} \tag{32.4}$$

11. **REASONING** The number of nuclei in the beam is equal to the energy absorbed by the tumor divided by the energy per nucleus (130 MeV). According to Equation 32.2, the energy (in joules) absorbed by the tumor is equal to the absorbed dose (expressed in grays) times the mass of the tumor. The absorbed dose (expressed in rads) is equal to the biologically equivalent dose divided by the RBE of the radiation (see Equation 32.4). We can use these concepts to determine the number of nuclei in the beam.

SOLUTION The number N of nuclei in the beam is equal to the energy E absorbed by the tumor divided by the energy per nucleus. Since the energy absorbed is equal to the absorbed dose (in Gy) times the mass m (see Equation 32.2) we have

$$N = \frac{E}{\text{Energy per nucleus}} = \frac{[\text{Absorbed dose (in Gy)}]m}{\text{Energy per nucleus}}$$

We can express the absorbed dose in terms of rad units, rather than Gy units, by noting that 1 rad = 0.01 Gy. Therefore,

$$\text{Absorbed dose (in Gy)} = \text{Absorbed dose (in rad)}\left(\frac{0.01\,\text{Gy}}{1\,\text{rad}}\right)$$

The number of nuclei can now be written as

$$N = \frac{E}{\text{Energy per nucleus}} = \frac{[\text{Absorbed dose (in rad)}]\left(\dfrac{0.01\,\text{Gy}}{1\,\text{rad}}\right)m}{\text{Energy per nucleus}}$$

We know from Equation 32.4 that the Absorbed dose (in rad) is equal to the biologically equivalent dose divided by the RBE, so that

$$N = \frac{E}{\text{Energy per nucleus}} = \frac{\left[\dfrac{\text{Biologically equivalent dose}}{\text{RBE}}\right]\left(\dfrac{0.01\,\text{Gy}}{1\,\text{rad}}\right)m}{\text{Energy per nucleus}}$$

$$= \frac{\left[\dfrac{180\,\text{rem}}{16}\right]\left(\dfrac{0.01\,\text{Gy}}{1\,\text{rad}}\right)(0.17\,\text{kg})}{\left(130\times10^{6}\,\text{eV}\right)\left(\dfrac{1.60\times10^{-19}\,\text{J}}{1\,\text{eV}}\right)} = \boxed{9.2\times10^{8}}$$

13. **REASONING** The reaction given in the problem statement is written in the shorthand form: $^{17}_{8}\text{O}\,(\gamma,\,\alpha n)\,^{12}_{6}\text{C}$. The first and last symbols represent the initial and final nuclei,

respectively. The symbols inside the parentheses denote the incident particles or rays (left side of the comma) and the emitted particles or rays (right side of the comma).

SOLUTION In the shorthand form of the reaction, we note that the designation an refers to an α particle (which is a helium nucleus ^4_2He) and a neutron (^1_0n). Thus, the reaction is

$$\boxed{\gamma + {}^{17}_{8}\text{O} \rightarrow {}^{12}_{6}\text{C} + {}^{4}_{2}\text{He} + {}^{1}_{0}\text{n}}$$

---

21. **REASONING** The rest energy of the uranium nucleus can be found by taking the atomic mass of the $^{235}_{92}\text{U}$ atom, subtracting the mass of the 92 electrons, and then using the fact that 1 u is equivalent to 931.5 MeV (see Section 31.3). According to Table 31.1, the mass of an electron is $5.485\,799 \times 10^{-4}$ u. Once the rest energy of the uranium nucleus is found, the desired ratio can be calculated.

SOLUTION The mass of $^{235}_{92}\text{U}$ is 235.043 924 u. Therefore, subtracting the mass of the 92 electrons, we have

$$\text{Mass of } {}^{235}_{92}\text{U nucleus} = 235.043\,924 \text{ u} - 92(5.485\,799 \times 10^{-4}\text{u}) = 234.993\,455 \text{ u}$$

The energy equivalent of this mass is

$$(234.993\,455 \text{ u}) \left(\frac{931.5 \text{ MeV}}{1 \text{ u}} \right) = 2.189 \times 10^5 \text{ MeV}$$

Therefore, the ratio is

$$\frac{200 \text{ MeV}}{2.189 \times 10^5 \text{ MeV}} = \boxed{9.0 \times 10^{-4}}$$

---

27. **REASONING AND SOLUTION** The binding energy per nucleon for a nucleus with $A = 239$ is about 7.6 MeV per nucleon, according to Figure 32.9 The nucleus fragments into two pieces of mass ratio 0.32 : 0.68. These fragments thus have nucleon numbers

$$A_1 = (0.32)(A) = 76 \qquad \text{and} \qquad A_2 = (0.68)(A) = 163$$

Using Figure 32.9 we can estimate the binding energy per nucleon for A_1 and A_2. We find that the binding energy per nucleon for A_1 is about 8.8 MeV, representing an increase of 8.8 MeV – 7.6 MeV = 1.2 MeV per nucleon. Since there are 76 nucleons present, the energy released for A_1 is

$$76 \times 1.2 \text{ MeV} = 91 \text{ MeV}$$

Similarly, for A_2, we see that the binding energy increases to 8.0 MeV per nucleon, the difference being 8.0 MeV – 7.6 MeV = 0.4 MeV per nucleon. Since there are 163 nucleons present, the energy released for A_2 is

$$163 \times 0.4 \text{ MeV} = 70 \text{ MeV}$$

The energy released per fission is $E = 91$ MeV + 70 MeV = $\boxed{160 \text{ MeV}}$.

---

29. **REASONING** We first determine the total power generated (used and wasted) by the plant. Energy is power times the time, according to Equation 6.10, and given the energy, we can determine how many kilograms of $^{235}_{92}\text{U}$ are fissioned to produce this energy.

SOLUTION Since the power plant produces energy at a rate of 8.0×10^8 W when operating at 25 % efficiency, the total power produced by the power plant is

$$\left(8.0 \times 10^8 \text{ W}\right) 4 = 3.2 \times 10^9 \text{ W}$$

The energy equivalent of one atomic mass unit is given in the text (see Section 31.3) as

$$1 \text{ u} = 1.4924 \times 10^{-10} \text{ J} = 931.5 \text{ MeV}$$

Since each fission produces 2.0×10^2 MeV of energy, the total mass of $^{235}_{92}\text{U}$ required to generate 3.2×10^9 W for a year (3.156×10^7 s) is

$$
\overbrace{\left(3.2 \times 10^9 \text{ J/s}\right)\left(3.156 \times 10^7 \text{ s}\right)}^{\substack{\text{Power times time gives} \\ \text{energy in joules}}}
$$

$$
\times \underbrace{\left(\frac{931.5 \text{ MeV}}{1.4924 \times 10^{-10} \text{ J}}\right)}_{\substack{\text{Converts joules to} \\ \text{MeV. See Sec. 31.3.}}} \underbrace{\left(\frac{1.0 \ ^{235}_{92}\text{U nucleus}}{2.0 \times 10^2 \text{ MeV}}\right)}_{\substack{\text{Converts MeV to} \\ \text{number of nuclei}}} \underbrace{\left(\frac{0.235 \text{ kg}}{6.022 \times 10^{23} \ ^{235}_{92}\text{U nuclei}}\right)}_{\substack{\text{Converts number of nuclei to} \\ \text{kilograms}}} = \boxed{1200 \text{ kg}}
$$

---

35. **REASONING** To find the energy released per reaction, we follow the usual procedure of determining how much the mass has decreased because of the fusion process. Once the energy released per reaction is determined, we can determine the mass of lithium ^6_3Li needed to produce 3.8×10^{10} J.

SOLUTION The reaction and the masses are shown as follows:

$$\underbrace{^{2}_{1}\text{H}}_{2.014\ u} + \underbrace{^{6}_{3}\text{Li}}_{6.015\ u} \rightarrow \underbrace{2\ ^{4}_{2}\text{He}}_{2(4.003\ u)}$$

The mass defect is, therefore, $2.014\ u + 6.015\ u - 2(4.003\ u) = 0.023\ u$. Since 1 u is equivalent to 931.5 MeV, the released energy is 21 MeV, or since the energy equivalent of one atomic mass unit is given in Section 31.3 as $1\ u = 1.4924 \times 10^{-10}\ \text{J} = 931.5\ \text{MeV}$,

$$(21\ \text{MeV}) \left(\frac{1.4924 \times 10^{-10}\ \text{J}}{931.5\ \text{MeV}} \right) = 3.4 \times 10^{-12}\ \text{J}$$

In 1.0 kg of lithium $^{6}_{3}\text{Li}$, there are

$$(1.0\ \text{kg of}\ ^{6}_{3}\text{Li}) \left(\frac{1.0 \times 10^{3}\ \text{g}}{1.0\ \text{kg}} \right) \left(\frac{6.022 \times 10^{23}\ \text{nuclei/mol}}{6.015\ \text{g/mol}} \right) = 1.0 \times 10^{26}\ \text{nuclei}$$

Therefore, 1.0 kg of lithium $^{6}_{3}\text{Li}$ would produce an energy of

$$(3.4 \times 10^{-12}\ \text{J/nuclei})(1.0 \times 10^{26}\ \text{nuclei}) = 3.4 \times 10^{14}\ \text{J}$$

If the energy needs of one household for a year is estimated to be 3.8×10^{10} J, then the amount of lithium required is

$$\frac{3.8 \times 10^{10}\ \text{J}}{3.4 \times 10^{14}\ \text{J/kg}} = \boxed{1.1 \times 10^{-4}\ \text{kg}}$$

---

43. **REASONING** The momentum of a photon is given in the text as $p = E/c$ (see the discussion leading to Equation 29.6). This expression applies to any massless particle that travels at the speed of light. In particular, assuming that the neutrino has no mass and travels at the speed of light, it applies to the neutrino. Once the momentum of the neutrino is determined, the de Broglie wavelength can be calculated from Equation 29.6 ($p = h/\lambda$).

SOLUTION
a. The momentum of the neutrino is, therefore,

$$p = \frac{E}{c} = \left(\frac{35\ \text{MeV}}{3.00 \times 10^{8}\ \text{m/s}} \right) \left(\frac{1.4924 \times 10^{-10}\ \text{J}}{931.5\ \text{MeV}} \right) = \boxed{1.9 \times 10^{-20}\ \text{kg} \cdot \text{m/s}}$$

where we have used the fact that 1.4924×10^{-10} J $= 931.5$ MeV (see Section 31.3).

b. According to Equation 29.6, the de Broglie wavelength of the neutrino is

$$\lambda = \frac{h}{p} = \frac{6.63 \times 10^{-34} \text{ J} \cdot \text{s}}{1.9 \times 10^{-20} \text{ kg} \cdot \text{m/s}} = \boxed{3.5 \times 10^{-14} \text{ m}}$$

47. **REASONING** The lambda particle contains three different quarks, one of which is the up quark u, and contains no antiquarks. Therefore, the remaining two quarks must be selected from the down quark d, the strange quark s, the charmed quark c, the top quark t, and the bottom quark b. Choosing among these possibilities must be done consistent with the fact that the lamda particle has an electric charge of zero.

SOLUTION Since the lambda particle has an electric charge of zero and since u has a charge of $+2e/3$, the charges of the remaining two quarks must add up to a total charge of $-2e/3$. This eliminates the quarks c and t as choices, because they each have a charge of $+2e/3$. We are left, then, with d, s, and b as choices for the remaining two quarks in the lambda particle. The three possibilities are as follows:

$$\boxed{(1) \ u,d,s \qquad (2) \ u,d,b \qquad (3) \ u,s,b}$$

51. **REASONING** To find the energy released per reaction, we follow the usual procedure of determining how much the mass has decreased because of the fusion process. Once the energy released per reaction is determined, we can determine the amount of gasoline that must be burned to produce the same amount of energy.

SOLUTION The reaction and the masses are shown below:

$$3 \, {}^{2}_{1}\text{H} \ \rightarrow \ {}^{4}_{2}\text{He} \ + \ {}^{1}_{1}\text{H} \ + \ {}^{1}_{0}\text{n}$$
$$\underbrace{\phantom{3\,{}^{2}_{1}\text{H}}}_{3(\,2.0141\text{ u})} \qquad \underbrace{\phantom{{}^{4}_{2}\text{He}}}_{4.0026\text{ u}} \qquad \underbrace{\phantom{{}^{1}_{1}\text{H}}}_{1.0078\text{ u}} \qquad \underbrace{\phantom{{}^{1}_{0}\text{n}}}_{1.0087\text{ u}}$$

The mass defect is, therefore,

$$3(2.0141 \text{ u}) - 4.0026 \text{ u} - 1.0078 \text{ u} - 1.0087 \text{ u} = 0.0232 \text{ u}$$

Since 1 u is equivalent to 931.5 MeV, the released energy is 21.6 MeV, or since it is shown in Section 31.3 that $931.5 \text{ MeV} = 1.4924 \times 10^{-10}$ J, the energy released per reaction is

$$(21.6 \text{ MeV}) \left(\frac{1.4924 \times 10^{-10} \text{ J}}{931.5 \text{ MeV}} \right) = 3.46 \times 10^{-12} \text{ J}$$

To find the total energy released by all the deuterium fuel, we need to know the number of deuterium nuclei present. Noting that 6.1×10^{-6} kg $= 6.1 \times 10^{-3}$ g, we find that the number of deuterium nuclei is

$$(6.1 \times 10^{-3} \text{ g}) \left(\frac{6.022 \times 10^{23} \text{ nuclei/mol}}{2.0141 \text{ g/mol}} \right) = 1.8 \times 10^{21} \text{ nuclei}$$

Since each reaction consumes three deuterium nuclei, the total energy released by the deuterium fuel is

$$\frac{1}{3} (3.46 \times 10^{-12} \text{ J/nuclei})(1.8 \times 10^{21} \text{ nuclei}) = 2.1 \times 10^{9} \text{ J}$$

If one gallon of gasoline produces 2.1×10^{9} J of energy, then the number of gallons of gasoline that would have to be burned to equal the energy released by all the deuterium fuel is

$$\left(2.1 \times 10^{9} \text{ J}\right) \left(\frac{1.0 \text{ gal}}{2.1 \times 10^{9} \text{ J}} \right) = \boxed{1.0 \text{ gal}}$$

---

55. *CONCEPTS*

(i) The power P is equal to the intensity I times the cross-sectional area A of the beam, as indicated by Equation 16.8 $(P = IA)$.

(ii) The energy E is equal to the product of the power P delivered by the beam and the time of exposure t, according to Equation 6.10b $(E = Pt)$.

(iii) The absorbed does is defined by Equation 32.2 as the energy E absorbed by the tissue divided by the mass m of the tissue, or Absorbed dose $= E/m$. The unit of the absorbed dose is the gray (Gy).

(iv) The biologically equivalent dose is equal to the product of the absorbed dose and the relative biological effectiveness (RBE), as indicated by Equation 32.4. However, the absorbed dose must be expressed is rad units, not in gray units. The relation between the two units is 1 rad = 0.01 Gy.

CALCULATIONS

The biologically equivalent dose is given by Equation 32.4 as

Biologically equivalent dose $= \left[\text{Absorbed dose (in rad)} \right](\text{RBE})$

At this point we need to convert the absorbed dose from Gy units to rad units:

$$\text{Absorbed dose (in rads)} = \left[\text{Absorbed dose (in Gy)}\right]\left(\frac{1 \text{ rad}}{0.01 \text{ Gy}}\right)$$

The biologically equivalent dose then becomes

$$\text{Biologically equivalent dose} = \left[\text{Absorbed dose (in Gy)}\right]\left(\frac{1 \text{ rad}}{0.01 \text{ Gy}}\right)(\text{RBE})$$

Equation 32.2 indicates that the absorbed dose (in Gy) is equal to the energy E divided by the mass m of the tissue, so that we have

$$\text{Biologically equivalent dose} = \left(\frac{E}{m}\right)\left(\frac{1 \text{ rad}}{0.01 \text{ Gy}}\right)(\text{RBE})$$

From the answers to the first and second Concept Questions, we can conclude that the absorbed energy E is related to the intensity I, the cross-sectional area A, and the time duration t of the beam via $E = Pt = IAt$. Substituting this result into the equation for the biologically equivalent dose gives

$$\left(\begin{array}{c}\text{Biologically}\\\text{equivalent dose}\end{array}\right) = \left(\frac{IAt}{m}\right)\left(\frac{1 \text{ rad}}{0.01 \text{ Gy}}\right)(\text{RBE})$$

$$= \frac{\left(0.40 \text{ W/m}^2\right)\left(0.072 \text{ m}^2\right)(0.20 \text{ s})}{3.6 \text{ kg}}\left(\frac{1 \text{ rad}}{0.01 \text{ Gy}}\right)(1.1) = \boxed{0.18 \text{ rem}}$$

CPSIA information can be obtained at www.ICGtesting.com
Printed in the USA
BVOW09s1701080815

412213BV00009B/11/P

9 781118 836903